Essential Cell Biology

Practical Approach Series

For full details of Practical Approach titles currently available, please go to
www.oup.co.uk/pas

The following titles may be of particular interest:

Enzyme Assays (second edition) (No. 257)
Edited by Robert Eisenthal and Michael Danson

Building upon the highly successful first edition, this book combines revised or
rewritten chapters with entirely new contributions. This second edition will be
valuable not only to biochemists, but to researchers in all areas of the life sciences.

Plant Cell Biology (second edition) (No. 250)
Edited by Chris Hawes and Beatrice Satiat Jeunemaitre

"Experienced plant cell biologists will find the book useful as a quick reference guide
and will be pleased to see some of their favourite methods, with updates and tips for
improvement. For anyone curious about cell biology, this book provides an excellent
introduction to the major relevant techniques and works as well as a laboratory
manual." *Trends in Plant Science*

Protein Purification Techniques and Applications
(2 volume set, Nos 244 & 245)
Edited by Simon Roe

This two volume set covers both protein purification techniques, focusing on unit
operations and analytical techniques, and protein purification applications which
describes strategies and detailed practical procedures for protein purification from
common sources.

Receptors—Structure and Function (No. 253)
Edited by Clare Stanford and Roger Horton

"...most chapters convey a good deal of practical advice that could save days or
months of effort on the part of seasoned investigators and could make the difference
between a revealing discovery and uninterpretable chaos." *Journal of Cell Science*

Membrane Transport (No. 230)
Edited by Stephen A. Baldwin

"Stephen Baldwin has performed a service to the scientific community in compiling
these technical approaches to membrane transport studies. He offers us a great
opportunity to have, within a single book, a comprehensive panorama of simple and
advanced procedures to investigate the membrane transporter structure and
function." *Journal of Cell Science*

No. 262

Essential Cell Biology
Volume 1: Cell Structure
A Practical Approach

Edited by

John Davey and Mike Lord
Department of Biological Sciences,
The University of Warwick,
Coventry CV4 7AL, UK

OXFORD
UNIVERSITY PRESS

OXFORD

UNIVERSITY PRESS

Great Clarendon Street, Oxford OX2 6DP

Oxford University Press is a department of the University of Oxford.
It furthers the University's objective of excellence in research, scholarship,
and education by publishing worldwide in

Oxford New York

Auckland Bangkok Buenos Aires Cape Town Chennai Dar es Salaam
Delhi Hong Kong Istanbul Karachi Kolkata Kuala Lumpur Madrid
Melbourne Mexico City Mumbai Nairobi São Paulo Shangai Taipei
Tokyo Toronto

Oxford is a registered trade mark of Oxford University Press in the UK and
in certain other countries

Published in the United States by Oxford University Press Inc., New York

© Oxford University Press, 2003

A catalogue record for this title is available from
the British Library

Library of Congress Cataloguing in Publication Data
Essential cell biology : a practical approach / edited by John Davey
and Mike Lord.
p. cm. (Practical approach series ; 262–)
Includes bibliographical references and index.
Contents: v. Cell structure – v. 2. Cell function.
1. Cytology–Laboratory manuals. I. Davey, John. II. Lord, Mike.
III. Practical approach series ; 262, etc.
QH583.2. E85 2002   571.6–dc21   2002029013

ISBN 0 19 963830 6 (v. 1 : hbk.)   ISBN 0 19 963832 2 (v. 2 : hbk.)
ISBN 0 19 963831 4 (v. 1 : pbk.)   ISBN 0 19 963833 0 (v. 2 : pbk.)
10 9 8 7 6 5 4 3 2 1

Typeset in Swift by Footnote Graphics Ltd, Warminster, Wilts
Printed in Great Britain on acid-free paper
by The Bath Press Ltd, Avon

Preface

Cell biology relies upon an integrated understanding of how molecules within cells interact to carry out and regulate the processes required for life. Obtaining such an integrated understanding imposes great demands on today's investigators. Being an expert in a relatively narrow area is no longer sufficient and researchers need to be able to call upon a battery of techniques in their quest for information. Unfortunately, a lack of familiarity with the experimental possibilities can often discourage diversification. This two-volume set is designed to try to help overcome some of these problems. With the help of experienced researchers, we have been able to gather together a compendium of protocols that covers most of the approaches available for studying cell biology. Inevitably in a project of this size it has not been possible to include every technique but with Volume 1 focusing on the techniques for studying cell structure and Volume 2 concentrating on understanding how the cell functions, we believe that all of the essential protocols are included. Both traditional and more modern techniques are covered and the theory and principles of each are described, together with detailed protocols and advice on trouble shooting. Directions to more specialized developments are also included. Although written by experts, each section is accessible to those new to science. We hope the result inspires readers to experience the challenges and rewards of entering a new area of cell biology for themselves.

We thank all of the authors for agreeing to undertake the time-consuming job of preparing their chapters, and the staff at Oxford University Press for their help at all stages of preparation and production. Finally, we would appreciate receiving any comments on the text and the correction of any errors that might have been missed.

Warwick   J. D.
November 2002   J. M. L.

Contents

CONTENTS

7 Gel electrophoresis of proteins   *197*

David E. Garfin

8 Biophysical methods in structural cell biology   *269*

Mavis Agbandje-McKenna, Arthur S. Edison, and Robert McKenna

11 Extracellular matrix protocols for the study of complex phenotypes   349

Calvin D. Roskelley and Shoukat Dedhar

12 The cytoskeleton   365

Theresia B. Stradal, Antonio S. Sechi, Jürgen Wehland, and Klemens Rottner

Protocol list

Basic biochemical analysis of actin and actin binding proteins

Abbreviations

A_{260}	absorption at 260 nm
AP	acid phosphatase
APS	ammonium persulfate
ATP	adenosine 5′-triphosphate
Bis	bis(hydroxymethyl)aminomethane
bp	base pairs
BSA-C	acetylated BSA
CBB	Coomassie Brilliant Blue
CCD	charge-coupled device
cDNA	complementary DNA
CFE	colony-forming efficiency
Chol	cholesterol
Con A	concanavalin A
COSY	correlation spectroscopy
1D	one-dimensional
2D	two-dimensional
DAPI	4′,6′-diamidino-2-phenylindole
dATP	2′-deoxyadenosine 5′-triphosphate
dCTP	2′-deoxycytidine 5′-triphosphate
ddA	2′,3′-deoxyadenosine 5′-triphosphate
ddC	2′,3′-deoxycytidine 5′-triphosphate
ddG	2′,3′-deoxyguanosine 5′-triphosphate
ddNTP	dideoxynucleotide
ddT	2′,3′-deoxythymidine 5′-triphosphate
DEPC	diethyl pyrocarbonate
DETAPAC	diethylenetriaminopentaacetic acid
dGTP	2′-deoxyguanosine 5′-triphosphate
DIC	differential interference contrast
DMEM	Dulbecco's modified Eagles medium
DMSO	dimethyl sulfoxide
DNA	deoxyribonucleic acid
dNTP	deoxyribonucleotide

d.p.m.	disintegrations per minute
DPPE	L-α-dipalmitoyl phosphatidylethanolamine
d.p.s.	disintegrations per second
dTTP	2'-deoxythymidine 5'-triphosphate
ECM	extracellular matrix
EDTA	ethylenediaminetetraacetic acid
EGTA	ethyleneglycolbis(aminoethyl)tetraacetic acid
EM	electron microscopy
EYPC	egg yolk PC
FITC	fluorescein isothiocyanate
FN	fibronectin
FSG	fish skin gelatin
GFP	green fluorescent protein
HBSS	Hanks balanced salt solution
HEPES	N-2-hydroxyethyl piperazine-N'-2-ethanesulfonic acid
HGF	hepatocyte growth factor
HIC	hydrophobic interaction chromatography
HPLC	high performance liquid chromatography
HPTLC	high performance thin-layer chromatography
IEF	isoelectric focusing
IF	intermediate filaments
IL	immunolabelling
ILK	integrin-linked kinase
IMAC	immobilized metal affinity chromatography
IPG	immobilized pH gradients
kb	kilobase pairs
LR	London resin
MAP kinase	mitogen-activated protein kinase
MC	methylcellulose
MEM	minimal essential medium
MMO	methane monooxygenase
mol. wt.	molecular weight
MOPS	3-morpholinopropanesulfonic acid
MT	microtubules
NA	numerical aperture
NBD	7-nitrobenz-2-oxa-1,3-diazole-4-yl
NC	nitrocellulose
NMR	nuclear magnetic resonance
NOESY	nuclear Overhauser effect spectroscopy
NR	neutral red
OD	optical density
PAGE	polyacrylamide gel electrophoresis
PBS	phosphate-buffered saline
PC	phosphatidylcholine
PCR	polymerase chain reaction

PE	phosphatidylethanolamine
PE-PDP	*N*-succinimidyl-3-(2-pyridyldithio) propionate-derivatized DPPE
PE-Lys	lysine-coupled PE
p*I*	isoelectric point
PIPES	piperazine-1,4-bis(2-ethanesulfonic acid)
PKB	protein kinase B
PLL	poly-L-lysine
PLT	progressive lowering of temperature
PMSF	phenylmethylsulfonyl fluoride
poly(A$^+$)	polyadenylated
polyHEMA	poly 2-hydroxyethylmethacrylate
ppm	parts per million
PS	phosphatidylserine
PVDF	polyvinylidene fluoride
PVP	polyvinylpyrrolidine
QE	quantum efficiency
RACE	random amplification of cDNA ends
Rh	Lissamine rhodamine B sulfonyl
RI	refractive index
RNA	ribonucleic acid
RPC	reverse phase chromatography
RT	room temperature
SDS	sodium dodecyl sulfate
SEM	scanning electron microscopy
SPDP	*N*-succinimidyl-3-(2-pyridyldithio)propionate
SSC	salt/sodium citrate
TAE	Tris/acetate/EDTA
TBE	Tris/borate/EDTA
TBS	Tris-buffered saline
TEM	transmission electron microscopy
TEMED	*N,N,N′,N′*-tetramethylethylenediamine
TFA	trifluoroacetic acid
TGF	transforming growth factor
TOCSY	total correlation spectroscopy
Tris	tris(hydroxymethyl)aminomethane
UV	ultraviolet
VSV	vesicular stomatitis virus
XRC	X-ray crystallography

Chapter 1
Cell culture

Samantha Touhey and Mary Heenan[*]
Previous address: National Cell and Tissue Culture Centre/National Institute for Cellular Biotechnology, Dublin City University, Glasnevin, Dublin 9, Ireland.
[*]Present address: Wyeth Bio Pharma, Grange Castle International Business Park, Clondalkin, Dublin 22, Ireland.

1 Introduction

Until animal cell culture emerged as a valuable research tool in the 1940s and 1950s, experimental approaches to the study of cellular physiology were limited to either studies in the whole animal or short-term studies with isolated cells or slices from various tissues. The development of cell culture led to a new experimental approach to cellular physiology in which isolated, functionally differentiated cells could be maintained in culture under conditions that allowed direct manipulations of the environment and measurement of the resulting changes in the function of a single cell type. Today many aspects of research and development involve the use of animal cells as *in vitro* model systems, substrates for virus replication, and in the production of diagnostic and therapeutic products.

2 Safety

In general, working with animal cells in culture is a low-risk procedure but care should be taken when working with human and other primate cells. Antibodies, sera, and cells (particularly, but not exclusively, those of human and non-human primate origin) pose a significant biological hazard. Many animal cells contain C-type particles, which may be retrovirus-related. All such materials may harbour pathogens and should be handled as potentially infectious material in accordance with local guidelines. A cell culture laboratory should be well ventilated, preferably with Hepa filters on the inlets. A Class II downflow recirculating laminar flow cabinet will provide a safe working environment and should be checked periodically for efficiency. A horizontal flow cabinet should never be used as it can, even in the absence of viruses, increase exposure to allergens. Operators must make sure that any cuts, especially on the hands, are covered. Wearing disposable gloves is recommended. Thorough washing of hands before and after cell work is imperative. It is advisable to obtain immunization against hepatitis B if working with primary human material. Laboratory coats and protective eyewear are essential. These cautionary procedures are for general 'non-hazardous' cells. However, if using viral-infected cell lines additional containment requirements may be necessary.

3 Good laboratory technique

A designated room should be reserved for cell culture work and equipment in this designated area should be kept to the minimum required. There should be proper entry facilities and surfaces must be easy to clean and dust free. When setting up a cell culture laboratory the following equipment is essential:

- Class II downflow recirculating laminar flow cabinet
- low speed centrifuge
- CO_2 incubator
- inverted microscope with phase contrast capabilities

  For a more specialized laboratory, extra equipment will be required.

Prevention of contamination by any extraneous organisms (bacteria, yeast, fungi, mycoplasma, and viruses) is a necessity in cell culture. Good laboratory practice in the cell culture laboratory requires that certain standard procedures are followed.

(a) It is important that cell lines are obtained from a reputable source—preferably the laboratory of origin or an established cell repository.

(b) Only sterile items (i.e. pipettes, culture flasks, etc.) should enter the rooms, and remove discarded media and waste each day. Dispose of all used materials safely, efficiently, and routinely, in accordance with local regulatory requirements.

(c) Stock cultures of two cell lines should never be used at the same time in a laminar flow cabinet. Between work with different cell lines a thorough cleaning of surfaces with a suitable disinfectant and a 15 min 'clearing' gap should occur. Use of pipettes, medium/waste bottles, etc. for more than one cell line is another possible source of cross-contamination and must be avoided.

(d) Cell lines can change properties (e.g. growth rate, antigen expression, enzyme profile) quite rapidly. It is essential to set up a frozen stock of each cell line and to work only within a defined number of cell doublings or passages at a defined dilution/split ratio. A ten-passage period is satisfactory for many cell lines and properties.

(e) When setting up frozen stocks of a cell line, aliquots should be thawed to test for viability, growth, and absence of contamination (including mycoplasma). They should also be characterized by some appropriate criteria (e.g. DNA fingerprinting, cytogenetic analysis) both for comparison to the parent cell line and as a standard for comparison of future stocks.

(f) Quality control all of the reagents used in tissue culture to ensure all materials used are free from contamination prior to use with cells.

(g) Ensure laboratory coats are changed regularly.

(h) A disinfectant (e.g. 70% isopropyl alcohol) should be used liberally to swab work surfaces, bottles, flasks, etc.

(i) Maintain facilities efficiently. Have Hepa filters tested at appropriate intervals (e.g. six months). Ensure that all incubators, microscopes, centrifuges, and other equipment are cleaned and serviced regularly.

4 Types of cell cultures

4.1 Primary cell cultures

The cultivation of animal tissues *in vitro* was first shown to be possible in 1907 and the first human cell line was obtained in 1952 (1). However it is only in the last 20–25 years that reproducible and reliable large scale culture of mammalian cells has been achieved. The process of initiating a culture from cells, tissue, or organs taken directly from an animal and before the first subculture is known as primary culture. Primary culture is the source of all cells in culture. There are a number of advantages and applications associated with primary culture:

(a) To provide tissue for surgery, e.g. primary culture of skin samples in skin grafting and plastic surgery.

(b) To optimize drug treatment, especially anti-cancer drug treatments, using samples from each individual patient.

(c) To produce cells that have the potential to become cell lines. This allows animals to be replaced as a model and improves the reproducibility of experiments.

(d) It is an essential component in the future technology of organ culture.

However even in primary culture loss of gene expression can occur quite rapidly, e.g. drug metabolism enzymes such as those responsible for cyclophosphamide activation in liver cells.

Explant cultures are useful as they provide cells with optimal cytodifferentiation and tissue cell interactions. They also provide insights into tissue structure development and pathological progression which are not possible *in situ*. A procedure for the preparation of tissue explants is described in Protocol 1. However, the main disadvantage of preparing tissue explants is the high rate of culture failure.

Protocol 1

Preparation of tissue explants

Equipment and reagents

- Sterile Petri dishes
- Vented 25 cm² tissue culture treated flasks
- Iredectomy scissors
- Sterilized stainless steel dissecting instruments
- Sterile tweezers
- Sterile surgical quality gloves
- Sterile surgical scalpels
- Eagles minimal essential medium (MEM) with penicillin and streptomycin (500 U/ml each), supplemented with 10% serum
- Hanks balanced salt solution (HBSS) without phenol red

Method

1. Carry out all work in a Class 2 laminar flow cabinet and use sterile surgical quality gloves.

2. Prior to the preparation of tissue explants, wash the samples with PBS to remove extraneous blood. Remove the fatty deposits, necrotic tissue, and blood vessels where possible using sterile surgical scalpels and scissors.

3. Cut the tissue fragments (>0.5 cm³) into smaller (2-4 mm³) fragments in HBSS (10 ml in a Petri dish).

4. Pre-wet the required number of 25 cm² flasks with 2.5 ml of MEM and remove the excess media. Transfer the small cubes prepared in step 3 to the growth surface of 25 cm² flasks (approx. 15–20 fragments of tissue/flask) using sterile tweezers or a scalpel.

5. Allow the tissue to adhere to the flask for approx. 30 min.

6. Taking care not to dislodge the explants, add 2.5 ml of growth medium.

7. Incubate the flasks at 37 °C and 5% CO_2 and periodically check for outgrowths of cells from the tissue explants.

To generate a primary culture, some disruption of the tissue architecture within which the cells exist must occur. Disruption of tissues can be performed with one of three methods: mechanical disaggregation, chemical disruption, and enzymatic digestion. Primary cultures are generally representative of the original tissue, with a mixture of cell types present. Such cell cultures can be subcultured to produce a cell line that may be cloned, characterized, and preserved. Subculturing allows greater uniformity of the cells and provides large amounts of consistent material for prolonged use.

4.1.1 Mechanical disaggregation

To mechanically disaggregate tissue, finely cut (to approx. 1 mm or less), mince, or shear the tissue through a sieve. Plate the resulting fragments into the desired culture media and examine, after a period of time, for the outgrowth of cells. This method is restricted to soft tissues such as the spleen.

4.1.2 Chemical disruption

To chemically disrupt the tissue, the divalent cations that are essential components of the adhesion junctions between the cells are removed by the addition of chelating agents such as EDTA. Chemical disruption is infrequently used alone and in cases where it is, it is limited to the dissociation of soft tissues that are loosely held together by a matrix. In general, chemical disruption is combined with enzymatic digestion of tissue (see Section 4.1.3).

4.1.3 Enzymatic digestion

Enzymatic digestion utilizes a range of different proteolytic enzymes to degrade the extracellular matrix, releasing cells from tissue. Generally it is necessary to perform some mechanical separation prior to the use of enzymes. Blood vessels, fatty and necrotic tissue is removed, and large tissue fragments are aseptically diced, in the presence of medium. In doing so, small tissue fragments (providing increased surface area for future enzymatic disruption) and single cells are released from the tissue into the medium. The medium, containing the single cells, may be transferred to a flask and incubated, where the cells may attach and grow. The smaller tissue fragments may by exposed to mixtures of enzymes to allow further dissociation of the tissue.

Enzymes used in tissue digestion include trypsin, collagenase, pronase, dispase, papain, deoxyribonuclease I, hyaluronidase, and elastase, alone or in combinations and possibly in association with EDTA. Trypsin, a pancreatic serine protease that acts on peptide bonds involving the carboxyl group of arginine and lysine, is considered the most potent of these enzymes. However, collagenase, which splits collagen's triple helix conformation, is reported as being the least traumatic of the enzymes for tissue disaggregation and is most commonly used today, usually in combination with another of the enzymes (2). Elastase is a serine protease with specificity for peptide bonds adjacent to neutral amino acids. The polysaccharidase, hyaluronidase, has specificity for the digestion of connective tissue. Papain is a sulfhydryl protease, dispase is a neutral metalloenzyme, and pronase has a mild enzymatic activity. Deoxyribonuclease I is used to degrade nucleic acid which may leak into the media as a result of cell damage during enzymatic digestion.

The choice of one enzymatic digestion over another is often arbitrary, although the nature of the tissue being digested must be considered. Epithelial tissue consists of closely packed and tightly bound epithelial cells. Generally, trypsin, collagenase, and hyaluronidase, used either individually or in combination, and/or in association with EDTA, is used in the enzymatic digestion of epithelial tissues. Connective tissue is composed of cells and extracellular fibres; collagenous, reticular, and elastic. Collagen is a major component of connective tissue, reticular and elastic fibres are generally less abundant. Careful consideration has to be given to the selection of an enzyme or combination of enzymes that will give the optimum yield and viability from the tissue type. The enzyme concentrations and ranges required for the disaggregation of tissues will vary according to tissue source. Preliminary assessment of the effect of various concentrations of each of the enzymes should be carried out before deciding on which to use. The release of most cells from tissue by enzymatic digestion may result in a problem with fibroblastic growth. It may be required to follow the digestion with cell separation or selective culture.

4.2 Finite life-span cell cultures

Certain primary cultures may be passaged for a finite number of population doublings before senescence occurs. The number of population doublings can vary significantly between different cell types. As high as 70 population doublings can be obtained with human diploid cell strains but usually the number of doublings is more limited in adult-derived or differentiated cell types. Normal tissues usually give rise to cultures with finite life-span.

4.3 Continuous cell lines

After a number of subcultures a cell line will either die or a population of cells can transform to become a continuous cell line. This occurs frequently in rodent cell cultures, but rarely in normal human cells (as opposed to cultures derived from tumours). The resulting cell lines will generally exhibit a combination of the following characteristics: smaller cell size, less adherence and more rounded, increased growth rate and higher cloning efficiency, increased tumorigenicity, a reduction in their serum dependency, variable chromosome complement, divergence from the donor phenotype, and loss of tissue-specific markers. It is not clear in all cases whether the stem line of a continuous culture pre-exists, masked by a finite population or arises during serial propagation. Lines of transformed cells can also be obtained from normal primary cell cultures by infecting them with oncogenic viruses or treating them with a carcinogenic chemical. It is very difficult to obtain a normal human cell line from a culture of normal tissue. In contrast, neoplasms from humans have been generated into many cell lines. It appears that the possession of a cancerous phenotype allows for the easier adaptation to cell culture, which may be due in part to the fact that cancer cells are aneuploid.

There are two main sources of cell lines:

(a) Cell lines established in one's own laboratory, from tissue cultures.

(b) Cell lines obtained commercially, e.g. from the ATCC (American Type Culture Collection) or the ECACC (European Collection of Animal Cell Culture).

Cultures may be supplied either growing in tissue culture flasks or frozen in a cryopreserved state.

5 Media requirements

In vitro growth of cell lines requires a sterile environment in which all the nutrients for cellular metabolism can be provided in a readily accessible form at the optimal pH, temperature, and atmospheric conditions for growth. The culture medium is produced so that it mimics the physiological conditions within tissues. Medium contains a number of essential elements required for the growth of the cells. These include an energy source, usually in the form of glucose, but fructose and galactose are also used. Inorganic salts provide osmotic balance and are involved in a physiological role (e.g. cofactors for enzymatic reactions, maintenance of membrane pumps). Other components include amino acids (essential and non-essential), lipids (essential fatty acids, glycerides, etc.), vitamins, cofactors, and a pH indicator such as phenol red.

Most cells grow at an optimum pH of 7.4. The pH of medium should therefore be maintained at this level. At a pH higher than 7.6, precipitation of components of the medium may occur. When phenol red is included in the medium, the medium turns purple at high pH and yellow in acidic conditions. During cell growth and metabolism, the pH of the medium will fall due to the build up of waste products such as lactic acid. Therefore, buffering of the culture medium is essential and is normally provided by sodium bicarbonate, which dissociates in solution releasing CO_2 and hydroxyl ions. Medium in which sodium bicarbonate is used as the sole buffer are generally used to grow cells in association with vented flasks and using CO_2 incubators. Cultures require a 5% CO_2/95% air environment for growth and this will also help to maintain the pH of the medium at the correct level. There are a number of synthetic buffer compounds available, the most popular of which is HEPES (*N*-2-hydroxyethyl piperazine-*N′*-2-ethanesulfonic acid). These are stronger buffering systems and may be used in association with closed, non-vented flasks. A particular buffering systems may not suit certain cells. HEPES may be toxic to certain cells, so it is advisable to compare the morphology and growth of a cell line using a number of different buffering systems before deciding on which system to apply to the medium. Alternatively, if cells are obtained from a culture collection or another research facility, the details for the optimum growth of the cells will, in general, be predetermined.

Serum is generally incorporated into the growth medium at concentrations varying from 5% to 20%, although certain production processes and experimental procedures require the use of serum-free conditions. Serum is a complex solution of albumins, globulins, attachment factors, growth promoters, and growth inhibitors. The majority of cell lines require the addition of serum to defined culture medium to stimulate growth and cell division but serum is subject to significant biological variation. The most common source of serum used in cell culture is of bovine origin and this may be of adult, newborn, or fetal origin. Serum-supplemented medium is the medium/complete medium referred to in all further protocols in this chapter, except where it specifically states that it is serum-free medium.

Media formulations vary in complexity and have been developed to support a wide variety of cell types (3), examples of which are:

(a) DMEM (Dulbecco's modified Eagles medium) supports the growth of a wide range of mammalian cell lines.

(b) RPMI 1640 is an enriched medium suitable for the growth of a wide range of mammalian cell lines. This medium is also suitable for the growth of human leukaemia cells.

(c) Ham's F12 was originally formulated for cloning Chinese hamster ovary cells.

(d) Eagles basal medium is suitable for primary mammalian cells and human diploid cells.

(e) Liebovitz's L-15 medium is used in carbon dioxide-free environments.

It is important to check the medium requirements of the cells (available when a stock is bought) against the nutrients present in the medium (data from supplier) to determine if the addition of extra nutrients is required. All components coming into contact with the cells must be sterile (by autoclaving or filter sterilization). Sterility checks must be carried out on medium prior to using with cells to ensure it is contaminant-free before use (see Section 9).

6 Serum-free medium

Due to significant biological variation in serum, the development and routine use of serum-free media is a high priority for cell culture from both an industrial and scientific point of view. There are a number of detailed reviews which outline the development of serum-free media (4–7). The preparation of serum-free media involves the selection of the best basal medium for a particular cell line and the addition of a selection of factors to the basal medium. These replace the complex mixture of factors present in serum while still adequately supporting the growth of the cells for which the media is required. These supplements include combinations of attachment factors (such as collagen, fibronectin, and laminin), growth promoters (such as epidermal growth factor, oestradiol, fibroblast growth factor, insulin, insulin-like growth factors, platelet-derived growth factor, and dexamethasone), transport proteins and detoxifying agents (such as albumin and transferrin), lipids (such as essential fatty acids, phospholipids, and triglycerides), and trace elements (such as Co, Cu, Fe, Mn, Mo, Ni, Se, Sn, Va, and Zn).

Serum-free medium is also commercially available from such companies as Sigma, Roche Biochemicals, Flow Laboratories, and HyClone. However, the constituents of these SFM are proprietary, which is not ideal when a defined medium is required for studying signal transduction and differentiation.

There are a number of different approaches to replacing fetal calf serum in the medium. The amount of the serum in the medium can be reduced and gradually replaced by defined components. This method of adaptation helps to maintain the original cell phenotype. Cells may also be cultured directly into SFM and any resulting clonal populations isolated.

There are a number of advantages to growing cells in serum-free conditions, including:

(a) Serum-free medium has a defined composition and so the interpretation of experimental data is unhampered by the effect(s) of the unknown factors present in serum.

(b) Selective growth of one cell type over another. Serum-free medium has been used for the selective growth of small cell carcinoma cells from specimens containing both malignant and non-malignant cells and other types of lung cancer (8).

(c) Less possibility of bacterial, viral, and mycoplasma contamination. Historically, serum has been a significant source of mycoplasma contamination of cell cultures.

(d) As it is a defined medium, there is minimal batch-to-batch variation, as compared to serum-supplemented medium.

(e) No growth inhibitors present in the medium, such as chalones, transforming growth factor-β, and glucocorticoids, as present in serum.

(f) Control of cellular differentiation. By changing the composition of the serum-free medium, proliferating cultures can be differentiated into cells possessing specific phenotypes and producing/expressing specific gene products (9).

(g) Lower protein levels present in the culture medium, resulting in fewer problems in purification of proteins/products from the cells.

However, there are also problems encountered with using serum-free media, including:

(a) Different cell types exhibit varying requirements for specific medium components (10). Generally, a specific serum-free medium formulation is not sufficient to support the growth of a broad range of cell lines.

(b) Cell growth is often slower in serum-free versus serum-supplemented medium (11).

(c) Cells are more susceptible to shearing damages and cannot be subcultured or frozen in the same way as cells grown in serum-supplemented medium.

(d) Adaptation of cells to serum-free media is often required.

(e) Changes in cellular phenotype may occur when cells are grown in serum-free medium. In general, phenotype alterations are associated with differentiation of the cells. For example, Najar *et al.* (12) found that human peripheral blood monocytes which were cultured in serum-free medium exhibited decreased expression of markers typical of monocytes and macrophages and adapted a phenotype of lymphoid dendritic cells. Under different experimental conditions, that ability of cells to differentiate/alter their phenotype may be advantageous.

7 Subculturing cells

7.1 Subculturing adherent cells

In order to maintain cell cultures in optimum conditions, it is essential to keep cells in the log phase of growth as far as practicable. Adherent cells will continue to grow *in vitro* until either they have covered the surface available for growth or they have depleted the nutrients in the medium. The usual doubling time for animal cells is 24–48 h. Cells kept for prolonged periods in the stationary phase of growth will lose plating efficiency and have a reduced overall quality of healthiness. Frequency of subculture is dependent on several factors, which will vary between cell lines. These include inoculation density, growth rate, plating efficiency, and saturation density. Once established in culture, some tumour cell lines do not produce attachment factors and remain in suspension either as single cells or clumps of cells. Examples of this cell type are lymphoblasts and cells of haemopoietic origin. There are two general types of culture method appropriate for adherent and non-adherent cells. The subculturing of adherent cells is described in Protocol 2.

Protocol 2

Subculturing of adherent cells

Equipment and reagents

- Sterile tissue culture flasks
- Trypsin/EDTA solution
- Growth medium

Method

1 Examine the condition of the cell monolayer using an inverted microscope with phase contrast capabilities and check the cells for healthiness and confluency. Examination of the media will also indicate if media requires changing to remove build up of waste products and to provide additional nutrients: a red–orange colour is normal but a change to orange–yellow indicates that the culture has become too acidic and is at risk of going into stationary phase and decline.

2 Discard the spent growth medium from the flask and wash the monolayer with a small volume (1–2 ml) of pre-warmed trypsin solution (0.25% trypsin, 0.02% EDTA in PBS).

3 Add an appropriate volume of the pre-warmed trypsin solution to the flask (enough to cover the base of the flask), and incubate at 37 °C, with periodic microscopic monitoring, to allow the cells to detach from the surface of the flask and from each other.

4 Deactivate the trypsin solution by adding an equal volume of complete media to the flask (trypsin inhibitors are present in the serum).

5   Remove the cell suspension from the flask and transfer to a sterile Universal container and centrifuge at 1000 g for 5 min.

6   Discard the supernatant from the Universal and resuspend the pellet in complete medium. Perform a viable cell count (see Protocol 5) and re-seed a flask with an aliquot of cells at the required density. The supplier of the cells generally provides information on the cells' growth requirements, including the plating density. A cell count may not always be necessary. If the cell line has a known split ratio (e.g. 1:2, 1:4, or 1:16) the appropriate dilution factor can be used. The volume of culture medium required in the tissue culture flasks is 5–7 ml for a 25 cm^2 flask, 15–30 ml for a 75 cm^2 flask, and 30–50 ml for a 175 cm^2 flask. The size of culture flask used depends on the number of cells required. Saturation densities for adherent cell lines range from about 10^5 cells/cm^2 for monolayers to 3–4 $\times$ 10^5 cells/cm^2 for cells forming multilayers.

7.2 Subculture of suspension cell systems

Lymphoblastoid and many other tumour cell lines do not require a surface on which to grow and therefore do not require proteases for subculturing. Non-adherent cells may be subcultured by a number of methods, including subculture by direct dilution, and subculture by sedimentation followed by dilution. The disadvantage of subculturing suspension cells by dilution is that cells produce metabolic by-products, which become toxic if allowed to accumulate in the medium. Initially, sufficient waste metabolites will be removed by diluting the cell cultures, allowing growth to continue, but the viability of the cells will gradually decline after a few subcultures. Surviving cells may also undergo selection pressures, resulting in altered characteristics. Subculture by sedimentation overcomes this problem of toxic metabolites but one must be careful not to over-dilute the cells, as this will increase the lag phase of the cells and may prevent them from dividing. It also removes any growth factors produced by the cells. The subculture of non-adherent cells is described in Protocol 3.

Protocol 3

Subculture of non-adherent cells by sedimentation

Equipment and reagents

- Sterile tissue culture flasks
- Growth medium

Method

1   Visually examine cultures for signs of cell deterioration and for cell density. Healthy cells will appear refractile, whereas dying cultures show cell lysis and appear shrunken. As in the case of adherent cell cultures, examination of the medium may also indicate the stage of growth.

Protocol 3 continued

2 Remove a volume of the cells for a viable cell count (see Protocol 5).

3 Centrifuge the suspension cells to form a pellet.

4 Discard the supernatant and resuspend the pellet using fresh medium.

5 Set up the required number of flasks. Add the appropriate volume of pre-warmed growth medium to the flask(s). Add the cells to the flasks and incubate the flasks at 37 °C. A typical growing density range for many suspension cells is 10^5 to 10^6 cells/ml, but again, consult the specific information supplied with each cell line.

6 The viability of suspension cells should not be allowed to fall below 90%. If the viability appears to be lower, this may indicate that the cells are being exposed to toxins in their growth medium. A combination of more frequent feeding (by dilution, as per the supplier's instructions or at an approximate dilution ratio of 1:3) or subculturing (by sedimentation) may be required, to enrich for a more viable culture.

7.3 Subculturing cells in serum-free medium

When trypsinizing cells in serum-free conditions, serum should not be used to inactivate trypsin as residues from the serum may be left in the cell suspension. To overcome this, trypsin inhibitor is used. For best results, cells should be in the log phase of growth when subcultured. The subculturing of cells in serum-free medium is described in Protocol 4.

Protocol 4

Subculturing cells in serum-free medium

Equipment and reagents

- Sterile tissue culture flasks
- Trypsin/EDTA solution: 0.25% trypsin, 0.01% EDTA in PBS
- Trypsin inhibitor
- Growth medium

Method

1 Examine the condition of the cell monolayer using an inverted microscope and ensure that the cells are healthy.

2 Remove the medium from the flask.

3 Rinse the cells with a small volume of 0.25% trypsin and remove this trypsin. The trypsin solution used in this procedure is not pre-warmed.

4 Add sufficient trypsin to cover the surface of the flask, approx. 3–5 ml for a 75 cm^2 flask.

Protocol 4 continued

5 Incubate the flask at room temperature, monitoring the state of detachment regularly. As soon as the cells have rounded up (the length of time required will depend on the cell type) gently tap the end of the flask to dislodge the cells.

6 Add a volume of trypsin inhibitor, which is sufficient to inhibit the trypsin. A volume of trypsin inhibitor at 1.388 mg/ml is sufficient to inactivate an equal volume of 0.25% trypsin solution. However, the concentration and volume require may vary depending on the manufacturer of the reagents, so refer to the manufacturer's instructions in all cases.

7 Transfer the cell suspension to a sterile Universal and centrifuge the cells at 1000 g for 5 min, or until a pellet of cells has formed.

8 Remove the supernatant and resuspend the cells in 1 ml of fresh serum-free medium or PBS. Add 9 ml medium. Centrifuge the cell suspension at 1000 g for a further 5 min.

9 Remove the supernatant and resuspend the cells in 5 ml serum-free medium.

10 Determine the cell count and re-seed the cells immediately at an appropriate density. The density may vary depending on the particular cell line and its ability to grow in serum-free medium, but in general, a cell density of 1×10^4 cell/cm^2 should suit most cells.

Protocol 5

Assessment of cell number

Equipment and reagents

- A Neubauer counting chamber
- 0.4% (w/v) trypan blue solution in PBS

Method

1 Cells are trypsinized, pelleted, and resuspended in medium. Add an aliquot of the cell suspension to trypan blue solution at a ratio of 5:1. Care must be taken when using and disposing of trypan blue as it is a possible carcinogen.

2 Incubate the mixture for 3–15 min at room temperature. Add an aliquot of the mixture to the chamber of a glass coverslip-enclosed cell counter. A Neubauer counting chamber is suitable, which has a depth of 0.1 mm and graduated squares, for ease of counting.

3 Count the cells in the 16 squares (1 mm^2 area) of the four corner grids of the chamber (see Figure 1). Multiply the average cell numbers per 16 squares by a factor of 10^4 (to convert to cell/ml) and the relevant dilution factor to determine the number of cells/ml in the original cell suspension.

$$\text{Number of cells/ml} = \frac{(\text{total number of cells counted} \times 10^4)}{\text{number of 1 mm}^2 \text{ areas counted}} \times \text{dilution factor}$$

Protocol 5 continued

4 Non-viable cells stain blue because the uptake of trypan blue is the result of membrane damage or low ATP levels. Viable cells exclude the trypan blue dye, as their membranes are intact, and remain unstained. On this basis, percentage viability can be calculated by dividing the number of viable cells by the total number of cells (i.e. viable and dead). Healthy cultures should be more than 90% viable. A lower viability indicates that either the cells are exposed to suboptimal growth conditions or have exhausted their nutrient supply.

8 Cryopreservation of cells

8.1 Cryopreservation of adherent cells

It is impractical for laboratories to maintain cell lines in culture indefinitely, as cell cultures will undergo genetic drift with continuous passage, and risk losing their differentiated characteristics. Therefore, it becomes necessary to store cell stocks for future use and as a back-up in case of contamination. There should be a limit placed on the number of passages any cell line undergoes before being discarded and replaced with new cells from the cryopreserved stock. Nearly all cell lines can be cryopreserved successfully in liquid nitrogen at $-196\,^{\circ}\mathrm{C}$. Cryopreservation is based on slow freezing and fast thawing, together with high

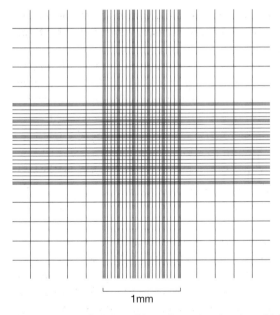

1mm

Figure 1 A magnified view of the cell counting chamber grid. The central 1 mm^2 square is surrounded by four corner squares, each of 1 mm^2 size, which are subdivided into 16 smaller squares. These corner squares are used when counting most cells. The central 1 mm^2 square is divided into 25 smaller squares, each 0.04 mm^2. These are enclosed by triple ruled lines and are further subdivided into 16 squares, each 0.0025 mm^2, to ensure accurate counting of very small cells (e.g. erythrocytes).

protein concentration (fetal calf serum), and the presence of a cryopreservative agent which increases membrane permeability. Cryopreservatives may be penetrative or non-penetrative. Penetrative cryopreservatives, such as dimethyl sulfoxide (DMSO) and glycerol, protect the cells against freezing damage caused by intracellular ice crystals and osmotic effects. Non-penetrative cryopreservatives, such as serum, protect the cells from damage by extracellular ice crystals. Cells should only be cryopreserved when healthy and in log phase of growth. It is important to check each batch of cryopreserved vials for viability, sterility, and maintenance of specific cell characteristics. Maintenance of accurate records of stored cryopreserved cells is very important and the cryovials should be clearly marked with the cell designation, passage number, and date of freezing. The cryopreservation of adherent cells is described in Protocol 6.

Protocol 6

Cryopreservation of cells

Equipment and reagents

- Cooling tray
- Liquid nitrogen
- Serum
- 10% DMSO/serum

Method

1 Cells for cryopreservation are harvested in the log phase of growth and counted as described in Protocol 5. If the cultures contain more than 20% of non-viable cells at this stage, it is advisable not to freeze them.

   Note: The final concentration of serum used in cryopreservation can vary from 20% to 100% and the concentration of DMSO can vary from 5% to 20%. The example described here uses a final concentration of 5% DMSO in 95% fetal calf serum (the serum routinely used when freezing cells), which is suitable for most cell types.

2 Resuspend the cells in a suitable volume of serum. Slowly, in a dropwise manner, add an equal volume of a 10% DMSO/serum solution to the cell suspension (this is toxic to cells if added too quickly), while at the same time gently swirling the cell suspension to allow the cells to adapt to the presence of DMSO.

3 Place 1 ml of this suspension (containing between 5×10^6 and 1×10^7 cells) into each cryovial.

4 Place the vials in the vapour phase of a liquid nitrogen container and allow to freeze for 3 h. An apparatus such as a sample tray attached to a liquid nitrogen container lid may be used for storing the cells in the vapour phase. The rate of cooling is determined by the depth of insertion of the sample tray in the neck of the liquid nitrogen container, controlled by a manually operated screw elevator. A cooling rate of approx. 1 °C per min is used for most cell types. Alternatively, the vials may be frozen by placing at -80 °C overnight.

5 Following freezing, transfer the vials to the liquid phase of liquid nitrogen (-196 °C) for long-term storage.

8.2 Cryopreservation of cells in serum-free medium

Cells routinely stored for indefinite periods of time in liquid nitrogen are normally frozen using a combination of the penetrative cryopreservative with serum. To freeze cells under serum-free conditions, other non-penetrative cryopreservatives are employed, e.g. polyvinylpyrrolidine (PVP) or methylcellulose (MC) (13, 14). A method for the cryopreservation of serum-free cells is described in Protocol 7.

Protocol 7

Cryopreservation of serum-free cells

Equipment and reagents

- Liquid nitrogen
- Serum-free medium
- 50% DMSO
- 10 × methylcellulose (MC): 1% (w/v) in PBS
- 10 × polyvinylpyrrolidine (PVP): 30% (w/v) in PBS

Method

1   Ensure that cells to be cryopreserved are not confluent and are in the log phase of growth.

2   Prepare 1% (w/v) MC (10 ×) or 30% (w/v) PVP (10 ×) in PBS and autoclave.

3   Dilute DMSO to 50% (w/v) in basal medium and filter sterilize.

4   Prepare 2 × freezing medium. Add 1 ml of 1% MC or 1 ml of 30% PVP to 2 ml serum-free medium, and add 2 ml of 50% DMSO.

5   Trypsinize cells as described in Protocol 4. Obtain a cell concentration of about 1×10^7 cells/ml.

6   Slowly, in a dropwise manner, add an equal volume of freezing medium to the cell suspension, while at the same time gently swirling the cell suspension to ensure even mixing. The required final concentrations of cryopreservatives are 0.1% MC or 3% PVP and 10% DMSO.

7   Aliquot cells into cryovials and slowly freeze in the vapour phase of liquid nitrogen for 3 h and then transfer the vials into the liquid phase ($-196\,°C$) for long-term storage.

The thawing of cryopreserved cells is described in Protocol 8.

Protocol 8

Thawing of cryopreserved cells

Equipment and reagents

- Centrifuge
- Tissue culture flasks
- Growth medium

Protocol 8 continued

Method

1   Add 9 ml of warm growth medium to a sterile Universal tube. Remove the cryo-preserved cells from the liquid nitrogen and thaw them at 37 °C. It is important to thaw the cryopreserved cells rapidly to ensure minimal damage to the cells by the thawing process and the cryopreservation agent, DMSO. Transfer the thawed cells to the aliquoted media.

2   Centrifuge the resulting cell suspension at 1000 g for 5 min to remove the cryo-protectant. Remove the supernatant and resuspend the cell pellet in fresh culture medium.

3   Carry out an assessment of cell viability on thawing (Protocol 5).

4   Add the thawed cells to a tissue culture flask with a suitable volume of growth medium. If the cells are anchorage-dependent, allow them to attach overnight and remove any dead or floating cells the next day. It is advisable to start cultures at between 30–50% of their final maximum cell density. This allows the cells to rapidly condition the medium and enter into the log phase of growth.

5   If thawing serum-free cells, the fresh culture medium used in step 2 should be the serum-free medium most appropriate for the particular cell line.

9 Monitoring of sterility of cell culture solutions

Sterility testing must be performed on all cell culture media and culture-related solutions. Samples of prepared basal media may be inoculated onto Colombia blood agar or tryptone soya agar (for the detection of aerobic bacteria, yeasts, and moulds), thioglycollate broths (for the detection of anaerobic bacteria), and Sabauraud dextrose agar (for the detection of yeasts and moulds). Details for their preparation come with the products and they are supplied by most suppliers of microbial media. The inoculated broths and agars are incubated at either 37 °C (aerobic and anaerobic bacteria) or 25 °C (yeasts and moulds). Complete cell culture media should be sterility tested at least four days prior to use.

10 Mycoplasma analysis of cell lines

Mycoplasma are common and very serious contaminants of cell cultures and remain one of the major problems encountered in biological research and diagnosis and in the biotechnology industry, where cell cultures are used. Mycoplasma contamination is widespread, with the incidence varying depending on the cell line, detection method used, and the quality control practised by the laboratory. Mycoplasma is the trivial name of the class *Mollicutes*, a group of minute, wall-less, self-replicating bacteria which can be as little as 0.3 μm in diameter. They are fastidious organisms, requiring complex growth medium. As a consequence of their small genome, mycoplasma lack many biosynthetic pathways and are dependent on their artificial culture media or host cells for the supply of many essential nutrients. An infection of cell cultures may persist for an extended time without apparent cell damage and can affect every parameter within a cell culture system such as changes in cell metabolism, alterations in cell karyotype, and rate of cell growth (15, 16).

As these organisms are not readily apparent, special techniques are required for their detection. A number of extensive reviews on mycoplasma contamination in cell cultures have been published, including detailed mycoplasma detection and identification protocols (17, 18). These methods include microbiological culture, fluorescent labelling of DNA, the detection of extranuclear DNA in a culture, immunodetection methods, electron microscope analysis, and molecular biology tests.

The choice of mycoplasma test method used and/or whether to perform the test in-house or to sub-contract the work will be determined by the facilities present in each organization. All mycoplasma testing, and handling of samples for mycoplasma testing, should be segregated from routing cell culture work and some procedures require specialized equipment. A number of PCR and ELISA kits are commercially available (from most companies that supply tissue culture and molecular biology reagents) (19) and should be used according to the manufacturer's instructions. Established cell repositories generally provide cell characterization services, including mycoplasma testing.

10.1 Collection of samples for mycoplasma testing

Samples of media, collected for mycoplasma testing, should be harvested and processed, so that they result in a maximum yield of viable mycoplasma and the absence of any growth inhibiting compounds. The sample of medium collected from the cell culture for mycoplasma testing should be taken from cells that are in the log phase of growth, near confluency, and from which the medium has not been removed for two or three days. The cells should be subcultured for three passages in antibiotic-free medium before mycoplasma testing as antibiotics may mask the mycoplasma infection. Samples should either be analysed immediately or stored at $-70\,°C$ until required.

10.2 Indirect staining procedure for mycoplasma analysis

This is a simple, inexpensive, and sensitive test for mycoplasma contamination, and is most effective in detecting the cytoadsorbing mycoplasma. The method, described in Protocol 9, involves growing a mycoplasma-free indicator cell line in medium which has previously been exposed to the cell line being investigated. The indicator cell line is then fixed and stained using a fluorescent DNA stain. The nucleus of the indicator cells fluoresce when viewed under UV light and any extranuclear fluorescence is an indication of possible mycoplasma contamination. In this procedure, the indicator cell line used is NRK, a rat kidney line. Vero cells (an African green monkey kidney cell line) and NIH 3T3 or 3T6 (mouse fibroblast cell lines) are also commonly used as indicator cells.

Protocol 9

Indirect staining procedure for the detection of mycoplasma

Equipment and reagents

- Fluorescent microscope
- Carnoy's fixative: 1 vol. glacial acetic acid, 3 vol. methanol
- Carnoy's/PBS: 1:1 (v/v) Carnoy's fixative/PBS
- Hoechst 33258[a]: 50 ng/ml

Method

1 Sterilize forceps by dipping in 70% alcohol and flaming, and place clean and sterile coverslips into 35 mm Petri dishes (one into each).

2 Trypsinize a growing culture of NRK cells. Resuspend at a concentration of 2×10^3 cells/ml in DMEM + 5% FCS + 1% L-glutamine. Add 1 ml of cell suspension to each coverslip. Incubate at 37 °C in a CO_2 incubator overnight to allow cell attachment.

3 When the cells have attached to the coverslips, add 1 ml of conditioned medium from the test samples to duplicate coverslips.

4 Thaw one aliquot of *Mycoplasma orale* (ATCC; 45539) and *Mycoplasma hyorhinis* (ATCC; 23234) positive cultures and dilute to the concentration required for the positive control (i.e. ensure the concentration used is such that the organism can be detected by the indirect test procedures). Add 1 ml of the mycoplasma cultures to duplicate coverslips. These are the positive controls. The NRK growth medium is used as the negative control.

5 Incubate the samples in a CO_2 incubator until the indicator cells are approx. 50% confluent (four to five days).

6 Remove the medium from the plates, wash the coverslips twice with sterile PBS, and once with Carnoy's/PBS.

7 Add 2 ml of ice-cold Carnoy's fixative to each coverslip and allow to fix for 10 min. Remove the fixative and allow the samples to air dry for at least 1 h.

8 When dry, add 2 ml Hoechst 33258 to each sample. Allow the samples to stain for 10 min. The stain is light-sensitive, so perform the staining in as little light as possible. Remove the stain and wash each coverslip thoroughly with 3×2 ml of sterile distilled water.

9 Mount the coverslips (top side down) on labelled slides using a drop of mounting medium. Store in the dark until ready to analyse.

10 Examine samples by fluorescent microscopy for mycoplasma contamination. Uncontaminated cells show only brightly fluorescing cell nuclei, mycoplasma appear as cocci or filaments in the cytoplasmic area.

[a] Hoechst is harmful; care should be exercised during use, and disposal should be in accordance with local regulations.

11 Cloning animal cells

The word 'clone' is used to describe a cell population that has been derived from a single cell. Such single cells are termed 'clonogenic' and only a small proportion of normal cells and probably a higher proportion of tumour cells have the proliferative capabilities required to give rise to clones. Both normal and neoplastic stem cells may be considered clones as they have the ability to renew themselves and produce descendants which differentiate. Clonogenic cells are detected *in vitro* by their ability to form individual colonies following isolation from the host tissue or from an established cell culture.

Colony-forming efficiency (CFE) is a measure of the ability of a cell population to form colonies (colonies formed/total cells seeded × 100%) and is often used as a means of comparing the tumorigenic potential of different cell populations. Cytogenetic and phenotypic heterogeneity is an important feature of tumour development and progression and has significant implications for the diagnosis and treatment of cancer. The ability to study individual tumour cells and descendants derived from a heterogeneous population is invaluable to the understanding of tumour cell biology. Cloning cells from cell lines following exposure to agents such as mutagens and toxins allows the establishment of useful variant populations. Cloning is also carried out on cells which have been transfected with DNA to yield genetically homogeneous populations and on hybridoma cells after fusion.

11.1 Isolation of clonal populations using cloning rings

This procedure is suitable for adherent cells with a CFE < 5%. The isolation of clonal populations using cloning rings is achieved by segregating specific colonies of cells. The objective of the cloning procedure is to selectively remove individual colonies arising from single cells, and maintain them in an isolated environment. This cloning method, described in Protocol 11, can only be used with strongly adhering cells.

Protocol 11

Isolation of clonal populations using cloning rings

Equipment and reagents

- Cloning rings: stainless steel or glass rings of 0.5–0.8 cm inside diameter and approx. 1 cm high; the base of the ring must be smooth and flat
- Silicone grease
- Forceps
- Microscope
- Growth medium

Method

1 Sterilize silicone grease and the cloning rings by baking at 120 °C overnight in separate glass Petri dishes.

Protocol 11 continued

2 Subculture the cells of interest and resuspend in suitable growth medium. The cells must be in a single cell suspension at this stage.

3 Dilute the cell suspension to a concentration that results in cells forming isolated colonies on the Petri dish. This concentration varies for each cell line and is dependent on the CFE of the cells.

4 Incubate the plates at 37 °C, 5% CO_2 in a humidified atmosphere and allow the cells to attach overnight. Examine the plates and note the location of attached cells. Areas where cells have attached as two or more cells are also noted and subsequently ignored.

5 Allow the cells to grow, re-feeding if required. Ensure cells have not detached and reattached as this may result in the mixing of colonies and the growth of non-clonal colonies of cells.

6 When the colonies have reached approx. 50 cells per colony, individual colonies are subcultured and transferred to a well of a 96-well plate as follows.

7 Remove the medium from the Petri dish and wash the cells with 5 ml sterile PBS.

8 Using sterile forceps, dip one end of the cloning ring in the silicone grease, ensuring the grease is evenly dispersed. Place the cloning ring around the colony of interest and with the edge of the forceps, press the cloning ring down to ensure a good seal between the bottom of the cloning ring and the Petri dish. Monitor the plate at this stage to ensure that the cloning ring encircles only the colony of interest.

9 Add approx. 100 µl of trypsin solution (0.25% trypsin, 0.02% EDTA in PBS) to the cloning ring and incubate the dish at 37 °C until the cells begin to detach.

10 Gently pipette the trypsin solution up and down to detach the remaining cells and transfer the solution to a well of a 96-well plate. Add 100 µl of growth medium to the cloning ring, pipette up and down, and remove the medium to the same well in the 96-well plate.

11 Incubate the 96-well plate overnight to allow the cells to attach and re-feed the wells with 200 µl of fresh medium.

12 Allow the cells to grow until they are approaching confluency, then subculture the cells into a larger dish.

13 Generate frozen stocks of the individual cell lines.

11.2 Cloning procedure with the limiting dilution assay

This cloning procedure, described in Protocol 12, is suitable for cells with a CFE > 5%.

Protocol 12

Cloning procedure with the limiting dilution assay

Equipment and reagents

- Microscope
- Tissue culture flasks
- Growth medium

Method

1 Dilute a cell suspension to produce a density of approx. 10 cells/5 ml of medium.

2 Aliquot 100 µl of suspension into each well of a 96-well plate and incubate at 37 °C, 5% CO_2.

3 Inspect the plate microscopically, and mark those wells that contain single cells.

4 Continue to inspect the plate every two or three days to ensure marked wells contain single colonies. The cells may need to be fed occasionally (usually about once a week).

5 When the marked wells are almost confluent, harvest the cells by trypsinization and seed into 6-well plates and then into 25 cm^2 and larger flasks.

6 Create frozen stocks of the clones and store in liquid nitrogen.

11.3 Semi-solid media cloning

This technique of cloning, described in Protocol 13, may be used to clone cells that grow in suspension. The normal growth medium is supplemented with agar or methylcellulose to produce a semi-solid medium that supports growth in suspension. Colonies are detected microscopically, isolated using a micropipette, and then cultured individually in soft agar or as monolayer cultures.

Protocol 13

Semi-solid media cloning

Equipment and reagents

- 35 mm Petri dishes
- Water-bath at 44 °C
- 2% agar
- Medium
- FCS

Protocol 13 continued

Method

1 Prepare the agar, autoclave, and allow to equilibrate to 44 °C in a water-bath.

2 Prepare the medium–agar solution. Mix 60 ml of medium (supplemented with 30% FCS) and 12 ml of the 2% agar. Dispense into 2.5 ml aliquots and equilibrate at 44 °C.

3 Prepare a single cell suspension in medium. The cell density will depend on the CFE of the cells. In general, prepare a range of cell suspensions ranging from 10^4 cells/ml to 10^1 cells/ml.

4 Add 0.5 ml of the cell suspension to the 2.5 ml medium–agar solution, mix, and pour into a 35 mm Petri dish immediately.

5 Place the Petri dishes on trays, which contain a small volume of sterile water to prevent drying out of the agar. Incubate the trays at 37 °C in a 5% CO_2 atmosphere.

6 Inspect the Petri dishes microscopically for colony formation. Using a micropipette, remove the individual colonies from the dishes and culture them in soft agar or in monolayer cultures.

12 The application of cell culture to *in vitro* toxicity testing

There are several reasons why the use of cell cultures in *in vitro* toxicity testing is useful to researchers and the pharmaceutical industry:

(a) Cell cultures can theoretically be cultivated from any species and tissue.

(b) Ethical problems connected with human experimentation are avoided as human tissues and cells may be utilized in *in vitro* toxicity testing.

(c) Cellular and molecular mechanisms are easily explored. Studies may be carried out on cellular models *in vitro* to determine specific toxic mechanisms or to elucidate organ-specific toxic effects. *In vitro* cytotoxicity assays have been developed to screen synthetic and natural compounds with unknown mechanisms of action, in order to determine possible injurious effects before they are used in the human body. These assays are utilized to identify the range of doses of the test substance.

(d) Cell damage may be identified and investigated at an early stage.

(e) The need for animal experiments is reduced.

(f) The technology is relatively simple with the possibility of automation.

(g) They are generally less costly and time-consuming than *in vivo* methods.

(h) It is possible to carry out a large number of *in vitro* assays with relative ease and there is good reproducibility of results.

(i) They often require smaller quantities of test compounds.

There are, however, some disadvantages to *in vitro* toxicity testing:

(a) Isolated cells are a very simplified system with respect to the complexity of the complete organism.

(b) The results obtained from *in vitro* toxicity testing may not truly reflect the cellular mechanisms within the human or animal body, as cellular models cannot consider the effects of the extracellular environment.

(c) The cellular models can only be used to investigate very specific and single mechanisms.

(d) Cells in culture may lose a great deal of their detoxifying/activating properties. Therefore, *in vitro* methods can miss toxicity associated with a compound being processed from a pro-drug to the active form of the drug *in vivo*. Also, in cases where the metabolite of the compound is more toxic than the specific test compound, the *in vitro* methods will not be a true reflection of the activity of the compound *in vivo*.

(e) There are difficulties in correlating the *in vitro* situation to the *in vivo* one both in terms of experimental design and results.

12.1 Miniaturized *in vitro* methods for toxicity testing

Due to their simplicity and sensitivity, *in vitro* miniaturized colorimetric end-point assays are used extensively in the determination of a substance's ability to enhance cell growth or promote cell death. These *in vitro* assays involve the determination of cell number, the most common measure of cell growth, after the cells have been treated with the test substance for a specified period of time. The most commonly used measure of toxicity is impairment of cell growth, generally determined by reduced cell numbers, as compared to untreated controls. Impaired cell growth may be caused by either cytostatic agents, where growth may be affected in a temporary manner and is reversible upon removal of the cytostatic agent, or cytotoxic agents, which cause irreversible cell damage, leading to cell death by an apoptotic or necrotic pathway. Both agents result in impaired cell growth, and will exhibit similar effects in colorimetric end-point toxicity assays.

There are a number of points to be considered when carrying out *in vitro* toxicity testing including:

(a) The cells to be assayed must be healthy and in the log phase of growth. If cells are in the lag phase, toxins that affect the cell cycle can be missed and metabolism is naturally slowed down.

(b) In most toxicity testing protocols cell lines are used due to their relative stability, generally good characterization, and easy handling. Consideration, however, must be given to the fact that many detoxifying and/or activating systems may be missing from the cells chosen.

(c) The compatibility of the compound under study with the culture conditions should be investigated in advance to avoid unwanted alterations of the culture medium that may alter the exposure conditions.

(d) As some agents may be toxic to the cells at all stages of the cell cycle, while others are cell cycle-specific, it is necessary to design experiments so that the cells are exposed to the test substance for sufficient time for the toxic effect to occur.

(e) Incubation times must be considered to allow the consequences of the exposure to become apparent in the assay system.

(f) The methods of end-point determination must be considered when setting up a miniaturized colorimetric assay system. Choose the best system for the analysis of cell number under the particular experimental conditions. In the initial calibration of the assay system, it is important to confirm by an independent method that the assay system is giving a true and proportional reflection of the end-point.

(g) It is important to demonstrate that the cell line being used in the assay has a suitable level of sensitivity and that a sufficient range of cell number can be analysed. The sensitivity and linear range for the colorimetric assays may vary depending on both the cell line being analysed and the colorimetric end-point being employed. The sensitivity of the assay end-point is offset against a range of linearity of optical density (OD) to cell number (20).

The acid phosphatase (AP) assay, the protein staining assays, and the neutral red (NR) assays are more sensitive than the MTT in determining cell number (21).

All of the assays described below are set up in 96-well plates and cell number is evaluated colorimetrically using a microplate reader.

Protocol 14

Setting up miniaturized *in vitro* assays

Equipment and reagents

- Tissue culture plates
- Incubator
- Cell culture medium
- Trypsin solution

Method

1   Set up the cell line to be tested two days prior to setting up the assay to ensure that the cells are 70–80% confluent and in the log phase of growth on the day of the assay.

Day 1

2   Trypsinize one, subconfluent flask of cells. Resuspend the cells in medium and perform a cell count.

3   Prepare cell suspensions containing 1×10^4 cells/ml in cell culture medium. The assays are devised to last for seven days so the initial amount of cells seeded can be altered depending on the growth characteristics of the cell line being used. Add 100 µl aliquots of these cell suspensions to each well of the 96-well plate. Gently agitate the plate in order to evenly disperse the cells over each well. Incubate the cells overnight at 37 °C in an atmosphere containing 5% CO_2.

Protocol 14 continued

Day 2

4   Examine the plates to ensure that the cells have attached to the base of the wells and are viable. Add the toxin(s) to be tested. The final volume in each well is 200 μl. If adding one toxin only, the solutions are made to a concentration twice that of the final desired solution, to take account of the diluting effect of adding 100 μl of chemical to 100 μl of cell suspension already in the well. When all the solutions have been added, gently agitate the plates to ensure adequate mixing and incubate at 37 °C in an atmosphere containing 5% CO_2.

Day 7

5   Remove the plates from the incubator and process as indicated for the relevant assay.

12.2 MTT assay

MTT, a soluble yellow dye, is metabolized by the enzyme succinate dehydrogenase in the mitochondria of metabolically active cells to a dark blue insoluble formazan product, which can be solubilized and the colour produced measured (22). This is the most widely used end-point determinant of cell number/cell growth. However, the sensitivity of this assay (described in Protocol 15) is quite poor. Furthermore, MTT is reduced to a greater extent by growing cells (23), and there are differences in the activity of MTT reducing enzyme among cell lines, which may lead to differences in assay sensitivity in various cell lines.

Protocol 15

MTT assay

Reagents

- MTT (tetrazolium dye): 3-(4,5-dimethylthiazol-2-yl)-2,5-diphenyl tetrazolium bromide (Sigma) is dissolved in PBS at a concentration of 5 mg/ml
- PBS
- DMSO

Method

1   At the end of the incubation period add 20 μl of 5 mg/ml MTT to each well and incubate for a further 4 h.

2   Remove the medium from each well without disturbing the formazan crystals.

3   Rinse the wells with PBS.

4   Add 100 μl of DMSO to each well and agitate to give a homogeneous colour. Measure absorbance at 570 nm.

12.3 Acid phosphatase assay

The acid phosphatase (AP) assay, described in Protocol 16, is based on the ability of the AP enzyme in the lysosomes of cells to hydrolyse p-nitrophenyl phosphate to yield the p-nitrophenyl chromophore (20). The AP assay has high sensitivity but a low range of linearity between OD and cell number.

Protocol 16

Acid phosphatase assay

Equipment and reagents

- Microplate reader
- PBS
- 1 M NaOH
- Substrate-containing buffer: 10 mM p-nitrophenyl phosphate (Sigma), in 0.1 M sodium acetate pH 5.5, containing 0.1% Triton X-100

Method

1 Remove the medium from the cells and rinse the cells once with 100 μl of PBS.

2 Add 100 μl of freshly prepared substrate to each well and incubate plates at 37 °C for 2 h.

3 Add 50 μl of 1 M NaOH to each well. This causes an electrophilic shift in the p-nitrophenyl chromophore and develops the yellow colour. Read the absorbance at 405 nm on a microplate reader.

12.4 Neutral red assay

This assay, described in Protocol 17, is based on the accumulation of neutral red (NR) (3-amino-7-dimethyl-2-methylphenazine hydrochloride) in the lysosomes of viable cells. The NR dye does not accumulate in the lysosomes of dead or damaged cells (24, 25). This assay is generally more sensitive than MTT, but less sensitive than crystal violet dye elution, acid phosphatase, or sulforhodamine B.

Protocol 17

Neutral red assay

Reagents

- NR dye: dissolve 50 μg/ml in medium, centrifuge at 1500 g for 10 min, and retain the supernatant
- Acetic acid/ethanol mixture: 1% glacial acetic acid in 50% ethanol
- Formol/calcium mixture: 4% formaldehyde, 1% anhydrous calcium chloride in water

---

Protocol 17 continued

Method

1 Remove the medium from the plates and rinse each well with 200 μl of PBS.

2 Add 200 μl of freshly prepared NR solution and incubate for a further 3 h.

3 Remove the NR solution and wash the wells with 50 μl of the formol/calcium mixture.

4 Add 200 μl of the acetic acid/ethanol mixture to elute the dye, agitate, and measure absorbance at 570 nm.

12.5 Protein staining assays

The crystal violet dye elution assay, described in Protocol 18, is a protein staining assay in which cells are fixed with formalin and stained with crystal violet (26). The sulforhodamine B assay, described in Protocol 19, is also a protein binding dye assay in which cells are fixed in trichloroacetic acid before staining with the dye (27). Both these methods are very sensitive but exhibit a loss of linearity of OD versus cell number at higher densities.

Protocol 18

Crystal violet dye elution

Reagents

- PBS
- 10% formalin
- 0.25% aqueous crystal violet (Sigma)
- 33% glacial acetic acid

Method

1 Remove the medium from the 96-well plates and rinse each well with 100 μl of PBS.

2 Fix the cells by adding 100 μl of 10% formalin for 10 min. Decant and allow the plates to air dry.

3 Add 100 μl of 0.25% aqueous crystal violet for 10 min. Rinse four times with water and allow to dry.

4 Solubilize the colour by addition of 100 μl of 33% glacial acetic acid and measure absorbance at 570 nm.

Protocol 19

Sulforhodamine B assay

Reagents

- 50% trichloroacetic acid (TCA) (toxic, handle with care)
- Sulforhodamine B (SRB): 0.4% solution dissolved in 1% acetic acid (Sigma)
- 1% glacial acetic acid
- 10 mM Tris buffer pH 10.5

Protocol 19 continued

Method

1 Fix the cells by layering 50 µl of 50% TCA directly on top of the incubation medium. Incubate the plates for 1 h at 2 °C.

2 Rinse the plates with water to remove solutes and allow to dry.

3 Stained with 200 µl of SRB for 30 min and rinse in 1% glacial acetic acid.

4 When dry, add 200 µl of 10 mM Tris buffer pH 10.5 to release the bound dye. Mix the resulting solution and measure absorbance at 570 nm.

12.6 Growth stimulation

The above protocols will test for the ability of compounds to inhibit cell growth or kill cells. The procedure (Protocol 14) may be modified to test for growth stimulation properties of test compounds, using the same colorimetric end-point determination. In a case where growth stimulation is being investigated, the cell plating density may need to be reduced and the control wells of the 96-well plate (no test compound added) will have suboptimal growth, with higher levels of growth, and therefore colour intensity, being present in the test wells. There is great variation in the growth stimulation properties of different batches of serum to particular cell lines. As described in Protocol 20, the miniaturized *in vitro* assays may be used to test batches of serum, allowing selection of the optimum serum for a particular cell line

Protocol 20

Serum screen, using miniaturized *in vitro* assays

Equipment and reagents

- Tissue culture plates
- Incubator
- Cell culture medium
- Trypsin solution

Method

1 Dilute the serum batches to twice the final concentration required. Use medium that has been supplemented with all additives except the serum for the dilutions. Include, as a control, the currently used batch of serum.

2 Aliquot 0.1 ml of the diluted serum into eight replica wells of the multiwell plate.

3 Trypsinize one, subconfluent flask of cells. Resuspend the cells in medium lacking serum and perform a cell count. Prepare cell suspensions containing 1×10^4 cells/ml in medium lacking serum. Add 0.1 ml of the cell suspensions to each well of the 96-well plate. Agitate the plate gently in order to ensure even dispersion of cells over a given well. Incubate the cells at 37 °C in an atmosphere containing 5% CO_2.

Protocol 20 continued

4 Microscopically examine the plate frequently. When any particular serum has resulted in approx. 90% cell confluency in a well, stop the assay.

5 Remove the plates from the incubator and monitor cell growth by any of the colorimetric end-point determinants (Protocol 15–19).

Additional reading

1. Freshney, R. I. (1986). *Animal cell culture: a practical approach*. IRL Press, Oxford.
2. Doyle, A., Griffiths, J. B., and Newell, D. G. (1998). *Cell and tissue culture: laboratory procedures*. John Wiley & Sons.
3. Spier, R. E. and Griffiths, J. B. (1985). *Animal cell biotechnology*. Academic Press Inc.
4. Clynes, M. (1998). *Animal cell culture techniques*. Springer.
5. Griffiths, B. and Doyle, A. (1997). *Cell culture: essential techniques*. John Wiley & Sons.

References

1. Gey, G., Coffman, W. D., and Kubiech, M. T. (1952). *Cancer Res.*, **12**, 264.
2. Leibovitz, A. (1986). *Cancer Genet. Cytogenet.*, **19**, 11.
3. Ham, R. G. and McKeenan, W. L. (1979). In *Methods in enzymology* (ed. W. B. Jaxoby and I. H. Pastan), Vol. 58, pp. 44–93. Academic Press, New York.
4. Barnes, D. and Sato, G. (1980). *Anal. Biochem.*, **102**, 255.
5. Hewlett, G. (1991). *Cytotechnology*, **5**, 3.
6. Bjare, U. (1992). *Pharmacol. Ther.*, **53**, 355.
7. Sandstrom, C., Miller, W., and Papoutsakis, E. (1994). *Biotech. Bioeng.*, **43**, 706.
8. Carney, D. N., Bunn, P. A., Gazdar, A. F., Pagan, J. F., and Minna, J. D. (1981). *Proc. Natl. Acad. Sci. USA*, **78**, 3185.
9. Rahemyulla, F., Moorer, C. M., and Wille, J. N. (1989). *J. Cell. Physiol.*, **140**, 98.
10. Zirvi, K. A., Chee, D. O., and Hill, G. J. (1986). *In Vitro Cell Dev. Biol.*, **22**, 369.
11. Ahearn, G. S., Daehler, C. C., and Majumdar, S. K. (1992). *In Vitro Cell Dev. Biol.*, **28A**, 227.
12. Najar, H. M., Bru-Capdeville, A. C., Gieseler, R. K., and Peters, J. H. (1990). *Eur. J. Biol.*, **51**, 339.
13. Merten, O. W., Peters, S., and Couve, E. (1995). *Biologicals*, **23**, 185.
14. Ohno, T., Kurita, K., Abe, S., Eimori, N., and Ikawa, Y. (1988). *Cytotechnology*, **1**, 257.
15. Boyle, J. M., Hopkins, J., Fox, M., Allen, T. D., and Leach, R. H. (1981). *Exp. Cell Res.*, **132**, 67.
16. Sasaki, T., Shintani, M., and Kihara, K. (1984). *In Vitro*, **20**, 369.
17. Kehane, I. and Adoni, A. (ed.) (1993). *Rapid diagnosis of mycoplasma*. Kluwer.
18. Tully, J. and Razin, S. (ed.) (1983). *Methods in mycoplasma*, Vol. IV. Academic Press.
19. Raab, L. S. (1999). *Scientist*, **13**, 21.
20. Martin, A. and Clynes, M. (1991). *In Vitro Cell Dev. Biol.*, **27A**, 183.
21. Martin, A. and Clynes, M. (1993). *Cytotechnology*, **11**, 49.
22. Vistica, D. T., Skehan, P., Scudiero, D., Monks, A., Pittman, A., and Boyd, M. R. (1991). *Cancer Res.*, **51**, 2515.
23. Mosmann, T. (1983). *J. Immunol. Methods*, **65**, 55.

24. Fiennes, A. G., Walton, J., Winterbourne, D., McGlashan, D., and Hermon-Taylor, J. (1987). *Cell Biol. Int. Rep.*, **11**, 373.
25. Triglia, D., Braa, S. S., Yonan, C., and Naughton, G. K. (1991). *In Vitro Cell Dev. Biol.*, **27A**, 239.
26. Scragg, M. A. and Ferreira, L. R. (1991). *Anal. Biochem.*, **198**, 80.
27. Skehan, P., Storeng, R., Scuderio, D., Monks, A., McMahon, J., Vistica, D., *et al.* (1990). *J. Natl. Cancer Inst.*, **82**, 1107.

Chapter 2
Basic light microscopy

Timo Zimmermann, Rainer Pepperkok,
David Stephens, Andreas Girod, and
Jens Rietdorf
Advanced Light Microscopy Facility, EMBL Heidelberg,
Cell Biology/Cell Biophysics Program, Meyerhofstr. 1,
69117 Heidelberg, Germany.

1 Introduction

There are a multitude of introductions to the basic principles of light microscopy reflecting on one hand the importance of the subject and on the other the fact that light microscopy has been used as a tool in different areas of research for a long time, and is constantly being developed further. The general approach for explaining light microscopy begins with considerations of the nature of light, properties of waves and lenses, the laws of diffraction, etc. and ends on the actual technical details of microscope usage.

However, we decided to approach this text from the side of the sample, for the principal reason that the direction of thinking of the researcher usually comes from this. Biological scientists will usually raise questions such as where the cell/protein localizes, where and when it moves, whether it co-localizes or interacts with something, rather than asking what he or she could do with a certain type of lens.

Given that biological microscopy does not start at the microscope but on the preparation of the sample, and does not end at the microscope but on the image in a manuscript or presentation, the overall idea of this article is to help biologists or clinical scientists to optimally employ microscopy techniques by:

(a) Emphasizing the importance of good sample preparation.

(b) Giving practical advice and 'rules of thumb' as to the use and limitations of different microscopy techniques.

(c) To address image acquisition and image processing topics.

2 The sample

In this section we briefly outline the importance of the sample preparation for the overall success of microscopic imaging. We therefore start with the important question of what we wish to observe (Section 2.1). This paragraph aims to provide an understanding of some of the underlying principles of microscopy without understanding each and every part of the microscope. We also give some protocols for optimized fixation and staining of samples.

We restrict ourselves to a discussion of the use of fluorescent dyes, since these allow for a highly specific detection of small amounts of dye molecules and thus are best suited to address the typical questions of a modern day microscopist. There is the possibility of obtaining high contrast images without the addition of dyes but by making use of the optical properties of the unstained specimen which is covered in Section 3.

2.1 Volume and length measurements and principal problems of sample preparation

In the biomedical research, light microscopy techniques are used to investigate objects of different sizes ranging from macromolecules in the nanometre range, such as ribosomes, up to whole embryos or tissues of almost 1 mm diameter. Some examples are listed in Figure 1. Even larger than the difference in lengths, in these examples spanning six orders of magnitude, is the difference in volumes, which is 18 orders of magnitude. It is important to estimate the volume of the

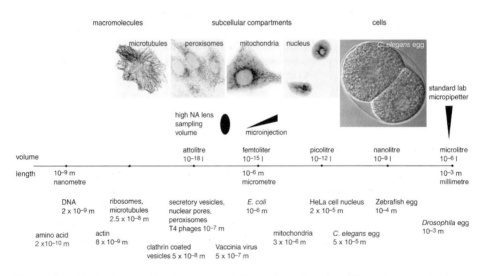

Figure 1 Important volume and length measures in biomedical microscopy. The entire span of objects for biomedical microscopy from macromolecules to whole cells is shown to give a feeling for the differences in size and volume. The micrographs show typical subcellular structures of different size stained with specific dyes or dye-conjugated antibodies and a DIC image of an *C. elegans* embryo. All micrographs are of the same scale. The 'high NA lens sampling volume' indicates the approximate focus volume of lenses used for fluorescence microscopy. Typical volumes for microinjection and micropipetting are indicated as well. Underneath the bar (log scale) there is a collection of biological structures ordered by their length or diameters.

structure of interest since there are several implications regarding methods to visualize the structure and separate it from structures nearby.

If one assumes a physiological concentration of dye of 1 μM inside the labelled structure, and if the structure itself is only 200 nanometres in diameter, then there can only be at most 30 molecules of dye inside. Depending on the lens and the dye used, detection may therefore become difficult. Using lenses that allow for best discrimination of fine detail—lenses with high numerical apertures (NA) (see below)—only light emerging from a small volume of 20–50 attolitres (the 'high NA lens sampling volume' in Figure 1), would enter the lens and contribute to the corresponding point inside a sharp image. Light emitted from the molecules inside this small focal volume is emitted in all directions and only a small part of it enters the objective lens and is detected. The final efficiency of detection after losses on optical surfaces inside the microscope, depending on the detector used, may be as small as 10% of the emitted light (1).

From the above points it becomes clear that it is critical to optimize staining in the first place to obtain a suitable micrograph. However, a very high concentration of dye may harm the function of the structure in question for live cell observation, or may not be targeted correctly causing a high background of non-specifically stained structures. Therefore it is important to choose a dye that has a high efficiency in absorbing light and re-emitting it. Among the best dyes currently available are the Alexa (www.probes.com) and the Cy- dyes (www.apbiotech.com). These outcompete 'classical' dyes like fluorescein in terms of photostability and efficiency by far, and are available in different variants that cover the entire spectrum of visible light. Other options beyond changing the fluorophor are to enlarge the detection volume by changing to a different lens, or sum the signal from larger areas of the object. This can only happen at the cost of the ability to resolve fine detail as we will see in Section 3. Increasing the exposure time or the illumination intensity on the other hand may harm the sample by photobleaching.

2.2 Sample preparation

With the need to investigate minute volumes inside the specimen in mind, and being aware of the fact that there are already tremendous losses from the light produced by the few dye molecules inside the small focal volume of the lens, it becomes critical to avoid additional shortcomings by an insufficient sample preparation. There are two fundamentally different situations to face: looking at fixed (Section 2.2.1) and non-fixed (Section 2.2.2) samples. While restricting the following section to cultured cells, the principles and considerations are similar for tissues or embryos.

2.2.1 Fixed samples

In order for the cellular structures to be adequately conserved, yet made readily available for labelling with specific dyes or antibodies, the reactivity of the cellular targets and epitopes should not be affected by the fixation procedure. To achieve this, an optimal fixation—and for immunocytochemistry, permeabilization—method should be selected. Conditions should be optimized for:

(a) The physical properties of the structure under study (for example actin filaments or microtubules).

(b) The specific cell type under study (some cells will not stay attached following light fixation).

(c) The particular antibody being used.

A positive reaction in the labelling may depend critically on the procedure used for fixation of the cells (2). The choice of a fixation protocol should also be made according to the subcellular localization of the structure to be labelled. Structures such as microtubules, which are in dynamic equilibrium with a soluble pool of tubulin, are best visualized after a brief extraction of the cells with a buffer containing detergent. This eliminates the soluble 'background'. On the other hand, antigens associated with membranes often cannot be visualized after pre-extraction with detergent.

In this section, we will describe several procedures including fixation with paraformaldehyde, glutaraldehyde, and methanol/acetone.

For all the protocols described below, cells should be grown on glass coverslips. Preparation of stock solutions is described in the Appendix to this chapter.

i. Paraformaldehyde fixation

Fixation of cells by paraformaldehyde is used when cell surface staining is required. It is also a good choice for labelling of membranes and microfilaments, although microtubules are not fixed well by it.

Protocol 1

Paraformaldehyde fixation

Reagents

- 3% paraformaldehyde: see Appendix for its preparation
- Phosphate-buffered saline (PBS): see Appendix for its preparation
- 30 mM NH_4Cl in PBS (or 30 mM glycine in PBS)

Method

1   Remove glass coverslips containing cells from a Petri dish and place (cells up) in 3% paraformaldehyde at RT (room temperature) for 20 min.

2   Aspirate the paraformaldehyde and wash with three changes of PBS at RT over ~15 min (when adding the PBS be careful not to pipette directly onto the coverslips).

3   A second wash with 30 mM NH_4Cl in PBS (or 30 mM glycine in PBS) for 10 min quenches the residual paraformaldehyde crosslinking activity.

For surface labelling do not permeabilize. In this way only antigen domains accessible from the outer side of the plasma membrane will be stained. Otherwise continue with Protocol 4. For the labelling of fixed cells with antibodies continue as described in Protocol 5.

ii. Glutaraldehyde fixation

Using glutaraldehyde for fixation typically preserves subcellular structures better than other fixation methods such as paraformaldehyde. It is however difficult to quench the autofluorescence of glutaraldehyde completely, which results in a high background fluorescence. As a consequence, antigens of low abundance will yield only weak fluorescent staining signals. In addition, many antigens lose their antigenic reactivity following glutaraldehyde fixation.

Protocol 2

Glutaraldehyde fixation

Reagents

- 1% glutaraldehyde (8% stock; Polyscience, EM grade)
- Phosphate-buffered saline (PBS): see Appendix for its preparation
- 0.5 mg/ml NaBH$_4$: made fresh in PBS
- 1 M glycine pH 8.5

Method

1 Remove coverslips from Petri dish and place (cells up) in 1% glutaraldehyde in PBS at RT for 20 min.

2 Aspirate the glutaraldehyde and wash with three changes of 0.5 mg/ml NaBH$_4$ (always made up fresh in PBS) for 4 min each to reduce Schiff bases (which are fluorescent) to secondary amines (which are not).

3 Wash three times with PBS (include two drops of 1 M glycine pH 8.5 in the second wash).

4 Continue with permeabilization and staining of cells as required (described in Protocols 4 and 5).

---

iii. Methanol and methanol/acetone fixation

These two protocols cannot be applied for surface staining since methanol fixation also simultaneously permeabilizes the cell membrane. Good results are obtained for nuclear antigens, microtubule and intermediate filament labelling, as well as for labelling of intracellular organelles like the Golgi apparatus or endoplasmic reticulum. Very often antigen accessibility is better with this fixation method, however it should be noted that the 3D architecture is slightly flattened by it.

Protocol 3

Methanol/acetone fixation

Reagents

- Methanol at −20 °C
- Acetone at −20 °C
- PBS

Method

1 Remove the coverslip from the Petri dish, dry off excess medium, and place in cold methanol at −20 °C for 4 min.

2 Remove coverslips and immediately dip in acetone at −20 °C. After 4 min transfer into PBS or air dry on filter paper at RT (these coverslips can be stored for several days at −20 °C). Extraction in acetone is not always necessary; therefore, this step may be omitted.

3 For the labelling of fixed cells with antibodies continue as described in Protocol 5.

2.2.2 Permeabilization

Protocol 4

Permeabilization

Reagents

- 0.1% Triton X-100 in PBS
- 0.05% saponin in PBS
- PBS

Method

1 Permeabilize the cells by incubation in 0.1% Triton X-100 in PBS for 4 min at RT.

2 Alternatively permeabilize cells using 0.05% saponin in PBS. Different antigen–antibody combinations will yield better results following one or other of these two treatments.

3 Wash with PBS (twice).

2.2.3 Specific labelling

A good source for dyes and labelled antibodies is Molecular Probes (www.probes.com) or Amersham (www.apbiotech.com) for Cy- dyes. For use of specific stains we recommend following the manufacturer's specific protocols. Whenever fluorescent molecules are used, keep in the dark. For a standard antibody labelling follow Protocol 5.

Protocol 5

Specific labelling

Equipment and reagents

- Parafilm
- PBS
- Antibodies
- Embedding media, e.g. Mowiol

Method

1   Tape a piece of Parafilm to the bench and pipette onto it a 50 μl drop of PBS containing the dilution of first antibodies for each coverslip. Remove the coverslips from the PBS wash buffer and dry off excess liquid by touching the edge of the coverslips on filter paper or tissue. Place the coverslips, cells downward, onto the drops and leave for 20–60 min at RT.

2   Gently squirt a little PBS at the edge of each coverslip to raise them up from the Parafilm. Transfer back into PBS and wash twice for 5 min.

3   If the first antibody was not labelled directly, place coverslips on Parafilm on 50 μl drops of PBS containing the secondary antibody. Incubate for 20–60 min.

4   Repeat step 2.

5   Remove the coverslips, dry off excess PBS with filter paper, and mount on glass slides by placing cell-side down on a 5 μl drop of Mowiol or other embedding media (see Section 2.2.4). If a polymer mounting is used allow polymerization before observation under the fluorescence microscope.

2.2.4 Mounting media

For longer-term storage and conservation of the structure during observation on the microscope there are a variety of different mounting media commercially available. An important consideration in the choice of the mounting medium is its refractive index (RI) a term explained in more detail in Section 3. Briefly, the RI of the mounting agent should match the RI of the immersion fluid of the objective to avoid distortions of the beam passing through the coverslip, sample, and objective. These distortions will lead to false estimations of the length of the sample along the z-(optical) axis and to losses of light. The effect will be more pronounced at a distance of more than 10 μm from the coverslip.

There are four categories of mounting media: aqueous, glycerol-based, natural resins/oils, and plastics. The refractive indices therefore cluster near water (1.34), glycerol (1.47), natural oils (1.53), and plastics (1.51).

A typical commercial medium includes one of the above as its base, plus a polymerizing compound that makes it solidify, plus an antifade compound such as DABCO, n-propyl gallate, or ppd. The base ingredients will largely determine the refractive index.

DABCO: 1,4-diazobicyclo-(2,2,2) octane.

ppd: para-phenylene diamine dihydrochlorate.

For Gel Mount and Fluorsave, the base ingredient is water. For Vectashield, ProLong, and Mowiol, the base ingredient is glycerol. For Permount, methyl salicylate, or Balsam oil, the base ingredient is natural oil. For glycol methacrylate, epoxy resins (Epon, Spurr's), or polyester resins (DPX), the base ingredient is a plastic.

2.2.5 Living samples

The preparation of living samples is rather easy even though there are a number of approaches to fluorescently label cytoplasmic and nuclear structures or organelles in living samples presently available. These vary considerably in terms of ease of use, the specificity with which they label their target molecules, and in how much they interfere with cellular function.

i. Cell-permeable markers

A simple approach to label cellular targets is to use commercially available fluorescent cell-permeable compounds. These are added to the cell culture medium to label the respective cellular structures. However, some of these also cross-react with structures different from the intended targets and therefore it is important to perform a stringent specificity test before they are used in real experiments. A short list of such compounds and their cellular targets is shown in Table 1. They have been extensively described in the literature, and the manufacturers supply details in their labelling protocols.

ii. Utilizing GFP and its variants in live cell fluorescence microscopy

The advent of the green fluorescent protein (GFP) has provided a remarkable opportunity to all areas of biological research (5); particularly live cell fluores-

Table 1 Selection of cell permeable fluorescent markers[a]

Cellular target	Marker	Comments
Plasma membrane	Con A	Is slowly internalized
Nucleus	SYTO dyes	Exist for several excitation/emission wavelengths
Endoplasmic reticulum	$DiOC_5$, $DiOC_6$	Cross-reaction with mitochondria
Golgi	NBD ceramide	Moves to the plasma membrane at 37 °C
Secretory vesicles	Acridine orange	Cross-reacts with any other acidic organelle
Endosomes	Rhodamine-transferrin	Labels early endosomes, is recycled to the plasma membrane
Endosomes	Fluorescein dextrans as fluid phase markers	Moves through the entire endocytic pathway
Lysosomes and acidic organelles	Lysotrackers	
Mitochondria	Rhodamine 123	

[a] This list of markers is far from being complete. All can be obtained from Molecular Probes (www.probes.com), which offers a whole range of fluorescent markers for cellular organelles and small regulatory molecules like calcium.

cence microscopy. Intensive mutagenesis of the primary sequence has expanded GFPs utility by altering the spectral and biochemical properties of the fluorescent protein (see Table 2). Numerous fusion proteins have now been generated and analysed by live cell imaging. GFP will more or less keep its fluorescence properties when fixed.

iii. Practical considerations in the use of GFP

Selection of GFP variants with the appropriate spectral properties allows for live visualization over a range of emission wavelengths and provides the potential for simultaneous analysis of multiple GFP chimeras inside the same single cell (2). Up to now the most useful, practical combination for dual labelling experiments has been CFP-YFP (see Table 2) (ref. 3).

Alternative approaches to the use of multiple GFP mutants in order to co-visualize more than one molecule in living cells is the combined use of a GFP chimera with cell-permeable fluorescent organelle markers (see above) or microinjection of fluorescently labelled marker proteins (4).

Besides the spectral properties of the GFP variants, there are a number of other considerations, and often limitations, which should be researched carefully before determining suitability for a specific application. It is of prime importance to ensure that the GFP chimera introduced into living cells mimics the properties of the parental protein. Unlike myc, HA, or FLAG tags, GFP is more than 200 amino acids in length and thus cannot be assumed to behave as a neutral addition to the protein of interest.

Moreover, the effect of ectopic expression of chimeras should be assessed for adverse effects on cell behaviour. For example, over-expression of the GFP chimera may compete with the endogenous counterpart for the binding of cellular factors and thus inhibit cell function. Similarly, expression of too high amounts of GFP in cells may become toxic due to the increased appearance of oxidative by-products generated during the folding of GFP after synthesis. These oxidative by-products may be counteracted by adding scavengers like oxyrase, ascorbic acid, or glucose oxidase into the culture medium during live cell imaging.

The fluorescence light emitted by most of the GFP mutants listed in Table 2 is also strongly pH sensitive.

Table 2 GFP mutants and their characteristics

GFP mutant	Excitation (nm)	Emission (nm)	Reference/source
EBFP	380	440	Clontech[a]
ECFP	433 (453)	475 (501)	Clontech
EGFP	488	507	Clontech
EYFP	513	527	Clontech
DsRed	558	583	(6), Clontech

[a] More information on available GFP expression vectors with multiple cloning sites are available under /www.gfp.clontech.com/ and from other suppliers.

More detailed information about creating GFP fusion proteins is beyond the scope of this chapter, but has been discussed elsewhere (5, 7, 8).

iv. Transient expression of GFP chimeras in cells

Many live cell fluorescence applications of GFP require the observation by microscopy of small numbers of cells, thus transient transfection strategies are likely to be sufficient. In addition to the standard transfection procedures, such as electroporation, calcium phosphate, or cationic lipid-based gene transfer (9), plasmids encoding GFP chimeras can be introduced directly into the nucleus using microinjection (see ref. 4). This procedure yields rapid, visible fluorescent protein expression, often within 0.5–1 h, and is particularly useful for quick analysis of a large number of different GFP chimeras.

2.2.6 Temperature and environmental control

One of the most crucial aspects in fluorescence microscopy of live specimens is that the environmental conditions are kept as close as possible to physiological. The key parameters in this respect are temperature, pH, and the composition of the incubation medium. A number of commercially available solutions exist to this problem and are available from most microscope manufacturers. In our experience, however the best results were achieved using a home-made Perspex box enclosing major parts of the microscope. With such a box the temperature is easily controlled, and once equilibrated, it remains stable for hours. An additional advantage of this approach is that it results in enhanced optical stability of the system and helps the sample to stay in the focal plane. This is important for experiments which require continuous imaging of the same cells for several hours, as is the case for studies on cell cycle regulation or apoptosis. The pH of the incubation medium may be controlled by using carbonate-free medium buffered with 20 mM HEPES pH 7.3. Alternatively, the Perspex box may be flooded with 5% CO_2 when carbonate-containing medium is used. An easy way to handle cultured cells and tissues is to use so-called glass-bottom dishes that come equipped with a cover glass glued in front of a hole in the bottom of a regular 3.5 cm culture dish.

3 The light microscope

The purpose of a microscope is to resolve objects of the size of single cells or smaller, and to create a suitably magnified image that can be comfortably viewed by the human eye. It therefore adapts the dimensions of the object to the limitations of the observing eye.

The basic set-up of a compound light microscope consists of a light source, a condenser, an objective, and an eyepiece, also called an ocular. The microscope can be considered in terms of illumination and detection components, with the specimen lying in between. Owing to the extreme magnification involved, the first requirement for microscopic imaging is sufficient illumination for the sample. For this purpose, the light created by the light source is focused in the plane of the specimen by the condenser. On the detection side, structures within the specimen are resolved by the objective and an intermediate magnified image is formed. The eyepiece then further magnifies this image.

Research microscopes can have an upright or an inverted architecture. Upright microscopes have the transmission light source at the bottom; the light passes through the sample from below to where it is collected by the objective above the specimen stage. An inverted microscope has the light source on top and the objective below the specimen. Both designs are equally well suited for the observation of slides but also have distinct properties. Inverted microscopes are used in cell culture because Petri dishes and multiwell plates can be observed directly through the base. On the other hand, upright microscopes work well with 'dip' lenses that can be lowered into the medium and image cells without a coverslip in between.

3.1 Objectives

Objectives of modern research microscopes are highly sophisticated multi-lens arrays with a wide variety of features. Lenses can produce imaging artefacts (spherical and chromatic errors, field curvature) so that high quality objectives are corrected for lens aberrations (Table 3).

Finite microscope objectives project an intermediate image at a fixed plane that is then magnified by the lenses of the eyepiece. Infinity-corrected objectives project an image of the sample to infinity and the intermediate image is formed by a tube lens in the light path. Infinity correction facilitates the introduction of further imaging components into the 'parallel' light path behind the objective, a prerequisite for many of the modern imaging techniques.

Although magnification is the most obvious characteristic of microscopes, the ability to resolve fine details of the specimen is the most crucial. The resolution of an objective is determined by three properties: the wavelength of the light, the refractive index of the medium between the coverslip and objective front lens, and the angular aperture which represents the angle of the cone of light

Table 3 Corrections for microscope objectives

Aberration correction	Description	Typical names
Achromatic	A lens doublet gives (not complete) chromatic correction	Achro, Achromat
Fluorite	Better spherical and chromatic aberration corrections	Fl, Fluar, Fluor, Neofluar, Fluotar
Apochromatic	Highest degree of correction for spherical and chromatic aberrations	Apo
Flat field optical correction	Field curvature correction	Plan, Pl, Plano
Apochromatic and flat field correction	Highest degree of correction for spherical and chromatic aberrations and field curvature correction	Plan Apo

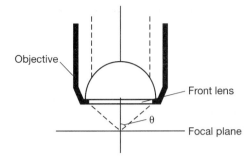

Figure 2 Angular aperture of a microscope objective.

collected by the objective. Angular aperture and refractive index can be combined into the numerical aperture of an objective:

$$\text{numerical aperture (NA)} = n \bullet \sin(\theta)$$

where n is the refractive index and θ represents half of the angular aperture of the objective (Figure 2).

The ability of a microscope objective to collect light for imaging is determined by its NA. It represents the number of highly diffracted light rays that are captured by the objective. Higher values of NA allow increasingly oblique rays to enter the objective and a more resolved image is obtained.

The minimum distance (d) at which two points can be distinguished as separate is given by the Rayleigh criterion as:

$$d = 0.61 \lambda / NA$$

where λ is the wavelength of the light. **Note:** This calculation of resolution is simplified and does not take into account the NA of the condenser in transmission imaging.

Objectives with high angular and numerical apertures generally have short working distances. The refractive index varies from 1.0 for air to 1.52 for some immersion oils. Oil immersion objectives with very short working distances have the highest NA and accordingly have the highest resolving power. As a direct consequence of the imaging properties, the depth of field of an objective decreases with increasing NA.

3.2 Magnification

The total magnification of a microscopic image consists of the product of the magnification of the objective and the eyepiece. The purpose is to adjust the details resolved by the objective to a size where they can be comfortably viewed by the eye. The total magnification should be adjusted so that all details resolvable by the objective are clearly detectable by eye or camera. Increasing magnification beyond this point adds no detail and leads to empty magnification.

As a rule of thumb, magnification should be at least 500 × the NA and not exceed 1000 × the NA.

3.3 Illumination

Setting up Köhler illumination on a microscope creates a uniformly bright and glare-free illumination of the specimen. It is also a prerequisite for the contrast enhancing methods described in the next section.

In Köhler illumination, the field diaphragm is opened just wide enough to illuminate the field of view. The field diaphragm protects the sample from unnecessary heat and prevents light not needed for imaging from entering the specimen. This leads to better contrast. When changing the objective, it is necessary to adjust the field diaphragm accordingly.

The aperture diaphragm of the condenser determines the contrast, depth of field, and resolution of the microscopic image. It generally should be adjusted to the NA of the objective. A lower value for the NA of the condenser (determined by the aperture diaphragm) than that of the objective degrades the image. An easy way to adjust it is to remove an eyepiece, to adjust the aperture diaphragm until it just appears in the pupil of the objective, and then to replace the eyepiece.

Protocol 6

Setting up Köhler illumination

1 Put the ×10 objective in the light path and focus on the specimen.

2 Set the condenser to bright field illumination.

3 Close the field diaphragm of the light source.

4 Adjust the height of the condenser until the edges of the field diaphragm are in-focus.

5 Centre the image of the field diaphragm by adjusting the condenser centring screws.

6 Open the field diaphragm until it just disappears out of the field of view.

3.4 Contrast methods for transmission imaging

Although a microscope can resolve and magnify small structures, these are often imaged with little contrast. The human eye is well suited to detect differences in brightness (amplitude) that are mainly created by absorption and the interference of direct and diffracted light in the sample. Very little contrast is created by the thin samples needed for transmission microscopy. Staining reagents are therefore used to enhance the visibility of specific structures by changing their intensity and/or by colouring them. Colour filters in the light path often enhance the contrast thus gained in stained samples.

Although staining works well with fixed samples, living material frequently has to be observed unstained. The visualization of the nearly transparent unstained cells is needed, for example, in microinjection experiments or for the handling of cell cultures. To overcome the problem of low detectable contrast, several solutions have been developed that are based on the wave characteristics of light.

3.4.1 Dark field illumination

Light passing through a sample is diffracted. These changes cannot be perceived because of the total amount of light in bright field illumination. In dark field microscopy the central light passing through the condenser is blocked out and only oblique illumination gets into the sample. Rays diffracted by the sample get into the objective and the sample appears as a bright structure against a dark background.

3.4.2 Phase contrast

Unstained objects that do not absorb light show no amplitude difference that is easily detectable by the human eye. Light diffracted by such objects does however have alterations in its phase by approx. ¼ of its wavelength (phase objects). Phase contrast transforms these phase differences into amplitude differences that are detectable by the eye (10). These amplitude differences are detected either by increasing this phase difference to ½ of its wavelength and thus creating destructive interference of diffracted and non-diffracted light (positive, dark phase contrast), or by reducing it for constructive interference (negative, bright phase contrast).

The non-diffracted direct light is separated from the diffracted light by inserting a ring annulus in the front focal plane of the condenser. This leads to a representation of the non-deviated light in the objective rear focal plane as a distinct ring while the fainter diffracted light is spread more widely. This distinct spatial separation of direct and diffracted light at the rear focal plane is used by a phase plate located there. The thickness of the phase plate is different for the direct (ring-like) and the diffracted light, thus further increasing (for dark phase contrast) the phase difference to approx. ½ of the wavelength. When the intermediate image is formed, the increased phase difference creates interference between the diffracted and non-diffracted light and the phase object is visualized in high (amplitude) contrast.

Phase contrast is a very powerful method to image unstained material, but there also some drawbacks associated with it. Optical artefacts are seen as halos around the outlines of structures. The ring annulus in the condenser decreases the working NA of the system to a certain extent. This also reduces the resolution. In thick specimens, phase shifts from above and below the plane of focus obscure the image.

For the details of setting up phase contrast on a specific microscope please refer to the manual of the microscope. Here, only a general guide can be given (Protocol 7).

Protocol 7

Setting up phase contrast

1   Focus the specimen.

2   Configure the microscope for Köhler illumination with the condenser set for bright field illumination (see Protocol 6).

3   Remove the specimen (this makes subsequent steps easier).

4   Rotate the proper ring annulus in the condenser into the light path. There are several annuli for different objectives. It is essential that the ring annulus is suited to the objective.

5   Remove one of the eyepieces and (optionally) insert the phase telescope. (There are different methods for this step.)

6   Centre the annulus to the ring of the phase using the annulus centring screws in the condenser.

7   Reinsert the eyepiece and put the specimen back on.

3.4.3 Differential interference contrast (DIC)

More recently, the polarization property of light has been used to gain contrast in unstained material (10).

In this technique, the light is polarized by passing through a polarizer before reaching the condenser. Here, a specialized Wollaston prism splits a polarized beam into two beams with slightly different paths and with 90 degrees of difference in polarization to each other. The path difference (shear) is very small, below the resolving power of the objective. The passage through the specimen on separate paths alters the characteristics of the beams and leads to differences between them. Above the rear focal plane of the objective, the shear of the beams is removed by a second prism, which is horizontally movable for adjusting optical path difference. The beams now travel along the same path but still with perpendicular polarization. An analyser in the light path subsequently brings the polarization of the beams into the same plane and thus causes interference between them that can be perceived as differences in brightness and colour. These variations reflect changes in refractive index, specimen thickness, or both. In DIC, one side of a structure appears brighter while the opposing side appears darker, giving the microscopic image a relief-like, pseudo three-dimensional effect.

Compared to phase contrast, fuller use of the NA of the microscope (condenser + objective) can be made because no ring annulus is in the light path. Resolution is therefore better and using the full objective aperture allows the visualization of a thinner section of the sample and 'optical sectioning' by focusing through the sample.

DIC is not suited for birefringent and some very thin specimens. The plastic in tissue culture vessels degrades the image, so other imaging methods should be used for this application.

For the details of setting up DIC on a specific microscope please refer to the manual of the microscope. Here, only a general guide can be given (Protocol 8).

In addition to phase contrast and DIC, other techniques for contrast enhancement in light microscopy exist. Presenting them all would exceed the limits of this chapter.

Protocol 8

Setting up DIC

1 Focus the sample.

2 Set the microscope up for Köhler illumination with the condenser set for bright field illumination (see Protocol 6).

3 Remove the sample.

4 Place the polarizer, the analyser, and the upper Wollaston prism in the light path.

5 Remove one of the eyepieces and (optionally) insert a phase telescope. Adjust the upper prism until you see a diagonal dark line at the centre. Rotate the polarizer slightly to make the line as dark as possible (polarizer and analyser are now crossed at 90°). Set the condenser aperture to 70–80% of the diameter of the objective at the back lens.

6 Replace the eyepiece and rotate the suitable prism on the condenser revolver into the light path. Put the sample back on and move the upper prism back and forth in its slot for the desired appearance of the specimen.

3.5 Fluorescence microscopy

More recent than transmission light microscopy, fluorescence microscopy is the most commonly used method of imaging in modern cell biology. In all modes of transmission microscopy, the light emitted by the light source carries the information along to the detector (eye or camera). In fluorescence imaging the light going into the sample and the light being detected are spectrally different.

Upon excitation of fluorescent molecules (fluorophores) with light of a shorter wavelength, fluorescence emission occurs at a longer wavelength that is determined by the Stokes shift of the molecule. The general principle is shown in Figure 3 for a fluorophore excited with green light at wavelengths around 550 nm and emitting in the red range around 610 nm. The spectra of the excitation and emission light are very specific for each fluorophore (Table 4). For fluorescence imaging, only light at the emission wavelengths is detected while

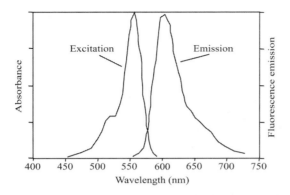

Figure 3 Excitation and emission spectrum of a fluorophore.

Table 4 Excitation and emission characteristics of common fluorophores

Derivative	Ex./Em. (nm)
DAPI	359/461
Flurescein isothiocyanate (FITC)	494/520
Alexa 488	495/519
Tetramethylrhodamine-5-isothiocyanate (TRITC)	550/573
Cy3	(514) 554/566
Alexa 546	557/572
Texas red-X conjugate	595/615
Cy5	649/666

light at the excitation wavelengths is rejected. The high amount of diffraction and refraction of the excitation light therefore does not obscure the image acquired with the microscope. In immunofluorescence microscopy, fluorophore-conjugated antibodies are used to specifically highlight intracellular structures that could not be detected using other microscopy methods. Fluorescence imaging is a very sensitive method that can detect objects below the resolution limit of the objective. Even single fluorophores can be detected; such objects are not optically resolved, but the light emitted by them is registered. To be able to detect even faint signals, as much of the emitted light as possible has to be collected, so that in fluorescence microscopy objectives with the highest possible numerical aperture are used.

Fluorescence microscopy differs from transmission imaging in some aspects discussed below.

In epifluorescence imaging, the excitation light is directed onto the sample through the objective instead of passing through a condenser on the opposite side of the sample. A condenser is therefore not needed (Figure 4a).

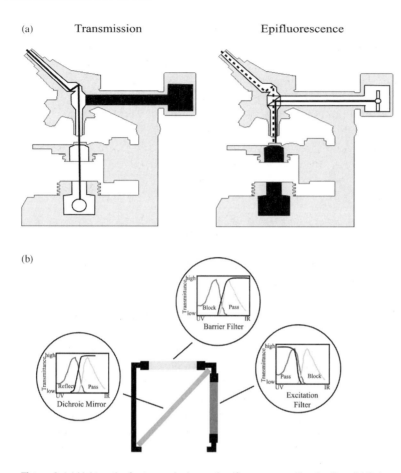

Figure 4 (a) Light paths for transmission and epifluorescence illumination. (b) Schematic of a fluorescence filter cube consisting of an excitation filter, dichroic mirror, and a barrier filter. The characteristics of the filters are displayed in the associated circles.

In illumination for transmission microscopy, a broad range of wavelengths is used and this light is directly detected after passing through the sample. Because of this, signal intensities are quite high in transmission imaging and tungsten-halide lamps are used as light sources. Fluorescence excitation uses only a very small part of the spectrum and the intensity of the emitted light is limited by the generally low quantum yield and the amount of the fluorophore in the sample. Therefore very bright light sources are needed in fluorescence microscopy to get a detectable signal. 50 or 100 W mercury arc lamps are commonly used for this purpose.

The light path for epifluorescence microscopy requires wavelength-sensitive filters for the excitation and emission light. Furthermore the excitation and emission light both pass through the objective, but while the excitation light is directed into the sample, the fluorescent signal passes into the microscope for detection so that a beamsplitter has to distinguish between these wavelengths.

For these purposes a filter cube, containing a combination of an excitation filter, a dichroic mirror for beam splitting, and a barrier filter for emission, is located behind the objective (Figure 4b). For different fluorophores, different combinations of filters and beamsplitters are installed in a multi-position holder and can be alternatively put into the light path in fluorescence microscopes.

By using filters and dichroics with several spectral 'windows' it is also possible to simultaneously image two to four distinct fluorophores.

Filters with all kinds of properties are commercially available. The proper combinations for common fluorophores are sold as packages. For emission filters, it is possible to detect all wavelengths above a threshold (longpass filters) or only a limited range (bandpass filters). The careful choice of the filter characteristics for the detection of a specific fluorophore is important when more than one kind of fluorescence is present, e.g. in double staining. A common problem here arises when the used fluorophores have partially overlapping spectra and can both be detected with the same filter. This leads to uncertainties in the assignment of the labels, especially when only the intensity and not the colour information is acquired, as is the case, e.g. with monochromatic CCD cameras. This problem can be overcome by using fluorophore combinations with little or no overlap and suitable bandpass filters instead of longpass filters.

In addition to the fluorescent label, a sample may also contain autofluorescence that interferes with the specific staining. Some autofluorescence can be overcome by treatment, e.g. glutaraldehyde fixation autofluorescence can be alleviated by treatment with sodium borohydride. In other cases it may be helpful to use a different fluorophore and a suitable bandpass filter. Autofluorescence often has a wide spectrum but low intensity at any wavelength and thus can be suppressed with a very narrow bandpass filter at the emission maximum of the fluorophore.

3.5.1 Practical imaging considerations for fluorescence microscopy

For a maximum amount of excitation light and an even illumination of the whole viewing field, the fluorescence lamp has to be properly centred and focused. The protocol for this procedure is supplied with the lamp and varies with the different designs, so it is not described here.

Exposure to the excitation light leads to loss of fluorescence in the sample. This is critical for faint signals like some that are obtained in immunofluorescence. Exposure therefore should be minimized to the times of observation and the shutter to the light source should be closed at any other time. Anti-bleaching reagents help alleviate this problem.

Changing of the filter cubes (including the dichroic mirror) for different fluorophores may distort the spatial relationship between differently labelled structures in the acquired images (pixel shifts).

4 3D microscopy

The central problem in imaging thick samples is out-of-focus light blurring the samples. The original biological samples for light microscopy consisted of thin sections, single cells, or films of material only a few microns in height. Current research, however, also requires the visualization of deeper and more complex samples. Sectioning microscopes have opened up views into tissues or even whole embryos. Researchers are now able to three-dimensionally resolve biological details smaller than a micrometre. There are two different approaches to remove out-of-focus light blurring.

4.1 3D sectioning microscopes

Wide field 3D sectioning fluorescence microscopes are capable of acquiring images with different excitation wavelengths and at different focal planes of the specimen. The resulting 3D data are reconstructed and possibly deconvolved with the help of specialized computer programs to remove the out-of-focus information contained in the individual images acquired at different focal planes. To ascertain reproducibility, image acquisition, filter changes, and z-focus control are automated in these systems. The sensitivity of these systems and the quality of the final images is very high, but they have to be extensively post-processed to remove the out-of-focus information. Filter changes between the acquisition of fluorophores may cause pixel shifts that have to be compensated in the final image. This problem can be minimized by using dichroic mirrors for the simultaneous acquisition of several colours and by just moving the excitation and emission filters. These dichroics are not quite as efficient for the single wavelenghts, though.

4.2 Confocal microscopes

In confocal laser scanning microscopy (CLSM), out-of-focus information is efficiently removed with a completely different set-up (11). All three items mentioned in the term CLSM contribute to this concept in the reverse order.

(a) Scanning. The enormous amount of scattered light created in the sample by the wide field illumination of conventional microscopes is drastically reduced by focusing the light source into a single point and thus illuminating only a tiny portion of the sample at any given time. This point light source is moved across the sample in a line-scanning pattern until every portion of the sample has been covered.

(b) Laser. Laser light is ideally suited for point illumination because it is very focused and very bright.

(c) Confocal. Using the laser and the point scanning approach does not get rid of the information from above and below the plane of focus. This is the role of the confocal pinhole. The confocal principle can only be applied in point scanning microscopes where no image is formed, but all the information at any given time resides in a single point. It consists of a pinhole aperture that is positioned before the light detector in such a way that only the light that is in-focus at this position can pass through the aperture (Figure 5). Light that is not yet or not anymore in-focus is rejected at the pinhole. This results in the representation of a single point at a defined focal plane. By combining the single points scanned along the sample into one image, a very defined representation of a thin slice of the specimen is created, with no contributing information from above and below the plane of focus.

By acquiring images at different focal planes with a confocal microscope, 3D data sets are created that can be immediately reconstructed without the need for time-consuming deconvolution.

Another advantage of confocal microscopes is that several fluorochromes can be imaged simultaneously by passing the emission signal through beamsplitters that deflect the different spectral components of the signal onto separate detectors.

However, the light budget of confocal microscopes is not as favourable as that of wide field microscopes. For reasonably fast image acquisition with a point scanning microscope, single points are illuminated only for a very short time in the range of microseconds. No signal integration over longer periods of time is possible. In a confocal microscope photomultiplier tubes are used for detection. These are not as light-efficient as some of the cameras available for wide field microscopes (see below).

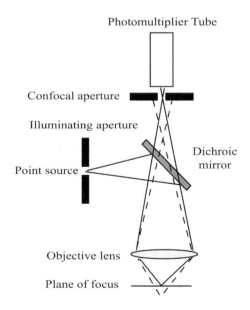

Figure 5 Schematic of the light path of a confocal microscope.

5 *In vivo* microscopy

Observation *in vivo* is an essential tool for the understanding of complex biological processes. Time-lapse microscopy allows the visualization of the dynamics of tissue development, cytoskeletal rearrangements, and vesicle trafficking in great spatial and temporal detail. Digital storage of image sequences makes these data easily accessible for further analysis of parameters such as speed and periodicity. In recent years, the field of *in vivo* microscopy has been substantially expanded by the use of variants of GFP to specifically tag and observe proteins or whole cells in living organisms (see Section 2.2.5ii).

The observation of living samples requires very efficient imaging and minimal exposure to excitation light to avoid stress, changes in behaviour, and/or photodamage that may be caused by the formation of free radicals. In planning an *in vivo* observation, one should consider the temporal resolution required for the experiment, the total time of observation needed, and the necessary image quality for analysis. These factors should be optimally balanced to avoid artefacts and the microscope and camera set-up should be chosen according to the required sensitivity and resolution.

With photosensitive organisms, care should be taken that the behaviour is not influenced by the observation light. This can be done by selecting parts of the spectrum that do not correspond to the sensitive regions.

A special form of *in vivo* microscopy is 4D microscopy, where 3D image stacks are repeatedly acquired over time. This can be done with the types of optical sectioning and confocal microscopes described above and creates very impressive visualizations of complex processes. The wide field set-up has advantages in image acquisition speed while confocal microscopes are better suited for complex samples and multichannel imaging. The excitation light is an even bigger consideration in 4D measurements since more images have to be made for any given time point. Also the rendering of these data sets requires sophisticated image processing software and, for data sets taken with wide field microscopes, requires a huge amount of deconvolution. As with other *in vivo* experiments, trade-offs and compromises between image quality, temporal resolution, and total length of the experiments are unavoidable.

6 Image acquisition

The final step in microscopy is the detection and acquisition of an image. While microscope set up and the selection of a suitable image is usually done by direct viewing through the eyepieces, the acquisition is done with a camera with different characteristics. In this section, only digital imaging will be discussed.

Digital imaging samples the analogue image into digital pixel values. Since the number of pixels is limited, image resolution may be lost when the pixel number is not matched to the resolving power of the objective (under-sampling). The Nyquist frequency (sampling frequency equals twice the maximum detected frequency) should be met or slightly exceeded (over-sampling) for imaging. In microscopic imaging this means that the minimal resolvable distance is covered by at least two pixels. Current cameras with approx. 1300×1000 pixels provide this resolution. This is demonstrated by the following simple example:

The region of a sample that can be viewed with a $\times100$ objective covers approx. 100×100 μm. With a NA of 1.2 and green light for viewing, the maximal resolvable distance is approx. 0.28 μm. The 100 μm of the viewing field are imaged in 1000 pixels. This gives a pixel resolution of 0.1 μm. 0.28 μm are therefore represented by 2.8 pixels and the Nyquist frequency is slightly exceeded.

Acquisition of a fluorescence image inevitably constitutes a compromise between sensitivity and resolution. This compromise has to be made according to the requirements for the image. The camera used is therefore a very important consideration and there is no camera design that suits all needs of any experiment. In the following sections three cameras with different properties will be discussed.

6.1 CCD cameras

The image detector in charge-coupled device (CCD) cameras consists of a CCD semiconductor chip comprised of an array of distinct wells, which can be addressed individually. Each well corresponds to one pixel in the image, so that the size and density of the wells determine the resolution of the camera. CCD cameras have excellent linearity over a broad signal range and little geometric distortions at the camera edges. They are therefore suited for quantitative imaging.

Many of the current CCD cameras allow the integration of the incoming signals on the chip before it is read out. This enables them to image signals over a wide range of brightness by varying the camera integration times. The ability to integrate gives these cameras very high detection sensitivities in comparison to video rate CCD cameras. 'Binning' is used to group adjacent wells and combine the signals collected by them. This increases the sensitivity of the camera but reduces the image resolution accordingly. Cooling the CCD chip, usually below 0 °C, reduces the thermal noise of the CCD chip and thus leads to a better signal-to-noise ratio of the image and allows higher quality imaging.

Further parameters for consideration are the quantum efficiency (QE), well capacity, and the overall geometry of the CCD chip.

Quantum efficiency is defined by the percentage of incoming photons that are converted into a useful electronic signal. The QE depends on the wavelength of the incoming light and usually does not exceed 40% for wavelengths in the visible range of the spectrum. QE is very poor below 400 nm for most CCD chips, so some chips are equipped with a special coating to overcome this problem. 'Back illuminated' cameras offer up to 80% QE. In comparison to 'front illuminated' cameras they are however more expensive and the increased well size results in decreased image resolution.

The size of the CCD wells determines the image resolution since one well corresponds to one pixel in the final image. Cameras with CCD chips that cover the resolution of the microscope are available.

The capacity of the CCD wells determines the amount of incoming signal that can be integrated before the well is physically saturated. At this point part of the signal will spill over to neighbouring wells and distort the image. The linearity of the signal and the resolution of the camera will be lost. Cameras with a high dynamic range (12-bit = 4096 grey levels, 16-bit = 65 536 grey levels) use larger well sizes. It is worthwhile to check the well capacity of a camera because some systems use a 12- or 16-bit data digitization protocol although the well capacity of the chip may be too low for a true 12- or 16-bit dynamic range.

6.2 Intensified CCD cameras

Intensified CCD cameras consist of a CCD chip coupled to an image intensifier. This significantly increases the sensitivity compared to normal CCD cameras. Light exposure, and consequently photobleaching, can be attenuated to a minimum with these cameras. On the other hand, image resolution, dynamic range, and the geometric distortion are determined by the image intensifier. The performance in these categories is considerably better for non-intensified CCD cameras. Intensified CCD cameras are therefore suited for experiments where less spatial resolution is required but where the detectable fluorescence is very weak.

6.3 Colour cameras

The cameras described above can only acquire monochromatic images. The colour information of the sample is lost. In cases of significant fluorescence crossover or autofluorescence, colour information is helpful in identifying the fluorescence signal. For the acquisition of colour images, three-colour, on chip integrating CCD cameras (red, green, and blue detectors) may be used. By combining the appropriate markers, up to three fluorescent labels can be visualized simultaneously. Some very efficient fluorescent dyes (e.g. Cy5) however, do not emit fluorescent light that can be properly separated into the spectral components of the camera. A quantitative association of the detected signals to the individual markers is therefore not possible.

7 Data analysis, image processing, and data presentation

It is far beyond the scope of this article to give a comprehensive introduction about image processing. It is however important to note that between the image as it is acquired with the microscope and the image that we see in the final publication, a lot of post-processing may have taken place. Many companies are active in this field and a lot of improvements have been made in recent years in the field of image processing as compared to image acquisition. Almost every microscope that is equipped with a camera comes with its own software for image acquisition and usually contains some functions for image analysis as well.

Image processing is a very important tool, but it requires an understanding of the subject to avoid introducing artefacts into the data.

While there is a variety of software that is commercially available, one of the most popular software packages for scientific image analysis and documentation still is the public domain NIH Image program (available from: http://rsb.info. nih.gov/nih-image/). The major advantages of this program are that it has existed for a long time, it has been continuously updated, and that it can be downloaded from the web for free. Therefore it is widely distributed which makes it is easy to find somebody to ask for help. It covers most basic features in image processing like basic filtering to smooth noisy images or contrast stretching. The generation of simple macros is possible without the user having a background in computer programming and it thus allows the automated acquisition and evaluation of image data. Created for Macintosh Computers it has been ported to PCs and is now also available as a platform independent JAVA application (Image J).

In general, the software package of choice for the more sophisticated analysis and documentation of time-lapse image data should have the possibilities to automatically analyse geometric and densitometric parameters on multiple colour images in 4D (time and space). Also, tracking of individual objects and visualization of their trajectories should be possible. An effective way of presenting data of live cell experiments is to convert the individual images into Quicktime movies which can be easily placed onto a World Wide Web server for general accessibility.

8 Online resources

The Internet offers a lot of information about microscopy and we would like to conclude this article by mentioning two proficient sources of information:

(a) Molecular Expressions Microscopy Primer is an excellently designed web site with information mainly on wide field microscopy. Microscopic concepts can be studied there with the help of interactive Java tutorials. The URL is http://micro.magnet.fsu.edu/primer/index.html

(b) The confocal newsgroup can be subscribed to at the listserver of the University of Buffalo. It is an up-to-date source of information on new techniques and some of the finer points of confocal imaging. The archive provides answers to a lot of the problems encountered in microscopy. The URL of the listserver is http://LISTSERV.ACSU.BUFFALO.EDU/

References

1. Wihelm, S., Gröbler, B., Gluch, M., and Heinz, H. (2000). *Confocal laser scanning microscopy. Principles*. Carl Zeiss micro special 40–617 e/02.00.
2. Allen, V. J. (ed.) (2000). *Protein localization by fluorescence microscopy: a practical approach*. Oxford University Press.

3. Ellenberg, J., Lippincott-Schwartz, J., and Presley, J. F. (1998). *Biotechniques*, **25**, 838.

4. Pepperkok, R., Saffrich, R., and Ansorge, W. (1998). In *Cell biology: a laboratory handbook* (ed. J. Celis), pp. 23–30. Academic Press, London.

5. Tsien, R. Y. (1998). *Annu. Rev. Biochem.*, **67**, 509.

6. Baird, G. S., Zacharias, D. A., and Tsien, R. Y. (2000). *Proc. Natl. Acad. Sci. USA*, **97**, 11984.

7. Sullivan, K. F. and Kay, S. A. (ed.) (1999). *Methods in cell biology. Green fluorescent proteins.* Vol. 58. Academic Press, San Diego, CA.

8. Patterson, S., Day, R. N., and Piston, D. (2001). *J. Cell Sci.*, **114**, 837.

9. Celis, J. (ed.) (1998). *Cell biology: a laboratory handbook*. Academic Press, London.

10. Abramowitz, M. (1987). *Contrast methods in microscopy: transmitted light*. Vol. 2. Olympus Corporation Publishing, New York.

11. Pawley, J. B. (ed.) (1995). *Handbook of biological confocal microscopy*. Plenum Press, New York.

Appendix

Preparation of stock solutions

Preparation of paraformaldehyde

1   Heat about 80 ml of PBS to 80 °C and add 3 g paraformaldehyde while stirring; mix until clear.

2   Add 100 μl of 100 mM $CaCl_2$ and 100 μl of 100 mM $MgCl_2$ (to give a final concentration of 0.1 mM) with stirring whilst the solution is warm, to prevent precipitation.

3   Allow the solution to cool; make up to 100 ml with PBS and adjust to pH 7.4.

4   Store in aliquots at −20 °C and use a fresh aliquot each time.

Preparation of Dulbecco's phosphate-buffered saline (D-PBS) (minus calcium and magnesium)

Composition	mg/litre
NaCl	8000
KCl	200
Na_2HPO_4 (anhydr.)	1150
KH_2PO_4	200

Made up in double distilled water (dH_2O). The pH should be 7.4, and therefore pH adjustment of the solution ought not to be necessary.

Preparation of Mowiol mounting medium

Composition	
Mowiol 4-88	2.4 g
Glycerol	6 g
DW	6 ml
0.2 M Tris pH 8.5	12 ml

1   Place glycerol in 50 ml disposable conical centrifuge tube.

2   Add Mowiol and stir thoroughly. Add dH_2O and leave for 2 h at RT.

3   Add Tris and incubate at approx. 53 °C until the Mowiol has dissolved. Stir occasionally.

4   Clarify by centrifugation at 3000–4000 g for 20 min.

5   Transfer the supernatant into glass vials with screw caps (about 1 ml in each).

6   Store at −20 °C. The solution is stable at this temperature for up to 12 months. Once defrosted, it is stable at RT for at least one month.

Chapter 3

Electron microscopy in cell biology

John Lucocq

Department of Anatomy and Physiology, Medical Science Institute,
The University of Dundee, Dundee DD1 4HN, UK.

1 Introduction

Electron microscopy is undergoing a strong revival as biologists strive to understand how molecular elements within cells are spatially and temporally integrated into functioning structures. The newer EM approaches have superseded the descriptive EM studies of the 1950s and 1960s by providing more quantitative characterizations of structure and molecular composition that more closely correspond to the *in situ* 'native' state. This chapter illustrates examples of important techniques currently used in EM study of structures in cell biology. In particular I have included descriptions of methods for immunoelectron microscopy and low temperature techniques of thawed cryosections, cryofixation, and embedding in resin. New approaches such as field emission scanning EM and SDS-cryofracture labelling along with quantitation of immunolabelling using stereology are also included. The protocols are presented as basic starter methods and for additional in-depth discussion see the more extensive treatises on EM methods (1, 2). Electron microscopy is a large subject to cover in a short chapter and I have out of necessity been very selective. The choice of protocols has been largely dictated by current trends in EM use, by personal preference, and the advice of colleagues. The chapter concentrates mainly on transmission EM (TEM) with a brief mention of scanning EM (SEM). Important but more specialized approaches such as electron tomography and cryoelectron microscopy of frozen hydrated specimens are not discussed here.

Many of the problems addressed in cell biology are suited to the TEM readout from ultrathin sections because this gives information about internal structure and allows quantitative immunogold labelling of tissues, intact cells, or organelles. TEM is also suited for viewing cryofracture/cryoetched replicas in which surface relief of cell structures and membranes can be determined, and also to high magnification display of macromolecules studied using freeze etch, metal shadowed, negative staining, or frozen hydrated samples. SEM provides information on surface morphology of cells, organelles, or molecules but its use in cell biology has historically been limited because of lower resolution. However, the newer high resolution field emission SEMs are particularly suited to the study of surfaces in combination with immunogold labelling and have already been used effectively, for example, in studies of the cell membrane, nuclear structures, and nuclear envelope.

2 Overview of sectioning methods for transmission EM (see Figure 1)

Ultrathin sections of 50–100 nm are thin enough to reveal details of the internal structure of cells and organelles when viewed in the electron microscope. The sectioning process relies on initial fixation of the specimen either by chemical crosslinking using aldehydes or by freezing rapidly at either ambient or high pressure followed by inclusion/embedding in solid media (see Figure 1). Aldehyde fixation remains the most frequently used starting point followed by either conventional embedding in epoxy resin for structural studies, or cryoprotection, freezing, and cryosectioning for antigen localization using immunoelectron microscopy. If the cryosectioning equipment is not available then an alternative approach is to embed in resin using progressively lowered temperature or by freeze substitution. Rapid freezing methods or freezing at high pressure are designed to immobilize structures without inducing structure-damaging ice crystal formation and thereby avoid the use of aldehydes. Once frozen the specimens can be embedded in resin by freeze substitution (suited to immuno-electron microscopy, and structural analysis), or sectioned to produce frozen hydrated sections on a cold EM stage for structural analysis. The basic principles

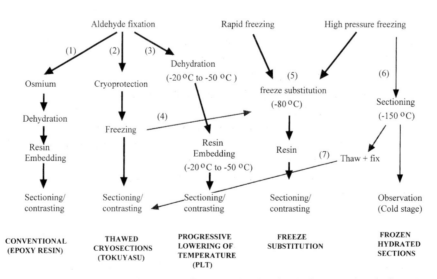

Figure 1 Processing biological specimens for sectioning. Numbers refer to sections in the text.

of all these sectioning techniques are described below and outlined in Figure 1 that shows the lines of processing leading to the production of sections for use in cell biology.

2.1 Methods using aldehyde fixation

The following are short descriptions of the methods numbered in Figure 1.

(a) **Conventional TEM (method 1)**. Generally comprises fixation in aldehyde followed by post-fixation in osmium tetroxide, solvent dehydration, and embedding in epoxy resin. This method is suited to structural analysis and quantitation of structure because it yields high contrast and also because serial and oriented sections can be prepared. The deleterious effects of osmium tetroxide, solvent dehydration at ambient temperature, and polymerization in resin at high temperatures generally dissuade against the use of this approach for immunolabelling studies (Protocols 5–9).

(b) **Thawed frozen sections (method 2)**. Involves aldehyde fixation, cryoprotection, freezing in liquid nitrogen, and sectioning at approx. −100 °C. The sections are then thawed, mounted on EM grids, immunogold labelled, and contrasted. This is the thawed frozen section or 'Tokuyasu' technique and contains fewer denaturation steps than for conventional epoxy embedding methods, and is a first choice for immunolabelling of proteins and also lipids. Usually this approach affords better membrane display and higher labelling densities compared to resin embedding after progressive lowering of temperature or freeze substitution (Protocols 10–13).

(c) **Progressive lowering of temperature (method 3) (PLT)**. Works by dehydrating aldehyde fixed material in liquid solvent at progressively lower temperatures followed by embedding in resin at low temperature. This technique can be done in a conventional laboratory freezer and does not require specialized cryomicrotomes. Dehydration at lowered rather than ambient temperature may preserve structure and preserve antigens better. It is therefore an alternative to thawed frozen sections and offers excellent visualization of cytoskeleton, desmosomes, and nuclear structures (Protocols 14–16).

(d) **Freeze substitution (method 4)**. Aldehyde fixed specimens can be cryoprotected and frozen in liquid nitrogen. The solid water in the frozen specimen is then exchanged for solvent and then resin at approx. −80 °C (see below). The resin is polymerized at low temperature and sections cut at ambient conditions. Freeze substitution also allows the processing and resin embedding of rapidly frozen or high pressure frozen specimens avoiding aldehyde fixation (see next section) (Protocol 17).

2.2 Methods using rapid/high pressure freezing

Freezing methods allow aldehyde crosslinking fixatives to be avoided and act faster. Rapid freezing allows adequate freezing (lacking obvious ice crystal effects) of a rather shallow thickness of 10–20 μm. Application of high pressure allows adequate freezing of thicker 200–300 μm specimens (Protocol 18).

(a) **Freeze substitution (method 5)**. This is the principal technique for embedding specimens physically fixed by freezing. After rapid freezing or high pressure freezing solid water in the specimen is exchanged for liquid solvent and then resin. For structural studies fixatives/contrasting agents can be included in the freeze substitution mixture and embedding completed at ambient temperature. For localization studies embedding is generally carried out in resin at low temperature (Protocols 19 and 20).

(b) **Frozen hydrated sections (method 6)**. A technique for examining structure in frozen material after sectioning at −150 °C and viewing structures on a cold stage in the electron microscope. An adaptation of this technique (**method 7**) (3) is one in which the frozen hydrated section is thawed on a droplet containing fixative and cryoprotectant. The sections are then further processed for immunoelectron microscopy.

3 Aldehyde fixation and preliminary specimen preparation (applicable to methods 1, 2, 3, and 4 in Figure 1)

The following methods describe the fixation of tissues and cells along with preliminary processing prior to embedding and sectioning. The mechanisms of aldehyde fixation are still poorly understood and most fixation conditions using these chemicals have been determined empirically.

Glutaraldehyde crosslinks amino groups present in most proteins and some lipids (most lipids are unlikely to be fixed with this aldehyde). This aldehyde can react with protein and free amino acids in the cells producing pyridine polymer crosslinks within seconds. For structural analysis the concentrations used vary between 0.5–2% (w/v). When combined with post-fixation in osmium, glutaraldehyde fixation of animal cells in phosphate or cacodylate buffers yields good visualization of cellular membranes at the expense of some extraction of the cytosolic proteins. On the other hand fixation in zwitterionic buffers such as PIPES and HEPES may allow maximum retention of soluble protein components. It is the norm to carry out fixation at approx. pH 7. Glutaraldehyde can be purchased as a 25% aqueous solution, stored as aliquots at −20 °C, and diluted immediately before use.

Formaldehyde has a more complex mode of action than glutaraldehyde and can exist as oligomers in solution at higher concentrations (>8%). The extent of crosslinking, and therefore the degree of structural preservation, is greatest at high concentrations and for thawed frozen sections 8% is an often used as a starting point. For structural studies using conventional TEM formaldehyde is rarely used on its own and can be combined with glutaraldehyde fixation in mixtures such as 0.1% glutaraldehyde/4%formaldehyde or 4% formaldehyde followed by post-fixation in glutaraldehyde. Formaldehyde solution is prepared from paraformaldehyde powder as a 16% stock solution in water and stored at 4 °C or at −20 °C. The powder is dissolved by heating to 80 °C with the addition of sodium hydroxide solution to adjust the pH to near-neutral and thereby facilitate solubilization of the paraformaldehyde.

The following protocols illustrate methods for fixation and subsequent pelleting of cells prior to further processing.

Protocol 1

Fixation and pelleting of cells in suspension

Equipment and reagents

- Centrifuge
- 1% glutaraldehyde in 0.4 M PIPES, 0.2 M phosphate, or 0.2 M cacodylate pH 7.2
- Phosphate-buffered saline (PBS)

Method

1 Prepare fixative at twice final required strength, e.g. 1% glutaraldehyde in 0.4 M PIPES, 0.2 M phosphate, or 0.2 M cacodylate at pH 7.2 and adjust to required temperature.

2 Add an equal volume of fixative to the cell suspension in growth medium.

3 After 2 min pellet by centrifugation at approx. 10000 g for 5 min.

4 Dislodge the pellet and wait for a total of 30 min. Wash carefully in PBS and store at 4 °C.

Protocol 2

Fixation and preparation of pellets from monolayer cell cultures

Equipment and reagents

- Centrifuge
- 0.5% glutaraldehyde in 0.2 M PIPES, HEPES, or cacodylate buffer pH 7.2
- PBS

Method

1 Drain off the medium in which the cells have grown (there is no need to wash with cold buffer before applying fixative).

2 Add fixative and incubate for the required fixation time, e.g. 30 min in 0.5% glutaraldehyde in 0.2 M PIPES, HEPES, or cacodylate buffer pH 7.2.

Protocol 2 continued

3 If necessary wash and store in PBS at 4°C. Cells may be sent through the post at this stage by topping up culture flasks with buffer or fixative as appropriate.

4 Drain off the buffer/fix and add 1 ml of fixative (using fixative prevents the cells sticking to plastic centrifuge tubes).

5 Scrape the cells into the fixative with a rubber scraper (quantitative studies in our laboratory have shown that scraping does not in general cause significant disruption to cell structures) (J. Lucocq, unpublished results).

6 Break up any sheets of cells by successive aspiration into a 1 ml plastic pipette tip (if necessary control using phase contrast light microscopy that the cell sheets have been broken-up).

7 Centrifuge for at least 15 min at 10000 g in fixative (note that the total fixation time is the fixation on the dish plus the centrifugation time).

Dispersion of cells from cell pellets during further processing is a common problem, especially when the pellets are further incubated in sucrose for the thawed frozen section method. In our laboratory we use a method to consolidate pellets (Protocol 3).

Protocol 3

Consolidation of pellets using fixation in bovine serum albumin (BSA)

Equipment and reagents

- Centrifuge
- 1.5 ml Eppendorf tubes
- PBS
- BSA
- Fixative, e.g. 0.5–2% glutaraldehyde in PBS

Method

1 Pellet cells in a 1.5 ml Eppendorf tube and wash in PBS by repeated centrifugation to remove the fixative.

2 Remove the supernatant from the pellet and add 0.5 ml of 5% (w/v) BSA in PBS.

3 Resuspend and centrifuge to make a pellet.

4 Remove the supernatant and overlay the pellet with the same fixative used for fixing the cells (e.g. 0.5–2% glutaraldehyde in PBS).

5 Dislodge the pellet and continue fixing for as long as is compatible with preservation of structure or antigenicity.

This procedure may significantly improve the cutting qualities of the pellet in thawed frozen sections.

3.1 Perfusion fixation of tissue

Perfusion fixation is strongly recommended for tissues (especially nervous tissue; see Figure 2) and is discussed in more detail elsewhere (1, 2). A sample protocol for brain fixation through the heart is given below.

Protocol 4

Perfusion fixation of the brain through the heart (ref. 4)

Equipment and reagents

- 25 gauge microlance needle
- 0.1 M sodium phosphate buffer pH 7.4 containing 0.1% (w/v) sodium nitrite
- PBS
- 4% paraformaldehyde, 0.05% glutaraldehyde in PBS

Method

1 The deeply anaesthetized mouse[a] is pinned out with its ventral surface uppermost. The ribcage is exposed and retracted.

2 A 25 gauge microlance is pushed into the left ventricle through which is perfused, via a peristaltic pump, ice-cold 0.1 M sodium phosphate buffer pH 7.4 containing 0.1% (w/v) sodium nitrite. The right atrium is cut under slight positive pressure from the perfusate. The animal is exsanguinated in this manner for 5 min (~25 ml buffer).

3 The perfusate is exchanged for freshly prepared, ice-cold fixative (normally 4% paraformaldehyde, 0.05% glutaraldehyde in PBS or other suitable buffer).

4 The fixation procedure is continued for 20 min (100 ml) after which the brains are dissected out and immersed in fixative, at 4°C overnight.

5 Fixed, PBS-washed brains are further processed for embedding and sectioning.

[a] Most countries require a special license to perform such procedures.

Embedding protocols

4 Conventional epoxy resin embedding (method 1, Figure 1)

Conventional TEM uses aldehyde fixation followed by post-fixation in osmium tetroxide and dehydration/embedding in epoxy or similar resin. Because of the emphasis on processing of monolayer cell cultures and cell suspensions in cell biology these protocols deal with the embedding of cells *in situ* on culture dishes as well as processing of microinjected cells.

Protocol 5

Fixation and embedding of cell pellets prepared from monolayer cultures

Equipment and reagents

- Centrifuge
- 0.5% glutaraldehyde in 0.2 M PIPES, PBS, or 0.1 M cacodylate buffer pH 7.4
- 0.1 M PIPES buffer pH 7.4
- 1% OsO_4 in 0.1 M cacodylate buffer pH 7.2
- 1% OsO_4/1.5% potassium ferrocyanide
- Cacodylate buffer
- Ethanol
- Epoxy resin
- Propylene oxide

Method

1 Fix in 0.5% glutaraldehyde in 0.2 M PIPES, PBS, or 0.1 M cacodylate buffer pH 7.4 for at least 30 min.

2 Wash three times in PBS buffer pH 7.4 for 10 min (and store at 4°C if necessary).

3 Scrape the cells in 1 ml fixative and break up sheets by pipetting through a plastic 1 ml pipette tip.

4 Centrifuge at 15000–20000 g for 15 min.

5 Cut pellet into 0.5 mm (or smaller) blocks and wash in 0.1 M cacodylate buffer pH 7.2.

6 Post-fix pellets in 1% OsO_4 in 0.1 M cacodylate buffer pH 7.2 for 30 min. If enhanced membrane contrast is needed then use reduced osmium tetroxide: 1% OsO_4/1.5% potassium ferrocyanide (made by mixing 2 parts of 2% OsO_4, 1 part 0.4 M cacodylate buffer pH 7.2, and 1 part 6% KFeCN).[a]

7 Wash in cacodylate buffer three times for a total of 10 min.

8 Wash in distilled water three times for a total of 10 min. (Can be done in plastic tubes up to this point but should be transferred into glass vials for the dehydration steps.)

9 Ethanol washes: 70, 90, 100, 100, and 100% each for 10 min.

10 Propylene oxide washes twice each for 10 min.

11 Epoxy resin/propylene oxide 50:50 (v/v) for at least 1 h.

12 100% epoxy resin for 2 h or longer.

13 Epoxy resin embedding in beam or gelatin capsules.

14 Polymerization at 60°C for one day/overnight.

[a] Further increases in contrast may be achieved by following osmium tetroxide with (a) 0.5% (w/v) magnesium uranyl acetate in water (1 h) (so-called *en bloc* staining) or, by (b) 1% (w/v) tannic acid (30 min) followed by a 10 min rinse in 1% (w/v) sodium sulfate both in 0.1 M cacodylate buffer pH 7.2, followed by washes in distilled water, and incubation in 0.5% (w/v) magnesium uranyl acetate in water (1 h). These modifications enhance the contrast of structures such as cytoplasmic filaments, ribosomes, and vesicle coats.

Some plastic Petri dishes used for culture are soluble in the solvents that are used for epoxy resin embedding, particularly propylene oxide. Protocol 6 avoids the use of propylene oxide allowing the cells to be embedded *in situ*. Once the resin is polymerized in the culture dish the blocks can be cut either parallel or perpendicular to the plane of the culture dish.

Protocol 6

Embedding monolayer cultures in dishes (avoiding propylene oxide)

Equipment and reagents

- Petri dishes
- PBS, 0.1 M cacodylate, 0.2 M PIPES, or HEPES pH 7.2
- 1% osmium tetroxide in cacodylate buffer
- Ethanol
- Uranyl acetate
- Epoxy resin

Method

1 Aldehyde fixation in PBS, 0.1 M cacodylate, 0.2 M PIPES, or HEPES pH 7.2 as required.

2 Rinse in 0.1 M cacodylate buffer pH 7.2.

3 Post-fix in 1% osmium tetroxide in cacodylate buffer or alternative reduced osmium tetroxide detailed in Protocol 5 for 30 min.

4 Wash in 50% ethanol.

5 Optional uranyl acetate *en bloc* staining 1% (w/v) in 50% ethanol in the dark.

6 Wash in 70% ethanol and then in 90% ethanol.

7 Wash in 100% ethanol twice for 30 min.

8 Ethanol/epoxy resin (Epon 812 equivalent) ratio 1:1 (v/v) for 30 min.

9 Epoxy resin (Epon 812 equivalent) 100% for 30 min.

10 Epoxy resin (Epon 812 equivalent) 100% for 30 min.

11 Epoxy resin (Epon 812 equivalent) 100% to fill the Petri dishes, at 45°C for 2.5 days, and then at 60°C for 2.5 days.

It is also possible to use the solubility of the plastic culture dish in propylene oxide to aid in the removal of the monolayer from the dish immediately prior to infiltration with resin (Protocol 7).

Protocol 7

Fixation and embedding of cells grown on plastic Petri dishes (using propylene oxide to detach the cells)

Equipment and reagents

- Centrifuge
- Petri dishes
- PBS, 0.1 M cacodylate, 0.2 M PIPES, or HEPES pH 7.2
- 1% (w/v) OsO_4 in cacodylate buffer

- Ethanol
- Propylene oxide
- Epoxy resin

Method

1 Remove the medium.

2 Aldehyde fixation in PBS, 0.1 M cacodylate, 0.2 M PIPES, or HEPES pH 7.2 for 30 min.

3 Wash in 0.1 M cacodylate buffer three times over 15 min.

4 1% (w/v) OsO_4 in cacodylate buffer (2 ml per 5.5 cm dish) over 30 min or alternative post fixative described in Protocol 5.

5 Wash in cacodylate buffer three times over 15 min.

6 Wash in distilled water three times over 15 min.

7 Dehydrate in 70, 90, 100, 100, 100% ethanol each for 10 min.

8 Remove the cells from the dish with propylene oxide (pipette the solvent over the cells until they come off in sheets). At this point the sheets can either be left intact[a] or broken into small pieces by pipetting before centrifugation and pelleting.

9 Wash cells once with propylene oxide to remove solubilized plastic.

10 Centrifuge at 14000 g for 3 min.

11 Resuspend pellet in propylene oxide/epoxy resin ratio 1:1 (v/v) in a beam capsule pushed into the top of a plastic Eppendorf tube.

12 Pellet at 14000 g for 2 min and resuspend in epoxy resin.

13 Repeat centrifugation as short 10 sec bursts at 14000 g to place evenly at the bottom of the beam capsule.

14 Polymerize the resin.

[a] If cell sheets are left intact they can be handled with a glass rod through the dehydration and infiltration procedure and embedded flat for sectioning.

Low melting point agarose or agar are convenient alternatives to gelatin (Protocol 8).

Protocol 8

Embedding of suspensions or cell pellets in agarose

Equipment and reagents

- Water-bath
- Razor blade
- Agar or agarose
- Osmium tetroxide
- Uranyl acetate
- Epoxy resin
- PBS

Method

1 Fix in aldehydes as described in Protocols 1–3 and wash the cells.

2 Pellet the cells, remove the supernatant, and warm the tube to the temperature of molten agar or agarose.

3 Add 2% solution (w/v in distilled water) of agar or agarose pre-heated (molten), and gently resuspend the cells.

4 Leave for 10 min in a water-bath.

5 Pellet the cells.

6 Cool on ice to solidify gel and cut pieces containing the cells with a razor blade.

7 Post-fix in osmium tetroxide (Protocol 5, 6, or 7), dehydrate, and embed in epoxy resin.

Note: An alternative technique for embedding in agarose is by the pipette method as follows: suck up a small amount of agarose, add to the pellet, and resuspend. Expel into ice-cold PBS. The agarose will solidify and produce a ribbon of agarose containing cells which is ready to cut up.

Microinjection of cells on coverslips is increasingly used to analyse the effects of proteins or expressed proteins from DNA constructs on cellular morphology and cell function. The cells may be identified with the use of a locator grid on which the cells are cultured and by using microinjected fluorescent markers.

Protocol 9

Embedding and sectioning of microinjected cells (ref. 5)

Equipment and reagents

- CELLocate coverslips (Eppendorf)
- CCD camera
- 4% paraformaldehyde in PBS
- 1% glutaraldehyde in PBS
- 1% OsO_4 (with or without 1.5% potassium ferrocyanide, see Protocol 5) in 0.1 M cacodylate buffer
- Ethanol
- Epoxy resin

Protocol 9 continued

Method

1 Cells are seeded onto CELLocate coverslips and microinjected with a fluorescent marker and protein/DNA of interest.

2 After required incubations cells are fixed in 4% paraformaldehyde in PBS and the location of the injected cells on the locater grid is recorded by fluorescence and phase/DIC imaging, using water immersible objectives and a CCD camera. Up to this point the cells can be used for either immunofluorescence or EM.

3 For EM the cells are then post-fixed with 1% glutaraldehyde in PBS followed by 1% OsO_4 (with or without 1.5% potassium ferrocyanide) in 0.1 M cacodylate buffer, dehydrated in graded ethanols, and embedded in epoxy resin. To infiltrate/embed the coverslips overfill plastic tubes with resin and invert the coverslip cell-side down onto the top of the tubes avoiding trapped bubbles. Before polymerization invert the tube and coverslip to allow any bubbles to rise from the coverslip.

4 Dip the coverslip briefly into liquid nitrogen and split off from the resin leaving the cells and an impression of the grid on the block. The location of the cells can be found by comparing the phase contrast and fluorescence CCD images with the grid impression prior to trimming and sectioning.

5 The region of injected cells is serially sectioned and sections collected on plastic/carbon coated slot grids and contrasted with uranyl acetate and lead citrate (see Section 4.1).

6 Cells are located at the EM by comparing the position of cell profiles in the section with the injected cells on the CCD images. Note that the arrangement of the cells on the block face may be a mirror image of that in the CCD image. So for comparison it may be helpful to invert the CCD image before examining the sections at the EM. The cells are then located by comparing the pattern of sectioned cell profiles with the phase contrast/fluorescence images.

4.1 Contrasting epoxy resin sections

The most frequently used sequence is aqueous uranyl acetate followed by lead salts. Typically grids are placed on drops of 2% or 3% (w/v) uranyl acetate for 15 min followed by rinses in distilled water. The grids are air dried and stained on drops of lead citrate or lead acetate for 1–3 min in a covered dish containing pellets of sodium hydroxide, washed in distilled water, and air dried. When ferrocyanide/osmium (Protocol 5) is used to enhance membrane staining, uranyl acetate staining can be omitted. Lead acetate is prepared as follows:

(a) Dissolve 20 g of NaOH and 1 g of sodium potassium tartrate in water to a volume of 50 ml.

(b) Add 1 ml of stock to 5 ml of 20% (w/v) lead acetate.

(c) Stir, filter, and dilute 15 times in distilled water.

(d) Store in a stoppered flask or under paraffin oil.

Lead citrate is prepared as follows:

(a) Put 30 ml of distilled water, 1.76 g of trisodium citrate ($2H_2O$), and 1.33 g of lead nitrate in a sealable 50 ml measuring flask.

(b) Shake vigorously for 1 min.

(c) Leave to stand for 30 min shaking from time to time.

(d) Add 8 ml of freshly made 1 N NaOH and make up to 50 ml.

5 Thawed ultrathin cryosections (see method 2, Figure 1)

Preparation of ultrathin cryosections and their contrasting has been perceived as one of the most technically demanding techniques in cell biology. This impression dates from the early days when cryomicrotomy was an art in itself. However the recent advent of high precision microtomes and reliable methods for section contrasting and embedding now make this a routine technique that can be applied in any laboratory. Excellent descriptions and movies of the technique are available at the following web site: http://pathcuric1.swmed.edu/Research/HaglerEMlab.htm.

5.1 Principle

Fragments of aldehyde fixed cells or tissues are soaked in cryoprotectant and mounted on metal stubs that fit into the specimen arm of the microtome and frozen in liquid nitrogen. The cryoprotectant prevents formation of structure-damaging ice crystals in the specimen. The stubs are transferred to the cryo-sectioning chamber and sections cut on glass or diamond knives at low temperature (-100 to $-120\,°C$). The sections are kept dry on the knife but picked up and thawed on drops of cryoprotectant introduced into the chamber. The sections are then adsorbed to EM support grids coated with a film of carbon coated plastic. The following protocols describe the important steps for cutting cryo-sections.

5.2 Preparing the blocks of pellet, tissue, or gelatin embedded material

The overall aim is to produce blocks with rectangular block faces, which cut much better than irregular shapes. Tissues fixed by perfusion or immersion, or pellets/gelatin embedded cells (see below) can be cut with a scalpel blade into thin slabs approx. 0.5 mm thick. Lay these slabs out and cut strips approx. 0.5 mm wide, then chop across these to make bricks which when mounted will present elongated block faces to the knife. Incubate in cryoprotectant (2.1 M or 2.3 M sucrose in PBS) for at least 30 min on ice, or preferably overnight at 4 °C.

5.3 Gelatin embedding

In some cases fixed cells do not form a closely adherent pellet and embedding in pig skin gelatin, agarose, or agar is required (Protocol 8 and 10). Gelatin embedding is well suited to preparation of cryosections because it improves the sectioning properties of the blocks, especially when cells have been fixed in formaldehyde. Note that it is important to wash out fixatives well before embedding in gelatin.

Protocol 10

Gelatin embedding of cell suspensions

Equipment and reagents

- Centrifuge
- Water-bath
- 12% (w/v) pig skin gelatin in PBS

Method

1 Make a solution of 12% (w/v) pig skin gelatin in PBS by heating in a water-bath. Cool and store in aliquots at $-20\,°C$.

2 Before use, warm the gelatin solution to $37\,°C$ in a plastic Eppendorf centrifuge tube. Disperse fixed cells or fragments of cell pellet in this solution and leave to equilibrate for 10–30 min.

3 Centrifuge to form a pellet of cells and put on ice for 30–60 min to solidify the gelatin. Cut off the ends of the tube and cut blocks of gelatin and incubate in cryoprotectant. A convenient way of cutting blocks is to slice the tube across the end to produce discs of gelatin containing fragments of pellet which can then be cut further into blocks suitable for mounting (see Section 5.2). An alternative method is to pour the molten gelatin and pellet onto a plastic Petri dish that has been inverted on ice to cool it down. A second Petri dish is placed over the molten gelatin and pressed down. Leave for 30–60 min on ice before preparing blocks and cryoprotection (see Section 5.2). Gelatin may be removed after sectioning by placing the sections on drops of warm ($37\,°C$) distilled water or PBS.

A recently introduced method allows *in situ* gelatin embedding of monolayer cultures in plastic dishes or on coverslips (6). The advantage here is that the position and orientation of the cells is preserved which can greatly aid interpretation and may allow certain quantitative techniques to be employed.

Protocol 11

The preparation of cell monolayers for *in situ* cryosectioning (ref. 6)

Equipment and reagents

- Scalpel
- Aldehyde fixative in appropriate buffer
- PBS containing 0.02 M glycine
- PBS containing 0.1% (w/v) BSA
- 12% solution of gelatin in PBS
- 2.3 M sucrose

Method

1   Pour off the culture medium and add aldehyde fixative in appropriate buffer.

2   Wash cells in PBS containing 0.02 M glycine and then in PBS containing 0.1% (w/v) BSA.

3   Cover the fixed monolayer with a 12% solution of gelatin in PBS (pre-warmed to 37°C).

4   Allow the gelatin to solidify at room temperature and rinse once more with PBS containing 0.1% BSA.

5   Apply a second thin layer of 12% (w/v) gelatin containing aldehyde fixed erythrocytes and allow to solidify at 4°C. This layer helps to prevent breaking of the block in liquid nitrogen and its colour helps to orient the block for cutting sections.

6   Score the gelatin layers with a scalpel and infiltrate with 2.3 M sucrose for two to three days at 4°C with agitation.

7   Detach pieces of gelatin slab using a thin spatula and if necessary stain with 1% Toluidine blue to aid detection of the cells.

8   Mount the slabs either flat on a cryostub (erythrocyte layer down), or sideways with the erythrocyte layer at one side to cut across the monolayer.

For tissues that are rather inhomogeneous in consistency improved sectioning qualities may be obtained by using a mixture of polyvinylpyrrolidone (PVP) and sucrose cryoprotectant.

Protocol 12

Preparation of PVP/sucrose (ref. 7)

Reagents

- Sucrose
- PVP: mol. wt. 10#000 (Sigma Chemical Co)
- PBS
- Na_2CO_3

Method

1   Make a 100 ml solution that is nearly saturated with sucrose and contains 25% of PVP as follows: 25 g of PVP is dissolved in a mixture of 5 ml of 1.1 M Na_2CO_3 in PBS and 75 ml of 2.3 M sucrose in PBS.

5.4 Mounting the specimen block on the specimen holder

To mount the specimen, clamp the specimen stub in forceps (e.g. arterial clamps) and under binocular microscope control, transfer a specimen 'brick' onto the stub using forceps, taking care not to crush the specimen. Stubs supplied by the microtome companies can be used but alternatively metal panel pins from the local do-it-yourself store, with the appropriate shaft diameter to fit the microtome chuck, can also be used (it is important to roughen the head of the pin using abrasive cloth and degrease them using acetone or ethanol). Remove excess cryoprotectant, orient the block, and leave enough cryoprotectant to form a frozen base to support the specimen after freezing. Plunge the stub into liquid nitrogen and store specimen upwards in cryovials vented for safety by making a hole in the lid.

5.5 Sectioning

Preparation of ultrathin sections is one process for which experienced help is needed. Detailed discussion of this process is given in key texts on this subject (1) and on the following website (http://pathcuric1.swmed.edu/Research/HaglerEMlab.htm). There are internationally recognized courses dedicated to training in this method. Diamond knives are increasingly in reach of the budget of most laboratories but good quality glass knives may be prepared as detailed elsewhere (8). The key to producing good cryosections is to trim a rectangular block face of 0.2 mm side or less. This can be done by first trimming a flat face on the block, and then successively each side of the rectangle using the edge of a glass knife or using a commercially available trimming tool. Once each side of the block has been trimmed the block is aligned with the knife edge and the sections cut at approx. 1 mm per sec, at nominal 50–90 nm thickness. Trimming a square or rectangular block face makes ribbons of sections much easier to prepare. The sections should appear clear and may display an interference colour. The sections can be manipulated using an eye-lash or hair probe. The build-up of static is often a problem making it difficult to retrieve the sections. Placing an antistatic device close to the knife during sectioning will usually help to reduce compression (e.g. Diatome Biel Switzerland at http://www.diatome.ch/). Static can also be reduced by humidifying the room prior to, and during, the sectioning procedure (Lucocq and Hug, unpublished observations).

5.6 Picking up the sections

If immunocytochemical labelling is required the sections need to be thawed. The most popular method for doing this is to use a drop of cryoprotectant on a metal wire loop (2–3 mm diameter). One suitable cryoprotectant is 2.3 M sucrose in PBS but increasingly mixtures of methylcellulose and sucrose are used (3), for example 1:1 or 1:3 (v/v) mixture of 2% methylcellulose and 2.3 M sucrose PBS. The methylcellulose reduces the stretching of the sections as they hit the cryoprotectant drop and significantly improves the structural display of membrane bound structures such as endo-lysosomes and Golgi apparatus. The sections are then transferred to a plastic/carbon coated EM support grid and the grid stored on distilled water, PBS, or PBS/gelatin at 4 °C until labelling. Alternatively sections can be transferred to the support grid that has not been freed from its plastic/carbon coating, left covered with their drop of methylcellulose/sucrose, and stored until use at 4 °C.

Protocol 13

Contrasting thawed cryosections using methylcellulose/uranyl acetate

Equipment and reagents

- Metal wire loops
- Whatman No. 1 filter paper
- 3% (w/v) uranyl acetate
- 2% (w/v) methylcellulose

Method

1 Wash grids on four 1 ml drops of distilled water to remove the phosphate ions that can precipitate with uranyl acetate stain.

2 Transfer grids to 1.5 parts 3% (w/v) uranyl acetate + 9 parts 2% (w/v) methylcellulose (25 centipoise) on ice (see ref. 2). Wash on two drops of this solution first before incubating for 10 min.

3 Make films needed to support and stain the sections using metal wire loops that have a diameter wider than the grids. Collect the grid on the loop taking care not to get methylcellulose/uranyl acetate on the back of the grid. Remove excess solution by streaking the edge of the loop over Whatman No. 1 filter paper at an angle of 45° and allow to air dry. Alternatively the staining solution can be removed by touching the filter paper on the side of the drop and controlling the level of solution on the grid. This allows more control over the final film thickness. Films are too thick if the sections and films appear dark and lack contrast under the EM, and too thin if membrane structures are not well seen and structures have a mottled appearance due to air drying artefacts.

---

6 Progressive lowering of temperature (see method 3, Figure 1)

This method is suited to immunolabelling of specimens fixed in aldehydes and requires only a commercial freezer and long wave ultraviolet (UV) lights as specialized equipment. It provides a suitable alternative for immunoelectron microscopy if cryosectioning is not available. In progressive lowering of temperature (PLT) the fixed tissue is dehydrated at low temperatures but at all times the solvents exchange for liquid (not solid) water in the specimen at progressively lower temperatures (this distinguishes this technique from freeze substitution). After dehydration the solvent is exchanged for resin (still at low temperature) and polymerization is achieved using UV light. Dehydration at

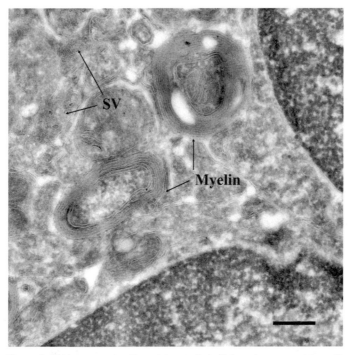

Figure 2 Picking up cryosections using methylcellulose/sucrose mixture. Structural display of membrane structures such as myelin figures and synaptic vesicles (SV) in aldehyde fixed rat cerebellum granular layer is improved. (Tissue supplied courtesy of Chris Thompson, University of Durham, UK). Bar 250 nm.

low temperatures particularly improves the structural display of membranes. Specialized methacrylate/acrylate resin formulations have been developed for use with this method. The embedding can be done in a regular freezer but apparatus developed for freeze substitution can be used (see Leica at http://www.leica-microsystems.com or BAL-TEC at http://www.bal-tec.com). Aldehydes are generally used alone as the colour imparted by osmium prevents UV-induced polymerization. A wide range of solvents have been used including ethanol, methanol, dimethylformamide, and acetone. The resins suitable for use at $-20\,°C$ or lower are Lowicryl K4M (hydrophilic), Lowicryl HM20 (hydrophobic), LR gold™, and Unicryl™. The following protocols use Lowicryl resins.

Protocol 14

Progressive lowering of temperature K4M (hydrophilic resin)

Reagents

- Pig skin gelatin or low melting point agarose
- Ethanol
- Lowicryl K4M resin

Protocol 14 continued

Method

Cut pellets of fixed cells or fixed tissue into blocks 0.5 mm or less in size. If necessary embed in either pig skin gelatin or low melting point agarose and cut to size.

1   Cool on ice in buffer.

2   Transfer to 30% ethanol on ice for 30 min.

3   Continue dehydration as follows in pre-cooled solvent solutions:
    - 50% ethanol at $-20\,°C$ for 1 h.
    - 70% ethanol at $-35\,°C$ for 1 h.
    - 100% ethanol at $-35\,°C$ for 1 h.
    - 100% ethanol at $-35\,°C$ for 1 h.

4   Exchange of resin for solvent. Ethanol/Lowicryl resin, 50:50 (v/v) at $-35\,°C$ for 1 h.

5   Infiltrate with resin as follows:
    - Pure Lowicryl K4M resin at $-35\,°C$ for 1 h.
    - Pure Lowicryl K4M resin at $-35\,°C$ overnight.
    - Pure Lowicryl K4M resin at $-35\,°C$ for 1 h.

Some workers prefer the hydrophobic resins because they do not show the swelling/spreading effects that occur when sections of the hydrophilic resins are floating on the water in the knife boat (see below). The low viscosity of the hydrophobic resin HM20 allows dehydration, infiltration, and embedding to be done down to $-50\,°C$.

Protocol 15

Progressive lowering of temperature HM20 (hydrophobic resin)

Reagents

- Pig skin gelatin or low melting point agarose
- Ethanol
- Lowicryl HM20 resin

Method

Cut pellet or fixed tissue into blocks 0.5 mm or less in size. If necessary embed in either pig skin gelatin or low melting point agarose and cut to size.

1   Cool on ice in buffer.

2   Transfer to 30% ethanol on ice for 30 min.

Protocol 15 continued

3   Continue dehydration as follows in pre-cooled solvent solutions:
- 50% ethanol at $-20\,°C$ for 1 h.
- 70% ethanol at $-35\,°C$ for 1 h.
- 100% ethanol at $-50\,°C$ for 1 h.
- 100% ethanol at $-50\,°C$ for 1 h.

4   Exchange of resin for solvent. Ethanol/Lowicryl resin, 50:50 (v/v) at $-50\,°C$ for 1 h.

5   Infiltrate with resin as follows:
- Pure Lowicryl HM20 resin at $-50\,°C$ for 1 h.
- Pure Lowicryl HM20 resin at $-50\,°C$ overnight.
- Pure Lowicryl HM20 resin at $-50\,°C$ for 1 h.

Embedding can be done in gelatin capsules, plastic vials, or plastic Eppendorf tubes and polymerization carried out under long wave UV according to the resin manufacturer's instructions. Heat is released during the polymerization and it is preferable to employ a heat sink such as an appropriately fashioned aluminium block or an ethanol bath (note there are potential fire risks associated with the use of solvents in freezers). After polymerization at low temperature the blocks may be 'cured' at ambient temperature either by exposure to daylight (a few days) or under UV light. This improves the cutting qualities of the resin.

6.1 Sectioning hydrophilic resins

The sectioning of hydrophilic low temperature methacrylate/acrylate resins such as Lowicryl K4M requires some special precautions. With hydrophobic Lowicryls (HM20 and HM23) sections can be cut at room temperature and floated onto a knife boat filled with water so that the water surface is flat and parallel up to the knife edge. However with hydrophilic resins (K4M and K11M) the water level should be lowered so that the surface forms a concave meniscus as the water approaches the knife edge. If the block surface does become wet then it should be dried immediately using a filter paper.

6.2 Contrasting of Lowicryl sections

The most frequently used contrasting methods employ heavy metal salts uranyl acetate and lead citrate or lead acetate.

Protocol 16

Contrasting Lowicryl sections

Reagents
- Uranyl acetate
- Lead acetate

Protocol 16 continued

Method

1   Float grids on drops of 3% (w/v) aqueous uranyl acetate for 5–10 min. Wash on drops of distilled water and air dry.

2   Float on drops of lead acetate for 45–60 sec for hydrophilic and 5 min for hydrophobic resins (lead citrate is less favoured because it may produce a more granular stain). Staining with lead salts is done in either a nitrogen atmosphere or in the presence of NaOH pellets to reduce carbon dioxide.

This technique in particular produces negative contrast of membranes and strong staining of cytoplasmic filaments such as keratin. Nuclear structures are particularly well contrasted compared to thawed cryosections.

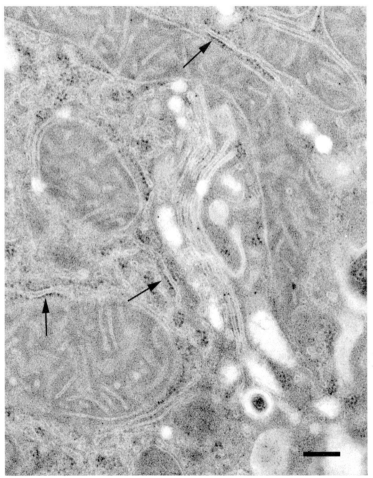

Figure 3 Perfusion fixed rat liver embedded in Lowicryl K4M by PLT. Immunogold labelled for a Golgi membrane protein and stained with uranyl acetate and lead acetate. Note the intense contrast of ribosomes studded along elements of rough ER (arrows). Bar 150 nm.

6.2.1 Adsorption staining

This is a method suitable for hydrophilic resins such as Lowicryl K4M.

(a) A mixture of 1.8% uranyl acetate and 0.2% (w/v) methylcellulose (25 centipoise) is made from a 2% stock of each.

(b) The sections are floated on drops of this solution and then picked up on a metal wire loop as described for thawed cryosections.

(c) The solution is dried down onto the grid. In tissues the contrast of glycocalyx and Golgi apparatus compartments can be especially prominent (9).

7 Freeze substitution of aldehyde fixed specimens (see method 5, Figure 1)

Freeze substitution differs from PLT in that the liquid solvent replaces ice in the frozen specimen rather than water. Accordingly the initial process of substitution occurs at lower temperatures. Small (less than 0.5 mm^2) sample blocks are first cryoprotected in 1.8 M sucrose in 5 mM PIPES buffer pH 7.2 (on ice for at least 1 h). They are then plunged into liquid nitrogen, and transferred to freeze substitution medium in appropriate solvent with or without a fixative/stain such as glutaraldehyde and/or uranyl acetate. After embedding at low temperature (Protocol 17), contrasting of sections can be achieved by using the staining procedures described for progressive lowering of temperature (see Protocol 16).

Protocol 17

Freeze substitution of aldehyde fixed samples in Lowicryl resins

Reagents

- Methanol
- Uranyl acetate
- Lowicryl HM20

Method

1 Methanol + 0.5% uranyl acetate at −90°C for 48 h.

2 Methanol + 0.5% uranyl acetate at −65°C for 6 h.

3 Methanol + 0.5% uranyl acetate at −45°C for 10 h.

4 Methanol at −45°C for 1 h.

5 Methanol/HM20 (1:1) at −45°C for 1 h.

6 Methanol/HM20 (1:2) at −45°C for 1 h.

7 Pure HM20 at −45°C for 12 h.

8 Pure HM20 at −45°C for 1 h.

Protocol 17 continued

9 Polymerization at −45 °C for one day.

10 Polymerization at room temperature for two days.

8 Cryofixed specimens

8.1 High pressure freezing (methods 5, 6, and 7, Figure 1)

High pressure freezing is a method by which ice crystal formation and therefore freezing damage can be avoided in relatively large specimens up to 200 μm thick. The high pressure is applied to the specimen during freezing with liquid nitrogen. High pressure frustrates the growth of ice crystals that would expand to occupy more space than the liquid water in the specimen. Previously, high pressure freezing machines were large and immobile but recent models are smaller, easier to use, and can be moved to sources of living cells/tissues. High pressure freezing is the only freezing technique that can be used to prepare relatively large specimens (approx. 200 μm) for frozen hydrated sectioning or freeze substitution. It is suited to tissues as well as suspensions of cells or organelles. Protocol 18 describes the use of capillary tubes as carriers of cell and organelle suspensions for freezing. This protocol is included because of its potentially general usefulness for handling suspensions in other techniques such as conventional resin embedding.

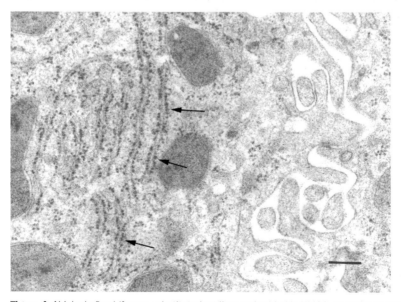

Figure 4 Aldehyde fixed/freeze substituted rat liver embedded in HM20. Note the positive contrast in plasma membrane and the prominent ribosome staining on the rough endoplasmic reticulum (arrows). Bar 200 nm.

8.1.1 Capillary tubes as carriers of suspension cells and organelle fractions

Cellulose capillary tubes from renal haemodialysis units can be used as carriers for freezing in high pressure freezing apparatus (10). They are approx. 200 μm in diameter and are an ideal size for freezing cells and organelles in suspensions. The tubes can also be used as carriers for fixed cells or cells can be fixed after insertion into the tube. Very small samples can be processed in this way. (High pressure freezing machines are available from Leica at http://www.leica-microsystems.com or BAL-TEC at http://www.bal-tec.com.)

Protocol 18

Loading capillary tubes

Equipment and reagents

- High pressure freezing machine
- Capillary tubes
- 1-Hexadecene

Method

1 Separate capillary tubes from the hollow fibre module of a haemodialysis unit and cut into 5 cm lengths.

2 Handle the tubes gently with gloves to avoid crushing and sealing the tubes.

3 Prepare a pellet of cells/organelle suspension.

4 Draw up the suspension by capillary action.

5 If the pellet is very viscous the tube can be embedded into a plastic pipette tip using dental wax, Parafilm, or gelatin. The viscous pellet can then be drawn up using the action of the pipette.

6 Cut into lengths appropriate for the model of the high pressure freezing machine being used if necessary under 1-hexadecene. Cutting the lengths with a scalpel will help to seal the ends of the tubes. (At this point the capillary tubes can be immersed in aldehyde fixative solutions if conventional embedding in epoxy resins is required.)

7 Freeze as directed by the high pressure freezing machine manufacturers and freeze substitute. Loading into a microtome for cutting frozen hydrated sections is facilitated by freezing within copper tubes available with the Leica high pressure freezing apparatus.

8.2 Rapid freezing (method 5, Figure 1)

Freezing of specimens without ice crystal damage is possible if the specimen is slammed on a cooled metal block (slam freezing) or plunged into coolant (plunge freezing). Other methods include spraying the coolant onto the specimen (spray freezing). The technical aspects of these procedures are outside the remit of this chapter but excellent descriptions of these methods are available (1).

9 Freeze substitution of cryofixed samples (method 5, Figure 1)

As already described, freeze substitution can be used to embed aldehyde fixed specimens that have been cryoprotected and frozen in liquid nitrogen. However this method is also a key approach for embedding rapidly frozen material in resin for sectioning and immunoelectron microscopy. The freeze substitution process is started below $-80\,°C$ and allows replacement of frozen water in the specimen, first with solvent and then resin. The infiltration and polymerization of resin may be done at low temperature (favoured for immunoelectron microscopy) or ambient temperature (favoured for structural studies). The two principal solvents used in these techniques are methanol and acetone. Contrasting/fixing agents are often included in the freeze substitution (see below). One of the advantages of freeze substitution over conventional resin techniques and progressive lowering of temperature is its potential ability to preserve lipids. Freeze substitution is conveniently carried out in one of the commercial thermostatically controlled freeze substitution chambers which are also adapted for UV light polymerization obtainable from Leica at http://www.leica-microsystems.com/ or BAL-TEC at http://www.bal-tec.com/. Otherwise it may be done using dry ice (1).

Protocol 19

Freeze substitution of cryofixed samples in Lowicryl resins

Reagents

- Acetone
- Stain
- Fixative

Method

	K4M	HM20	Time	K11M	HM23	K11M	HM23
	Temp.			Temp.		Time	
Acetone + fixative	$-85\,°C$	$-85\,°C$	72 h	$-85\,°C$	$-85\,°C$	72 h	72 h
Acetone + fixative	$-35\,°C$	$-50\,°C$	8 h	$-60\,°C$	–	7 h	–
Pure acetone[a]	$-35\,°C$	$-50\,°C$	1 h	–	–	–	–
Acetone/resin (1:1)	$-35\,°C$	$-50\,°C$	2 h	$-60\,°C$	$-80\,°C$	4 h	12 h
Acetone/resin (1:1)	$-35\,°C$	$-50\,°C$	2 h	$-60\,°C$	$-80\,°C$	4 h	12 h
Resin	$-35\,°C$	$-50\,°C$	2 h	$-60\,°C$	$-80\,°C$	18 h	8 h
Resin	$-35\,°C$	$-50\,°C$	14 h	–	–	–	–
Resin (embed)	$-35\,°C$	$-50\,°C$	1 h	$-60\,°C$	$-80\,°C$	1 h	1 h
Polymerization	$-35\,°C$	$-50\,°C$	24 h	$-60\,°C$	$-80\,°C$	24 h	5 days
Polymerization[b]	RT	RT	72 h	RT	RT	72 h	6 days

Methanol can be used in place of acetone. Added stains/fixatives can be uranyl acetate (0.5%), glutaraldehyde/osmium (see below).

[a] The specimen may be warmed to 0 °C and conventional embedding continued via ethanol as described in Protocol 5.

[b] RT = room temperature.

Protocol 20

Preparation of solvent/fixative mixtures suitable for freeze substitution

Reagents

- Glutaraldehyde/acetone
- Osmium/acetone

Method

1 Glutaraldehyde/acetone (suitable for immunolabelling). Mix 50% (w/v) glutaral-dehyde in water with acetone to give required final concentration. Add molecular sieve 0.4 nm to about 0.25 of the volume and leave for 4–8 h.

2 Osmium/acetone. Dissolve solid osmium tetroxide in pre-cooled analytical grade acetone (less than 0.1% water) at $-30\,^{\circ}$C. If there is more water in the acetone then add molecular sieve (0.4 nm).

3 The above fixatives can also be added to methanol instead of acetone (methanol can tolerate up to 10% water after addition of fixative).

9.1 Contrasting

Sections can be contrasted using the protocols described for the PLT method above. However a recently described technique for increasing contrast uses permanganate oxidation (11). Ultrathin sections are oxidized with 0.1% $KMnO_4$ in 0.1 N H_2SO_4 for 1 min at room temperature. After washing with distilled water sections are stained with a saturated aqueous solution of uranyl acetate (3 min) followed by lead citrate. This method can give improved membrane contrast with high pressure frozen/freeze substituted Lowicryl K4M embedded tissues.

10 Frozen hydrated sections (methods 6 and 7, Figure 1)

Frozen hydrated sections can be prepared from high pressure frozen material. The sections are cut at $-150\,^{\circ}$C and mounted in a TEM cold stage and viewed at low electron dosage. This is an important reference technique that displays near-native structure for comparison with other methods of fixation. The preparation and handling of frozen hydrated sections should be discussed with recognized experts in this procedure. Frozen hydrated sections can be thawed on drops of cryoprotectant containing fixatives for immunogold labelling if required (see method 7 in Figure 1).

11 Immunogold labelling

11.1 Preparation of reagents

Colloidal gold is the contrasting agent of choice for immunoelectron microscopy experiments because it is electron dense, particulate, and can be quantified easily. It can be prepared in a number of different sizes for double or triple labelling and can be complexed to most biological polymers including proteins which means a wide range of affinity reagents can be easily prepared. Colloidal gold can be prepared at low cost and colloidal gold–protein complexes may be stored indefinitely in liquid nitrogen or at −80 °C. The mainstay of colloidal gold preparation for use in electron microscopy is the method described in Protocol 21 (12). This method is important because it allows the experimenter to choose the particle size of the gold colloid with a high degree of precision. Colloidal gold labelling reagents are now available from a number of commercial sources.

Protocol 21

Preparation of colloidal gold

Reagents

- 1% (w/v) chloroauric acid (Sigma)
- 1% (w/v) tannic acid from nutgalls (Aleppo tannin from nutgalls; Mallinkrodt, St. Louis, MO)
- 1% (w/v) trisodium citrate·2H$_2$O

Method

Use clean glassware.

1 Make up a solution of 1% (w/v) chloroauric acid, 1% (w/v) tannic acid from nutgalls (Aleppo tannin from nutgalls—it is important to use this particular tannic acid), and 1% (w/v) trisodium citrate·2H$_2$O.

2 For 100 ml of colloid prepare first a reducing solution containing 4 ml of 1% (w/v) trisodium citrate·2H$_2$O and variable volumes of 1% (w/v) tannic acid. Volumes of tannic acid solution and appropriate pH adjustments are shown below

Reducing solution	Tannic acid
• 5 nm particles	1.00 ml
• 10 nm particles	0.08 ml
• 14 nm particles	0.025 ml

When 1 ml or more of tannic acid is needed, add an equal volume of 25 mM potassium carbonate.

3 Prepare a gold chloride solution with 1 ml of 1% (0.2 ml of 5%) chloroauric acid made up to 80 ml with distilled water.

4 Heat both solutions to 60 °C and add the reducing solution to the vigorously stirred gold chloride solution.

5 Continue heating until the resulting colloid boils gently. The colloids with the smallest particle sizes are formed rapidly but those with the largest particles may require up to 20 min gentle boiling (with reflux) to complete the formation of the preparation.

One of the most widely used colloidal gold reagents for detecting bound antibodies is the protein A–gold complex. Protein A and therefore by inference protein A–gold binds tightly to a large range of IgG antibodies from a diverse array of species (including rabbit, pig, guinea pig, human, and dog). When the primary antibody does not react with protein A (e.g. with most IgG from goat, rat, mouse, sheep, and horse) this reagent can also be used in combination with an intermediate antibody. Complexing of gold particles to protein A can be achieved using Protocol 22. The amount of protein A needed to stabilize the colloid depends on the size of the gold particles.

Protocol 22

Protein A gold

Equipment and reagents

- Polycarbonate centrifuge tubes
- Centrifuge
- Protein A: e.g. the Cowan strain of *Staphylococcus aureus* (Sigma)
- 20 mM Na phosphate buffer pH 6.05
- Colloidal gold

- BSA
- Sucrose
- PBS
- Sodium azide

Method

1 Protein A is dissolved in 20 mM Na phosphate buffer pH 6.05 to give a final concentration of 1 mg/ml.

2 The stabilizing amount for 5 nm gold (made using the tannic acid citrate procedure) is approximately 4–5 μg/ml of gold used (for 9–10 nm gold approximately 2.0–2.5 μg of protein A per ml of gold sol). A titration method for finding out the stabilizing amounts for different gold sizes is described in ref. 13.

3 Dissolve enough protein A to stabilize 60 ml of colloidal gold in 1 ml of sodium phosphate buffer at the bottom of a polycarbonate centrifuge tube (45Ti rotor tube) and pour in the colloidal gold pre-adjusted to pH 6.05 use pH paper and sodium hydroxide to adjust the pH).

Protocol 22 continued

4   2 min later 5% BSA is added to a final concentration of 0.1% (w/v).

5   The tubes are centrifuged at 210000 g max. for 30 min at 4°C (for 10 nm gold use 75000 g, and for 15 nm gold use 7000 g).

6   The loose sediments are then loaded on to 10–30% gradients of sucrose in PBS/0.01% BSA and centrifuged in a SW40 rotor for 45 min at 240000 g max. at 4°C (for 10 nm 650000 g, and 15 nm 5000 g); or until the main body of the gold has descended into the middle of the gradient.

7   The top 3 ml of the red coloured band on the gradient is collected, sodium azide added to 0.02%, and stored at 4°C. Preparations of protein A–gold remain active for more than one year when stored in this way. Alternatively, fractions from the sucrose gradient can be diluted 1:1 in 2.3 M sucrose/PBS and snap frozen in liquid nitrogen and either stored in liquid nitrogen or at −80°C (13). Protein A–gold concentrates are diluted immediately before use and the working dilutions determined by finding the maximum concentration that can be incubated with the sections without yielding significant background staining.

11.2 Immunolabelling

Protocol 23

Immunolabelling of ultrathin sections

Equipment and reagents

- Reusable porcelain or plastic blood typing plates
- Whatman No. 1 filter paper
- PBS/FSG: 0.5% fish skin gelatin in PBS
- 0.1 M ammonium chloride
- Antibody
- PBS
- Protein A–gold
- Uranyl acetate
- Methylcellulose
- Intermediate IgG if required

Method

Unless otherwise specified all steps are performed at room temperature (approx. 20°C).

1   Store grids on distilled water or PBS/FSG at 4°C.

2   Block free aldehyde groups by placing grids on drops 0.1 M ammonium chloride for 5 min.

3   Place on PBS/FSG for a further 5 min.

4   Incubate on 3–20 µl drops of antibody diluted in PBS/FSG (30 min) on Parafilm inside a humidified chamber.

11.3.2 Mixing two secondary antibody gold complexes

If the two primary antibodies have been obtained from distinct species then it is possible to label the antigen simultaneously using second step antibody gold reagents. In general, in our experience, antibody gold conjugates are variable in quality, generally less stable and more liable to aggregation than protein A–gold complexes.

Protocol 26

Using species-specific antibody gold complexes

Reagents

- 0.5% fish skin gelatin in PBS
- Antibody–gold complexes

Method

1 Block non-specific binding by incubating the sections on 0.5% fish skin gelatin in PBS, and incubate on diluted antibodies directed against the two antigens of interest.

2 Wash in PBS.

3 Incubate on antibody–gold of two sizes directed against the two primary antibody species, and wash.

4 Contrast according to the type of ultrathin section (cryosections/low temperature resins).

Note: It is prudent to try labelling with each of the second step reagents after each of the primary antibodies independently to ensure there is no cross-reactivity. It is also possible to use antibody gold complexes in sequential incubations instead of protein A gold in Protocol 25.

Finally resin sections can be double labelled by using both sides of the section. In this approach the sections are mounted on bare (uncoated) EM grids and labelled. The labelled side is then coated with plastic and carbon, the grids inverted, and the sections labelled on the other side with the second set of reagents (different gold size).

11.4 Adsorption of particulate/membrane organelle fractions to grids

Organelles adsorb spontaneously to carbon/plastic coated EM grids and these can then either be fixed or left unfixed before labelling using antibodies and gold probes. The organelles may be effectively supported and contrasted using the methylcellulose/uranyl acetate procedure described for cryosections.

Protocol 27

Gold labelling of organelle fractions adsorbed to EM grids

Equipment and reagents

- Plastic/carbon coated EM grids
- 0.5% glutaraldehyde in 0.2 M PIPES pH 7.4
- 0.1 M ammonium chloride in PBS

Method

1 Float plastic/carbon coated EM grids on unfixed membrane fractions.

2 Transfer directly to drops of fixative compatible with antibody/gold labelling of antigen of interest, e.g. 0.5% glutaraldehyde in 0.2 M PIPES pH 7.4 for 10 min. Do not drain or wash the grid during this transfer—this will retain a maximum amount of organelle attached to the grid film.*

3 Block aldehyde groups using 0.1 M ammonium chloride in PBS.

4 Label using antibodies/gold probes as detailed for thawed cryosections.

5 Contrast using methylcellulose again as detailed for thawed cryosections (Protocol 13).

* Fixation step may be omitted.

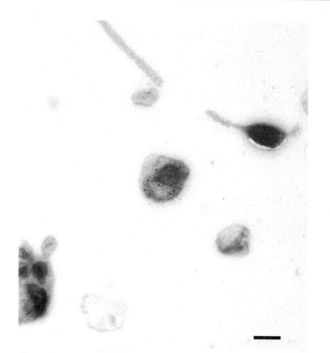

Figure 8 Immunolocalization of phospholipase D-1 in an isolated endosome fraction (ref. 15). Endosome fraction was adsorbed to carbon/pioloform coated grids, fixed in glutaraldehyde, and labelled using anti-PLD (15) before embedding and staining in a mixture of methylcellulose/uranyl acetate (Protocol 13).

11.5 Pre-embedding labelling

A formerly popular approach for immunolabelling was permeabilization of aldehyde fixed specimens using detergents, followed by immunolabelling using antibodies that were localized with horseradish peroxidase (HRP) conjugates. With the advent of cryosectioning, this approach has become less popular. However an adaptation using nanogold particles is worth mentioning. The gold conjugate is enlarged using silver enhancing techniques and the labelling can be correlated with the good structural contrast afforded by the post-fixation in osmium and the embedding in epoxy resin. This protocol was kindly provided by Alexander Mironov (Mario Negri Sud, Italy).

Protocol 28

Localization using small gold in tissues and whole cells

Reagents

- 4% formaldehyde, 0.05% glutaraldehyde in 0.15 M HEPES pH 7.3
- 4% formaldehyde in 0.15 M HEPES pH 7.3
- PBS
- 0.5% fish skin gelatin (FSG) in PBS
- 0.02 M glycine and 0.1% saponin in PBS pH 7.4
- Monovalent Fab fragments of anti-IgG conjugated with nanogold (from

Nanoprobes, NY, USA at http://www.nanoprobes.com/)

- 1% glutaraldehyde in 0.15 M HEPES pH 7.3
- 1% osmium tetroxide containing 1.5% potassium ferrocyanide
- Ethanol
- Epoxy resin

Method

1 Fix cells with 4% formaldehyde, 0.05% glutaraldehyde in 0.15 M HEPES pH 7.3 for 5 min at 37°C.

2 Replace this fixative with 4% formaldehyde in 0.15 M HEPES pH 7.3 for 30 min.

3 Wash with PBS.

4 Treat with blocking solution containing 0.5% FSG in PBS together with 0.02 M glycine and 0.1% saponin in PBS pH 7.4 for 20 min.

5 Incubate the cells with primary antibody in FSG/PBS solution for 60 min. As a guide use a dilution three- to four-fold lower than that used for immunofluorescence.

6 Wash with PBS.

7 Treat cells with monovalent Fab fragments of anti-IgG conjugated with nanogold for 30 min (dilution is about 1/50).

8 Wash with PBS.

9 Fix cells with 1% glutaraldehyde in 0.15 M HEPES pH 7.3 for 5 min.

Protocol 28 continued

10 Silver enhance the gold labelling according to instructions issued with the kit supplied by the manufacturer. It may be useful to test whether the final size of gold particles is optimal using a time course.

11 Wash in PBS and then distilled water.

12 Treat with 1% osmium tetroxide containing 1.5% potassium ferrocyanide in the dark on ice for 2 h.

13 Dehydrate in ethanol.

14 Embed in epoxy resin.

---

12 Quantitation of gold labelling on ultrathin sections

Ultrathin sections are very thin samples of a comparatively large 3D specimen. The problem is to obtain representative unbiased samples of gold labelling that give valid estimates for the whole specimen and ultimately for the population of animals/experiments. Random sampling gives each part of the specimen, or section, an equal chance of being examined. At the level of animals/experiments, this is simply done by selecting animals at random, and in tissue or large pellets by exhaustively slicing and/or chopping the specimen into pieces, and then selecting one or more at random. The piece(s) of specimen is embedded, sectioned, and labelled. Then it is crucial to collect data on gold labelling from the large expanse of section displayed at the electron microscope. Most often this is done in a series of micrographs but it would be a lengthy task to generate a large number of random positions for the micrographs. A better way is to distribute the micrographs evenly over the specimen profile in a regular array. This is easier and faster (and therefore more efficient) compared to random sampling alone. Randomness (and therefore unbiasedness) can still be ensured by placing the first field (position) of the array at a random position, e.g. at the corner of the EM support grid. This type of sampling is often termed 'systematic random'. Regular intervals of the array can be determined using the stage controls at the microscope (see below). It is essential that the array cover the whole specimen/pellet profile (or a randomly selected part of it) for the data to be representative of the specimen.

Experience has shown that it is not necessary to do huge amounts of work to get a useful estimate of gold labelling parameters for an experimental condition or animal. In general a total of 100–200 gold particles or point hits/line intersections (used for estimating structure sizes; see below) for each particular compartment or structure per experiment/animal is very likely to be sufficient. More detailed discussions of sampling schemes are found in refs 2, 16, and 17.

Figure 10 A systematic array of microscope fields are used to sample labelled profiles and count the fractions of gold label over different intracellular compartments. The arrow represents the random start point. Micrographs are taken systematically over the whole pellet. In this case pictures are only taken when the field falls over a cell profile and contains gold labelling. To ensure unbiasedness it may be advisable to apply two-dimensional unbiased counting rules to quadrats placed over the micrographs especially when gold particles are a significant diameter compared to the quadrat (see ref. 17).

(a) It can be used as a guide for labelling intensity when assessing labelling specificity controls.

(b) Because when appropriate controls have been done it may reflect the local concentration of antigen. The relationship between the amounts of gold labelling and antigen, better known as labelling efficiency, is discussed elsewhere (2, 17).

Labelling concentration can be obtained by relating gold counts to the size of labelled compartments estimated using stereological methods. Stereology uses statistical encounters of geometrical shapes/line/points with features in the specimen to obtain estimates of structural quantities in 2D and (more usually) 3D. For reasons of clarity and brevity only the estimators of profile area and profile length (membrane traces) are described here.

12.2.1 Labelling density over profile area

When antigens are distributed in a volume, i.e. they are soluble proteins, it is possible to relate the number of gold particles to the area of the organelle profile displayed on the section (see refs 16 and 17 for more details).

Protocol 30

Estimating the labelling density over cell compartments/structures

1   Select a magnification at which the gold particles and the compartment structure can be identified. Starting outside the pellet profile at a random location (EM grid corner), place fields at regular intervals over the whole pellet profile and check to see how many micrographs will include portions of the structure of interest. Aim for about 20 micrographs and adjust the spacing of the fields accordingly to increase or decrease the fields containing the structure (as described in Figure 10).

2   Either enlarge as prints or examine negatives with a 10× ocular and apply a transparency film with lines in a regular square lattice pattern. Count the number of grid square corners (point hits) and gold particles which fall over the organelle. Adjust the size of the grid to get about 100–200 points and 100–200 golds over all grids/blocks examined for each experimental value. Alternatively the negatives can be scanned on a flat bed scanner in grey scale transmissive mode, enlarged on-screen, and superimposed with a grid generated in Adobe Photoshop 5.5 or later version.

3   Sum the point hits and sum the gold particles and calculate the number of golds per unit area. The area of organelle is found from the product of the number of point hits and the real area of each grid square (the area associated with each point of the lattice).

12.2.2 Labelling density over membrane length area

If an antigen is distributed in membranes then the number of golds per unit length of membrane profile can be found by relating the number of golds to the number of intersections of a grid placed over the cell or tissue profile.

Protocol 31

Estimating the labelling density over cell membranes (e.g. plasma membrane)

1   Take micrographs of the membrane of interest at a magnification at which the gold particles and the membrane can be identified.

2   Starting outside the pellet profile at a random location, space fields at regular intervals over the whole pellet profile and check to see how many micrographs will include portions of plasma membrane. Aim for about 20 micrographs and adjust the spacing of the fields accordingly to increase or decrease the fields containing plasma membrane.

Protocol 31 continued

3   Either enlarge as prints or examine negatives with a 10× ocular and apply a square lattice grid on transparency film with spacing which will produce a total of 100–200 intersections with the plasma membrane on all grids/blocks for that experimental condition. Alternatively scan the negatives on a flat bed scanner in grey scale/ transmissive mode, enlarge on-screen, and apply a grid electronically using, for example, Adobe Photoshop version 5.5 or later versions.

4   Count the number of intersections of the grid lines with the plasma membrane and count the gold particles associated with this structure (count intersections with the grid lines in both line directions of the grid, and define the lines by one of their borders not the line itself, which has a finite thickness).

5   Finally from the total intersections over all micrographs calculate the length of profile trace examined from:

$$\frac{\pi}{4} \times I \times d$$

where I is sum of intersections and d is the real distance between the lines.

12.2.3 Counting when labelling is low or very dense

If labelling is low it may be necessary to scan all the cell profiles found in a square/hexagon of the support grid. A preliminary idea of the number of golds per grid hole can be obtained by counting the number of gold particles per cell profile and counting the cell profiles displayed in a grid square. As already mentioned it is generally useful to aim for 100–200 particles for a compartment over the whole experiment/animal. The grid holes should be selected at random or in a systematic random pattern. If the labelling is very dense then subsampling the gold on micrographs can be done on micrographs by applying a regular array of counting frames (for example quadrats) that together represent a known fraction ($1/P$) of the total area examined (this could be, for example, one in every four grid squares of a square lattice grid). The number of gold particles counted in any micrograph is then the product of the number of gold particles counted × P (see ref. 17 for how unbiased counting rules applied to these quadrats).

12.3 Controls for the specificity of labelling

The problem with fixed/resin embedded specimens is that components bound to antibodies within the section are not easily extracted to confirm their identity. Therefore more indirect tests of specificity are required. The most convincing of these are based on experimental introduction, withdrawal, or modulation of the antigen within the biological system of interest. Examples include temperature-sensitive mutations that abrogate expression of the target antigen, induction of antigen expression by using transient transfection techniques, induced expression in stable cell lines, or simply introducing the antigen by micro-injection or by uptake via a cellular pathway such as endocytosis. Other controls such as inhibition of labelling using the cognate purified antigen, the use of pre-immune sera, or independent evidence for specificity in immunoblots or immunoprecipitations, all increase confidence but unfortunately do not address directly the question of the identity of the antibody binding sites within the section.

13 Methods for visualizing molecules and macromolecular assemblies

The traditional method for visualizing individual proteins or complexes of proteins is negative staining. Newer cryomicroscopy methods are now becoming increasingly important for structure determination. These require specialized skills and equipment and are therefore not covered here (see for example refs 18–20).

13.1 Negative staining

Protocol 32

Preparation of samples for negative staining (general protocol)

Reagents
- Formvar, colloidion, or pioloform
- Uranyl acetate

Method

1 Prepare grids coated with formvar, colloidion, or pioloform.

2 Dilute sample in suitable buffer. Note that uranyl acetate is less soluble above pH 6 so if possible use water for dilution if it has no detrimental effect on your protein, etc.

3 Place a drop of protein solution on the coated grid and leave for 30–60 sec. Blot off excess.

4 Successively place three drops of aqueous 1–2% uranyl acetate onto grid, blotting off excess each time.

5 Blot dry and air dry for 5 min before putting into microscope.

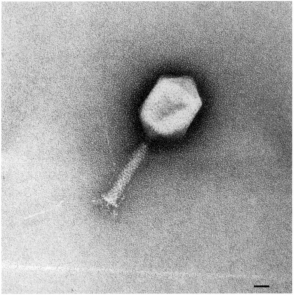

Figure 11 T4 bacteriophage negatively stained with 2% uranyl acetate. Bar 20 nm.

Protocol 33

Preparation of samples for negative staining (ref. 21)

Equipment and reagents

- Freshly split mica
- 1.5 ml Eppendorf tube
- Uranyl acetate

Method

1 Carbon coat (4 nm) 3 mm square pieces of freshly split mica. Prepare 1% uranyl acetate in water.

2 Dilute sample into 200 μl buffer or water in a 1.5 ml Eppendorf tube cap.

3 Put 200 μl of 1% aqueous uranyl acetate into a second cap.

4 Introduce mica piece into protein solution at an angle of 45° so that the carbon film floats on the surface of the solution; do not allow it to come off completely. Withdraw mica picking up the carbon film.

5 Introduce mica into the uranyl acetate solution at a 45° angle as in step 4, allowing the carbon film to float off onto the surface and the mica to drop to the bottom of the cap.

6 Pick up the carbon film on a small hole copper grid (we use 400 mesh).

7 Air dry for 5 min on filter paper before introducing into the microscope.

8 Several grids can be prepared from the protein solution but use fresh uranyl acetate each time.

Other staining solutions suitable for negative staining are:

(a) Ammonium molybdate, prepared as a 1% solution of ammonium molybdate in water. This has also been used for staining thawed cryosections.

(b) Neutral phosphotungstic acid, consisting of a 1–3% solution of phosphotungstic acid in water with pH adjusted to 7 using sodium hydroxide. This stain is useful when low pH of uranyl acetate must be avoided.

14 Cryofracture techniques: freeze fracture and freeze etching

This body of methods relies on cryofracturing frozen cells/organelles. The fractured surface can be coated with a metal replica that is then observed in the transmission electron microscope. To improve relief of intracellular structures water can be etched away prior to coating. Freeze fracture of aldehyde fixed cryoprotected specimens produces stunning images of cellular junctions and intramembrane particles as the plane of fracture passes between the lipid

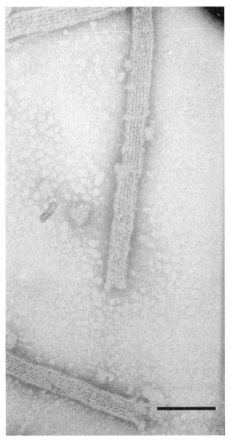

Figure 12 Microtubules negatively stained by the method of Valentine *et al.* (21). Micrograph supplied by Alan Prescott, Dundee University, UK. Bar 150 nm.

bilayer. However the popularity of the approach waned because it is not well suited to correlative immunolabelling. Recently the situation has changed with the advent of methods that allow cryofracture, coating, and labelling of replicas from rapidly frozen tissues, cells, or organelles.

14.1 Cryofracture labelling

Protocol 34 describes a procedure for combining freeze fracture with immuno-labelling. After rapid or high pressure freezing and fracture, a metal replica is shadowed onto the internal aspect of frozen membranes, membrane proteins and lipid adhere to the replica. Cell components are then digested away from the membrane using SDS allowing the replica to be labelled using antibodies and gold probes (22). This is the method of cryofracture after slam freezing and is suitable for use with suspension cells and organelle suspensions. Methods for slam freezing cells grown on coverslips are given in ref. 23.

Protocol 34

Cryofracture labelling

Equipment and reagents

- 10 μm thick copper foil
- Freeze fracture device
- 10% (v/v) nitric acid
- Ethanol
- 30% nitric acid
- SDS
- Tris pH 8.3
- Sucrose

Method

1 Cells or cell organelles are concentrated by centrifugation.

2 Cover foils are prepared using 5.5 mm diameter circles of 10 μm thick copper foil that have been punctured with many pinholes using a scalpel blade. Before use the copper circle is cleaned by dipping into 10% (v/v) nitric acid in distilled water. The circles are rinsed in distilled water and stored in ethanol until use.

3 Base plates are prepared from copper sheet of thickness 0.25 mm and cut into rectangles of appropriate size for the slam machine. These sheets are flattened out by pounding and dipped into 30% nitric acid in distilled water for 2 min, and then rinsed in distilled water, ethanol, and stored in ethanol until needed.

4 Cut a sheet of adhesive paper the same size as the copper base plate and punch a hole 5 mm diameter and adhere to the base plate.

5 Adhere the base plate of the slammer with recommended adhesive and place 2–5 μl of the cell or organelle suspension.

6 Apply the cover foil to the specimen by pressing gently with forceps and excess buffer/medium is removed making sure that there is no buffer/medium on the outside surface of the foil.

Protocol 34 continued

7 Mount the specimen holder onto the freezing machine and drop onto the copper block cooled with liquid helium.

8 The specimen is transferred to a freeze fracture device. The copper foil is separated from the frozen specimen and the fracture face is shadowed unidirectionally with platinum at an angle of 40–45°. Then it is backed with rotary deposition of carbon at 60°.

9 Place the specimen with attached replica into PBS. The replica will be released and can be transferred to 5 ml of 2.5% (w/v) SDS in 10 mM Tris containing 30 mM sucrose pH 8.3. Stir vigorously for 1–12 h; the replica will fragment into very small pieces.

10 The replica pieces are immunolabelled essentially as described for thawed cryo-sections (Protocol 23). After rinsing in distilled water they are brought to the water surface where they are transferred to plastic/carbon coated grids by touching the grid face onto the replicas.

11 Replicas are observed in the transmission electron microscope.

Note: Platinum wire (0.05 mm diameter) is used to make loops for manipulating the small fragments of replica. The loop can be formed around a needle approx. 0.5 mm in diameter.

15 Scanning electron microscopy

Newer field emission high resolution scanning electron microscopes have broken the resolution barrier that limited the usefulness of scanning EM in cell biology. The new microscopes now allow nanometre resolution of structure and visualization of correlative gold labelling.

Protocol 35

Sample preparation for high resolution SEM whole cell mounts (FESEM)

This protocol was provided by Alan Prescott, University of Dundee, UK.

Equipment and reagents

- Silicone wafer
- Glutaraldehyde
- Paraformaldehyde
- Osmium tetroxide
- Fish skin gelatin
- Antibody
- Ethanol
- Acetone

Method

1 Grow cells/stick samples to pieces of silicone wafer (sometimes requires poly-L-lysine).

> **Protocol 35** continued

2   Fix cells in suitable fixative; glutaraldehyde and paraformaldehyde for morphology, paraformaldehyde alone for immunolabelling.

3   Post-fix in osmium tetroxide if not immunolabelling. Permeabilize for antibody penetration.

4   Block with fish skin gelatin.

5   Label with primary antibody.

6   Wash, label with either gold-conjugated secondary antibody or intermediate rabbit antibody followed by protein A–gold. Use 8–15 nm gold for surface labelling; smaller for penetration.

7   Dehydrate in ethanol, change from 100% ethanol to 100% acetone (two changes, CO_2 is more soluble in acetone).

8   Critical point dry using at least 10 flushes with fresh CO_2. (We use the Baltec CPD 030.)

9   Coat samples with 2 nm chromium for immunogold or 5 nm chromium or gold/palladium (2–5 nm) for morphology. (We use a Cressington 208HR sputter coater.) Uncoated samples benefit from a thin carbon coating (5 nm) to prevent charging. (We use an Agar Turbo carbon coater.)

10 Image morphology at 5–15 kV for coated samples, use 1 kV on uncoated samples and 15 kV for immunogold. Use backscatter detector for gold and SE detector for morphology (e.g. we use the upper SE in-lens detector on the S4700, though the lower conventional detector may be better for low magnification and maximum depth of focus on 3D specimens). 10 nm gold can be imaged at $> \times 25$ K if the instrument is well aligned, $> \times 40$ K may be needed for difficult samples or smaller gold. Focus and astigmatism are critical for the BSE image (using analysis mode on the S4700 gives a better BSE signal).

Protocol 36

Immuno-FESEM on organelles

This protocol was provided by Martin Goldberg, Patterson Institute, Manchester, UK.

Equipment and reagents

- Silicon chip
- PBS
- 2% paraformaldehyde in PBS
- 100 mM glycine in PBS
- 1% fish skin gelatin in PBS
- Primary antibody in PBS
- Gold labelled secondary antibody in PBS
- 2% glutaraldehyde, 0.2% tannic acid, 0.1 M HEPES pH 7.4
- 0.2 M sodium cacodylate pH 7.2
- 0.1% osmium tetroxide in 0.2 M sodium cacodylate
- Ethanol
- 1,1,2-trichlorotrifluoroethane

Protocol 36 continued

A. Immunolabelling

1 Isolate organelle/cell and adhere to silicon chip.

2 Fix with 2% paraformaldehyde in PBS for 10 min.

3 Transfer to 100 mM glycine in PBS for 10 min.

4 Block in 1% fish skin gelatin in PBS for 1 h.

5 Incubate with primary antibody in PBS for 1 h.

6 Wash with PBS six times for 5 min.

7 Incubate gold labelled secondary antibody in PBS for 1 h (we use Amersham secondary antibodies).

8 Wash with PBS six times for 5 min.

B. Preparation for FESEM

1 Fix again with 2% glutaraldehyde, 0.2% tannic acid, 0.1 M HEPES pH 7.4 for 10 min.

2 Wash with 0.2 M sodium cacodylate pH 7.2 for 5 min.

3 Post-fix in 0.1% osmium tetroxide in 0.2 M sodium cacodylate for 10 min.

4 Wash two times in water for 5 min each.

5 Dehydrate in 30, 50, 70, 95, 100, 100% ethanol.

6 Transfer to 1,1,2-trichlorotrifluoroethane.

7 Critical point dry via 1,1,2-trichlorotrifluoroethane from CO_2.

8 Sputter coat with 2–4 nm chromium in a high vacuum coating unit.

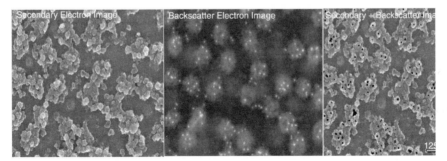

Figure 13 A *Xenopus* oocyte nuclear envelope labelled with an antibody called mAb414 which labels a family of nuclear pore proteins. The first image is a secondary electron image of the cytoplasmic side of the nuclear envelope. The second image is a backscatter electron image of the same area to show the position of the gold–secondary antibody. In the third image the backscatter image has been thresholded to show just the gold particles, contrast inverted, and superimposed onto the secondary image to enable comparison of the position of the gold together with the structural information. (Images kindly provided by Martin Goldberg.)

Acknowledgements

This article is dedicated to colleagues past and present whose work has contributed to this article and who have worked to promote the development of low temperature and quantitative EM methods. JML was supported by a Research Leave Fellowship from the Welcome Trust (059767/Z/99/Z) and by Tenovus Scotland. My thanks go to Dr Alan Prescott, John James, Callum Thomson and Stephen Watt for helpful comments.

References

1. Robards, A. W. and Wilson, A. J. (ed.) (1993). *Procedures in electron microscopy*.
2. Griffiths, G. (ed.) (1993). *Fine structure immunocytochemistry*, Springer–Verlag, Berlin.
3. Liou, W., Geuze, H. J., and Slot, J. W. (1996). *Histochem. Cell Biol.*, **106**, 41.
4. Thompson, C. L., Bodewitz, G., Stephenson, F. A., and Turner, J. D. (1992). *Neurosci. Lett.*, **144**, 53.
5. Prescott, A. R., Farmaki, T., Thomson, C., James, J., Paccaud, J. P., Tang, B. L., *et al.* (2001). *Traffic*, **2**, 321.
6. de Wit, H. (2000). PhD thesis. University of Utrecht.
7. Tokuyasu, K. T. (1989). *Histochem. J.*, **21**, 163.
8. Webster, P. (1993). *J. Microsc.*, **169**, 85.
9. Roth, J., Taatjes, D. J., and Tokuyasu, K. T. (1990). *Histochemistry*, 95, 123.
10. Hohenberg, H., Mannweiler, K., and Muller, M. (1994). *J. Microsc.*, **175**, 34.
11. Sawaguchi, A., Ide, S., Goto, Y., Kawano, J. I., Oinuma, T., and Suganuma, T. (2001). *J. Microsc.*, **201**, 77.
12. Slot and Geuze (1985). *Eur. J. Cell Biol.*, **38**: 87–93.
13. Lucocq, J. M. (1993). In *Fine structure immunocytochemistry* (ed. G. Griffiths), pp. 279–302. Springer–Verlag, Berlin.
14. Keller, G. A., Tokuyasu, K. T., Dutton, A. H., and Singer, S. J. (1984). *Proc. Natl. Acad. Sci. USA*, **81**, 5744.
15. Lucocq, J., Manifava, M., Bi, K., Roth, M. G., and Ktistakis, N. T. (2001). *Eur. J. Cell Biol.*, **80**, 508.
16. Lucocq, J. M. (1993). *Trends Cell Biol.*, **3**, 345.
17. Lucocq, J. M. (1994). *J. Anat.*, **184**, 1.
18. Unger, V. M. (2001). *Curr. Opin. Struct. Biol.*, **11**, 548.
19. Tao, Y. and Zhang, W. (2000). *Curr. Opin. Struct. Biol.*, **10**, 616.
20. Frank, J. (2002). *Annu. Rev. Biophys. Biomol. Struct.*, **31**, 303–319.
21. Valentine, R. C., Shapiro, B. M., and Stadtman, E. R. (1968). *Biochemistry*, **7**, 2143.
22. Fujimoto, K., Umeda, M., and Fujimoto, T. (1996). *J. Cell Sci.*, **109**, 2453.
23. Fujimoto, K. (1997). *Histochem. Cell Biol.*, **107**, 87.

Chapter 4
Subcellular fractionation

Alois Hodel

Department of Biology and Biochemistry, University of Bath,
Bath BA2 7AY, UK.

J. Michael Edwardson

Department of Pharmacology, University of Cambridge,
Tennis Court Road, Cambridge CB2 1QJ, UK.

1 Introduction

Subcellular fractionation has been crucial to the development of modern
cellular and molecular biology. In recent years, the major driving force behind
the isolation of organelles has been the requirement to study their function *in
vitro*. Naturally, if organelles are to be used in *in vitro* assays, native structure and
activity must be preserved during fractionation, and protocols are constantly
being improved with this in mind. Although separation techniques based on
non-specific physical properties are still widely used, additional methods based
on specific biochemical interactions, such as antigen–antibody recognition, have
been introduced. The principles behind these various techniques are discussed
in this chapter, and specific examples will be described in detail. Throughout,
the emphasis will be on the isolation of functionally competent organelles.

2 Basic principles

Subcellular fractionation protocols have been established for most organelles
and other cellular constituents, and the majority of these involve at least one
centrifugation step. Centrifugation exploits the fact that subcellular constituents
in suspension differ in size, density, and shape, and thus behave differently
under the influence of a centrifugal force. It also permits the convenient separation of large volumes of crude tissue homogenates into smaller volumes of
material for further purification. Centrifugation, however, suffers from several
drawbacks. For example, there is considerable heterogeneity in the size and
density of certain organelles, and these organelles may be influenced unpredict-
ably by the surrounding medium. Furthermore, organelles often vesiculate
during homogenization, so that the difference in the sizes of microsomes
derived from different cellular membranes may be eliminated. Consequently,

complete separation is seldom achieved using centrifugation alone, and more sophisticated methods have been introduced. These include size exclusion chromatography (1), partition between liquid phases (2), free-flow electrophoresis, and immunoisolation. Immunological methods, in particular, have led to considerable improvements in the resolving power of fractionation methods. Specifically, the presence of organelle-specific markers accessible on their surfaces has permitted the adsorption of the organelles to solid supports via specific antibodies, and subsequent purification to homogeneity.

3 Choice of starting material

The questions to be asked and the practicality of the methods to be used dictate the choice of starting material. For example, tissue culture cells are often used, either because they are the only pure source of a particular cell type or because an assay such as that for a transport step between two organelles may require the introduction of an exogenous protein by pre-incubation or transfection. Evidently, it is advisable to choose a tissue that is rich in the organelle to be purified. The tissue should also be repeatedly available in a high-quality form, and preferably in large quantities. Rat liver, which is easy to manipulate and readily obtainable, is often used as a source material, not least because numerous established fractionation procedures exist for this tissue. The nature of the starting material in turn determines which fractionation methods are feasible. Cells or tissues that can be disrupted using minimal shear forces are preferable because the organelles will consequently suffer a minimum amount of damage. The tissue to be homogenized ought to be fresh, although frozen tissue may also be used provided it has been frozen rapidly and in small pieces; ice crystal formation may disrupt organelles and thus lead to the leakage of luminal components (e.g. proteases). If possible, a cell type or tissue with little or easily suppressible protease activity should be chosen.

4 Homogenization of tissue and cultured cells

The major goal of homogenization is to achieve efficient cell breakage so that the organelles of interest are not significantly damaged during the process. A monodisperse suspension of intact organelles and other cellular constituents should be aimed for to allow subsequent separation by centrifugation or other techniques (3). To preserve the structural integrity and activity of organelles, homogenization conditions must be optimized for different tissues and cell lines, which will differ in structure and composition. Disruption can be achieved by osmotic shock, ultrasonic vibration, or by mechanical shearing. Devices used for mechanical disruption include Potter–Elvehjem (Teflon/glass) homogenizers, Dounce (glass/glass) homogenizers, French press cells, Polytron homogenizers, and ball-bearing homogenizers. The first two types of homogenizers are used most often, particularly for solid but soft tissues such as liver, kidney, or brain. Typically, 5–10 strokes (Potter–Elvehjem with a clearance of about 90 μm), or

15–20 strokes (Dounce homogenizer with loose-fitting pestle) are required to achieve complete disruption. Fibrous tissues such as skin, thymus, skeletal muscle, smooth muscle, uterus, or heart require some form of rotating-blade device (e.g. Waring blender, Polytron PT20, or Ultraturrax PT 18/2 tissue disintegrator), either alone or as a prelude to Dounce or Potter–Elvehjem homogenization. The severity of the shearing forces required to disrupt the cells dictates what kind of device is applicable. In a French press cell a cellular suspension is forced by a piston through a narrow orifice (often the annulus around a ball-bearing). The shear forces are controlled by the size of the orifice and the imposed pressure. The method is highly reproducible and can be used for bacteria as well as eukaryotic cells. Ultrasonication has relatively little application to eukaryotic cells, because it is difficult to achieve conditions that disrupt the plasma membrane but keep the organelles intact.

For efficient lysis of cultured cells it may be necessary to stress the cells osmotically in hypotonic buffer for up to five minutes before homogenization with either a Dounce or a Teflon/glass homogenizer. When doing so, it is important to use a high cell concentration, so that cytoplasmic proteins will protect the released organelles to some extent from osmotic stress. The isotonic milieu is restored subsequently by adding a concentrated sucrose solution. If homogenates are kept in hypotonic solution for longer, organelles will begin to lyse and lysosomal enzymes may degrade other cellular components. Altern-atively, cultured cells can be made hyperosmotic through glycerol treatment. When cells are subsequently returned to iso- or hypo-osmotic medium, most will lyse without further homogenization, and thus without shear forces (4). Again, conditions must be optimized for each particular cell type. Cultured cells may also be disrupted using a nitrogen cavitation device. In this technique, high pressure dissolves nitrogen in the cytoplasm and a subsequent pressure release leads to the formation of nitrogen gas bubbles and efficient cell rupture. This method has the advantage of being highly reproducible, although organelles are rendered fragile, endosomal and plasma membranes vesiculate, and ribosomes are often stripped from the rough endoplasmic reticulum.

Potential difficulties that occur during homogenization include the formation of aggregates of cytoskeletal elements, organelles, or genomic DNA released from ruptured nuclei. Aggregation may result in premature pelleting during centrifugation, as well as the loss of lighter subcellular components through co-sedimentation. Inclusion of $MgCl_2$ helps to preserve nuclear integrity, and DNase I (at 10 μg/ml) can prevent aggregation by liberated genomic DNA. Relatively large quantities of connective tissue or large blood vessels can hamper homogenization and it may be necessary to remove these by prior dissection. Also, many tissues are not homogeneous in terms of the cell types they contain, and in some cases contaminating cell types must be removed prior to homogenization in order to avoid artefacts. A further potential problem is the redistribution of macromolecules from the cytoplasmic surface of organelles to the cytosol or to other organelles when cells are disrupted; any protein dissolved in the medium can be trapped inappropriately during homogenization by vesiculation of membrane fragments. There also is a risk of overheating, and samples should be cooled during homogenization. Foaming should be kept to a minimum in order to prevent inactivation of enzymes through denaturation at the surface of bubbles. This can be achieved by filling the container to reduce the area of liquid surface exposed to air. Plasma membranes are particularly difficult to isolate because of vesiculation and fragmentation of the ruptured membranes. An isolation protocol designed to prevent this is referred to as 'density perturbation' (5). Cells are allowed to adhere to positively charged colloidal silica beads before disruption, and large sheets of plasma membrane remain attached to the beads and can subsequently be recovered.

Many organelles are surprisingly robust, and most retain their biological activities after homogenization if suspended in the appropriate buffer. Isotonic homogenization buffers typically contain 0.25 M sucrose, supplemented with mono- or divalent cations, protease inhibitors, chelating agents, and an organic buffer with a pK_a near the pH value of the working solution. Buffer components must not interfere with any of the enzymes under investigation (e.g. phosphate —being a metabolite—can affect phosphatases, act as an allosteric effector, or bind metal ions in the working solution). Dithiothreitol (prevents oxidation), EDTA (chelates heavy metals and inactivates metalloproteases), glycerols and other polyols are frequently added to stabilize enzymes and maintain protein–protein interactions. ATP is often added during the isolation of mitochondria, such as in the widely used Chapel–Perry medium (6). To protect organelles from proteases, it is common to include a cocktail of protease inhibitors in the homogenization buffer. A typical mixture is phenylmethylsulfonyl fluoride (PMSF), leupeptin, antipain, aprotinin, and α_2-macroglobulin. These are usually stored as 100 × concentrated stock solutions. PMSF is not stable and must be added repeatedly from a stock solution in dimethyl sulfoxide or isopropanol. Leupeptin and antipain stock solutions are usually kept in 10% dimethyl sulfoxide, whereas aprotinin is stored in water.

Protocol 1 is applicable to a single rat liver and will release all major organelles, although the endoplasmic reticulum will vesiculate. If nuclei are to be isolated, 25 mM KCl and 5 mM $MgCl_2$ instead of EGTA should be included in the homogenization medium. Intact Golgi stacks and contiguous sheets of plasma membranes, stabilized by junctional complexes and by desmosomes, respectively, can be purified under mild conditions (i.e. homogenization in hypotonic medium in the presence of stabilizing divalent cations). A Dounce homogenizer is used to apply a minimum amount of shearing forces, and it is essential to perfuse the liver prior to homogenization.

Protocol 1

Homogenization of rat liver in iso-osmotic sucrose buffer

Equipment and reagents

- Scissors and thick forceps
- Large Dounce homogenizer with medium-fit plunger

- Surgical gauze
- STE buffer: 250 mM sucrose, 5 mM Tris pH 7.4 , 2 mM EGTA at 4 °C

Method

1   Kill the rat by cervical dislocation.

2   Carry out all procedures at 4 °C. Open the abdominal cavity, and remove the liver.

3   Perfuse liver by injection of STE buffer through the portal vein and transfer to STE buffer in a beaker.

4   Chop the tissue into approx. 1 mm³ pieces.

5   Transfer the tissue to the homogenizer and add 10 vol. of STE buffer. Homogenize using 10 strokes of the pestle. Wear protective gloves—these will protect against injury in case the homogenizer shatters, and will also minimize heat transfer to the sample.

6   If the nuclear pellet is needed, filter the homogenate through three layers of surgical gauze to remove connective tissue and clumps of cells.

The cell cracking method employs a stainless steel or tungsten carbide ball homogenizer. Cells passed once through the chamber of the device are rendered permeable. However, multiple passages through the chamber are required to homogenize cultured cells thoroughly. Homogenization in isotonic medium results in the release of intact organelles (7). Protocol 2 describes the use of a ball-bearing homogenizer to homogenize cultured cells.

Protocol 2

Homogenization of cultured cells using a ball-bearing homogenizer

Equipment

- Ball homogenizer. The precise bore of the hole relative to the size of the ball needs to be determined empirically for each cell type. For most cultured cells (diameter approx. 15 μm), a clearance of 20–25 μm is optimal.

Method

1   Use the homogenizer at room temperature with a cold (4 °C) suspension of cells (pre-chilling the homogenizer results in suboptimal clearance, and consequently in an increase in cell debris and a decrease in reproducibility).

Protocol 2 continued

2   Scrape the cells from the culture dish using a rubber policeman, and resuspend in a suitable homogenization buffer.

3   Pass a suspension of 10^6 to 10^7 cells/ml through the homogenizer. The number of passes (typically 8–10) depends on the cell type and the degree of disruption required.

5 Fractionation by centrifugation

Centrifugation separates organelles and other subcellular components on the basis of differences in size, shape, and buoyant density (the buoyant density of a particle is related to its actual density and to the degree of hydration) (8). These parameters determine the velocity of a sedimenting particle in either uniform liquid (differential pelleting) or a liquid containing a continuous concentration gradient of a solute (velocity or isopycnic gradient centrifugation). When particles are to be pelleted in a gradient, a suspension is layered over the top of the gradient (low density). During centrifugation, particles travel at different velocities towards the bottom of the gradient (high density). It can be shown that the sedimentation velocity, v, of a particle in a centrifugal field is given by:

$$v = \frac{d^2(\rho_p - \rho_l)}{18\,\eta}\,\text{RCF}$$

where d is the diameter of the particle, ρ_p is the density of the particle, ρ_l is the density of the liquid, η is the viscosity of the liquid, and RCF is the relative centrifugal field. The equation is based on Stokes' law, which assumes that the sedimenting particle is spherical in shape. It shows that the sedimentation rate is proportional to the square of the diameter, and to the difference between the buoyant density of the particle and that of the surrounding medium. The relative centrifugal field is proportional to the distance of the particle from the centre of rotation and to the square of the rotor speed (9).

   In most centrifugation protocols, cell homogenates are fractionated initially by a series of differential pelleting steps, which reduce but do not eliminate the amount of contaminants in each fraction. Subsequently, centrifugation tech- niques with higher resolving power, such as velocity gradient centrifugation or isopycnic gradient centrifugation, are applied to the fractions obtained. Difficulties experienced during centrifugation can be caused by the non- homogeneity of cell types in many tissues. Furthermore, labile organelles that are ruptured during homogenization are likely to pellet in an inappropriate fraction. It is thus clear that the histology and cytology of the tissue of interest must be considered prior to fractionation. Essential equipment required for centrifugation includes centrifuges with compatible rotors, gradient-forming and gradient-collecting devices. Rotors are available in four basic types: vertical, fixed-angle, swing-out, and zonal rotors. If isolated fractions are to be used preparatively, one has to bear in mind the limits of the amounts of material that can be loaded onto a gradient. Fixed-angle and zonal rotors have a higher capacity than swing-out rotors and are frequently used in preparative centrifugation procedures.

5.1 Differential pelleting

Separation of subcellular particles by differential pelleting is based on differences in sedimentation rates that are largely a result of differences in size. Centrifugation of a cellular homogenate is continued for just long enough to pellet the largest particles. The supernatant is then spun at the next higher speed to pellet the next-largest remaining particles, and so on. The pellets obtained contain contaminating smaller particles. It is therefore important not to centrifuge longer than necessary and to wash the pellets at least once (i.e. to repeat the centrifugation step with the resuspended pellets). Differential pelleting can be used to prepare several subcellular fractions enriched in particular organelles. Protocol 3 illustrates this technique by describing the isolation of mitochondria from liver.

Protocol 3

Preparation of mitochondria from rat liver

Equipment and reagents

- Bench centrifuge
- 8 × 50 ml rotor, cooled to 4 °C, and suitable centrifuge tubes
- STE buffer: 250 mM sucrose, 5 mM Tris pH 7.4, 2 mM EGTA, at 4 °C

Method

1 Prepare a rat liver homogenate as described in Protocol 1.

2 Centrifuge for 3 min at 1000 g.

3 Carefully pour off the supernatant into another tube and centrifuge for 10 min at 12 000 g.

4 Discard the supernatant, and gently resuspend the pellet in a small volume of STE buffer, taking care not to disturb the blood spot.

5 Transfer the resuspended pellet into a centrifuge tube, and top up with STE. Centrifuge for 10 min at 12 000 g.

6 Resuspend and re-centrifuge the pellet as in steps 4 and 5.

7 Resuspend the pellet to the desired volume.

5.2 Velocity gradient centrifugation

As for differential pelleting, separation of particles in velocity gradient centrifugation (or rate-zonal density gradient centrifugation) is based upon different sedimentation rates. These are a function of particle size, shape, buoyant density, and the density and viscosity at each point in the gradient (10). Again, the time of centrifugation is crucial, since all of the particles will pellet eventually (the maximum density of the gradient is less than the buoyant density of the particles). Particles are layered on top of a continuous gradient in a small volume, and during centrifugation they form a series of bands according to the velocity of their migration into liquid of higher density. This approach does not work over a uniform liquid because the sedimenting bands of particles are not stable. Generally, the primary purposes of using a gradient are to ensure that each population of particles migrates as a band and to prevent convection.

Velocity gradient centrifugation should ideally be used after differential pelleting, since it has a higher resolving power. There is, however, a limit to the range of particle sizes that can be resolved, because each band containing a particular type of particle will be finite in width. Thus, if the range is too great, large particles will accumulate in the pellet before the small ones leave the starting zone. Normally, swing-out rotors are used to minimize wall effects and to maximize the length of gradients, and hence the size range of particles that can be separated. Generally, particles that differ in sedimentation rate by at least 20% can be distinguished in a velocity gradient. There is also an upper limit to the amount of material that can be loaded: if a gradient is overloaded, the sedimenting bands will be unstable.

Solutes used to form gradients must be water soluble and must not damage the structures of the organelles. They should not precipitate with calcium, magnesium, and other ions that occur in cell homogenates, and they must not interfere with enzyme assays to be performed after separation. The viscosity of the medium should be kept low, so as to minimize the time required for sedimentation and for the separation of individual bands of particles. This is an important consideration because the width of individual zones of particles increases with time (9). Solutes may increase the density of organelles either by drawing water out of the organelle, if they exert significant osmotic pressure, or by permeating the organelle membrane. The buoyant density of organelles thus depends not only on their intrinsic composition but also on the specific properties of the solute. Sucrose and glycerol have many of the desired properties of a gradient-forming solute, and both are widely used in velocity gradient sedimentation procedures, particularly for the separation of large proteins or protein complexes. Glycerol freely permeates membranes, while sucrose permeates only some organelle membranes. The relative buoyant densities of organelles suspended in glycerol or sucrose solutions are thus higher than those found in non-permeant solutes, which increases the times required to separate particles. Furthermore, for solutions of the same density, the glycerol solutions are more viscous. Hence, sucrose is more widely used than glycerol for the separation of organelles.

5.3 Isopycnic gradient centrifugation

Isopycnic gradient centrifugation is an equilibrium technique. It takes advantage of differences in buoyant density between subcellular particles, and provides a powerful complement to differential pelleting and velocity gradient centrifugation. A suspension of cellular components is layered over a density gradient and centrifuged to equilibrium. During centrifugation, particles migrate with velocities that depend on their size and on the difference between their buoyant density and that of the surrounding medium. As in velocity gradient centrifugation, gradients ensure stable sedimentation as bands along the gradient (10). Particles will eventually settle in a series of bands where their density equals that of the medium (the maximum density of the gradient must exceed that of the particles). At this point, no net forces act upon the particles and the bands will remain in place regardless of the time and force of centrifugation. Subcellular components can then be recovered from the appropriate zones (9). Alternatively, the sample may be placed at the bottom of the gradient, or be mixed throughout the gradient and, upon centrifugation, the particles will migrate to their isopycnic positions.

Density gradients may either be continuous or discontinuous. Two-chamber gradient makers of various volumes are commercially available and can be used reproducibly to prepare gradients. Alternatively, discontinuous step gradients may be used, which are particularly suitable for large scale separations and are easier to prepare. The degree of separation is limited by the volume of material loaded onto the gradient, because the sedimenting zones are at least as broad as the starting zone. Furthermore, the concentration of particles in the starting material is limited, because samples that are too concentrated will mix with the top of the gradient.

Advantages of isopycnic gradient centrifugation over velocity gradient centrifugation include the fact that once equilibrium is reached the process is insensitive to changes in centrifugation time. In addition, the banding densities differ markedly with the composition of the medium, making it possible to design a specific density range. For example, the density of the medium can be increased by using D_2O instead of H_2O as a solvent. A major disadvantage of isopycnic gradient centrifugation is that the components being separated are exposed to high concentrations of solute, which may damage subcellular organelles. Additional bands of damaged particles may thus form in the gradient. Furthermore, centrifugation times are longer, because the particles inevitably approach equilibrium slowly. It is thus advisable to use the fastest rotors available. Swing-out rotors are used predominantly, because the use of fixed-angle rotors can lead to mixing of the samples with the gradient during acceleration and deceleration. In vertical rotors, the liquid surface reorients during acceleration, forming a short path along which the gradient forms. Vertical tube rotors are therefore well suited for isopycnic gradient centrifugation because of their short sedimentation distance and consequent short banding times. They also avoid the build up of organelle-damaging hydrostatic pressure at the bottom of the gradient.

Because glycerol passes freely through membranes it increases the relative density of organelles to an extent that makes it impractical as a solute to sediment organelles in isopycnic gradient centrifugation. Sucrose, however, is widely used, although exposure of organelles to high sucrose concentrations may damage them and make them prone to lysis if the medium is diluted after separation. Furthermore, concentrated sucrose solutions are highly viscous, which increases the time taken to reach equilibrium. Also, the increased relative buoyant density of organelles in sucrose solutions leads to the banding of organelles at a greater density than in non-permeant solutes. Nycondez, derived from an X-ray contrast medium, and its dimer Iodixanol, do not permeate membranes. They are therefore especially suitable for isopycnic gradient centrifugation since they exert a low osmotic pressure, leading to lower equilibrium banding densities than sucrose. Because lower concentrations are required to produce a solution of a given density, the viscosity is kept low and thus banding times are relatively short (11).

Ficoll is a high molecular weight polymer made of sucrose and epichlorhydrin. Because it cannot be removed by dialysis, separated cellular components must be collected by differential pelleting, which may cause damage to organelles. However, Ficoll is non-permeant, exerts a low osmotic pressure, and can therefore be used for the separation of whole cells. Unfortunately, it is also extremely viscous and centrifugation to equilibrium is therefore relatively slow. Percoll is a dispersion of extremely small silica particles that are coated with polyvinylpyrrolidone. Particles are almost the size of microsomes, and centrifugation of Percoll (20–100 000 g for 20–30 min) results in the formation of stable gradients *during* centrifugation (centrifugal force is balanced by diffusion). Particles with diffusion coefficients lower than that of Percoll will band in such gradients. Self-forming gradients may be disadvantageous, however, if large particles are present in the sample, because these may pellet before the gradient has had time to form (10). As with sucrose, step gradients are often used instead of continuous gradients, because it is then not necessary to use gradient makers. The viscosity of Percoll is similar to that of water, and banding times are therefore short.

Protocol 4 describes the isolation of zymogen granules, granule membranes, and plasma membranes from rat pancreas using a combination of differential pelleting and density gradient centrifugation.

Protocol 4

Preparation of zymogen granules, granule membranes, and plasma membranes from rat pancreas

Equipment and reagents

- Dissection equipment and surgical gauze
- 20 ml hand-driven Dounce homogeniser (cooled to 4 °C)
- Homogenization buffer: 280 mM sucrose, 5 mM MES pH 6.0, containing 1 mM EGTA, 1 mM PMSF, 1 μg/ml pepstatin, 1 μg/ml

antipain, 1 μg/ml leupeptin, 17 μg/ml benzamidine, 50 μg/ml bacitracin, and 10 μg/ml soybean trypsin inhibitor, at 4 °C
- Granule lysis buffer: 170 mM NaCl/200 mM NaHCO$_3$ pH 8.0 (1:3, v/v), containing 1 mM EGTA and protease inhibitors (as above)

Protocol 4 continued

Method

1 Collect five rat pancreases. Carry out all of the following procedures at 4°C. Mince each pancreas in 10 ml of homogenization buffer by chopping with scissors. Homogenize using 10 strokes of the pestle. Filter the homogenate through surgical gauze.

2 Centrifuge at 400 g for 10 min. Keep both pellet and supernatant.

3 To prepare plasma membranes, resuspend the pellets in 15 ml of homogenization buffer.

4 Homogenize with one stroke in a Teflon/glass homogenizer, and bring the suspension slowly to 1.6 M sucrose by adding 48 ml of 2 M sucrose.

5 Overlay the suspension with 20 ml of 300 mM sucrose in four centrifuge tubes, and centrifuge at 100 000 g for 1 h.

6 Harvest the interface, dilute by adding 30 ml of 300 mM sucrose, and pellet the membranes by centrifugation at 70 000 g for 30 min.

7 Resuspend the pellet in homogenization buffer, to about 1 mg/ml protein. The suspension can be stored as aliquots at −20°C.

8 To prepare zymogen granules, centrifuge the supernatant from step 2 at 900 g for 10 min. This step produces a tight white pellet overlaid by a buff-coloured mitochondrial layer.

9 Discard the supernatant, and remove the mitochondrial layer with three surface washes with 1 ml of homogenization buffer. Resuspend the granules to the desired volume in the same buffer.

10 To purify the granules further (if required), resuspend the white pellet from step 9 in 50% Percoll in homogenization buffer. Centrifuge at 25 000 g for 90 min. Harvest the white band toward the bottom of the tube. Wash the granules three times in homogenization buffer.

11 To prepare granule membranes, lyse the granules by incubation for 1 h in lysis buffer, and recover the membranes by centrifugation at 70 000 g for 30 min.

12 Resuspend the membranes as required.

6 Free-flow electrophoresis

Subcellular fractionation by free-flow electrophoresis relies on differences in surface charge between organelles (12). The technique involves the introduction of the starting mixture of components as a fine jet into a thin sheet of buffer flowing 'downhill' perpendicular to an electric field. The sample inlet is near the cathode, and the organelles in the sample migrate towards the anode at different rates as the buffer flows down the flat bed of the apparatus. A linear array of tubes collects the various fractions at the downstream end. A major advantage of the method is that it is preparative, and can be operated continuously. Organelle recovery is high (up to 90%), and the organelles suffer little trauma during isolation. The starting material for this step is usually a partially purified sample containing the organelle of interest. Despite its obvious potential, free-flow electrophoresis is not widely used, perhaps because of the necessity for specialist equipment. However, it is currently used as a method for fractionating endosomal populations (13). In this case, efficient fractionation has been shown to depend on a brief treatment of the starting material with trypsin before the free-flow electrophoresis step. The molecular basis of this requirement is not defined.

7 Immunological methods of organelle purification

Subcellular fractionation based on specific biochemical criteria represents an attractive adjunct or alternative to fractionation through non-specific physical parameters. The most promising of these more specific methods is immuno-isolation using antigen–antibody recognition. The principles of immunoisolation are very simple: the antibody is bound to a solid support and used to extract a membrane fraction from a crude mixture through the formation of a complex with a membrane-bound antigen. As for free-flow electrophoresis, the starting material for this step is usually a fraction already partially purified using one or more of the conventional techniques described above. The success of this method depends on a careful choice of the antigen, the solid substrate, and the mode of coupling the antibody to the substrate (14).

7.1 Selection of antigen

If an endogenous antigen is used, it should be restricted to the appropriate subcellular organelle. Furthermore, to obtain high yields of the organelle, the antigen must be present at reasonably high density. This is particularly important if the organelle becomes vesiculated during homogenization. Once a suitable antigen has been chosen, antibodies must be raised against it. Polyclonal antibodies should be affinity purified on the antigen before use, to provide a high enough density of the appropriate antibody on the solid support, and also to improve the specificity of the reagent. Monoclonal antibodies are in general preferable, not only because of their epitope specificity but also because of the availability of an inexhaustible supply, which is helpful if the procedure is to be used routinely. If a membrane-spanning protein is chosen as the antigen, monoclonal antibodies, by binding to defined epitopes, also make it possible to immunoisolate either 'right-side-out' or 'inside-out vesicles', which may be an important consideration for some *in vitro* assays.

Exogenous antigens may also be introduced into cells for the purpose of immunoisolation of particular organelles. Viral envelope proteins have been used most widely, because their intracellular trafficking pathways are well characterized, and also because suitable monoclonal antibodies are available. For example, the membrane-spanning glycoprotein of vesicular stomatitis virus (VSV G) has been implanted into the plasma membrane of cultured cells by low pH-mediated membrane fusion. If this procedure is carried out by a brief incubation at 37 °C followed by a rapid return to 4 °C, then the protein remains in the plasma membrane, and antibodies to the cytoplasmic domain of the protein can be used to isolate 'inside-out' plasma membrane vesicles. On the other hand, if the cells are warmed to 37 °C, membrane traffic causes the protein to move through the endocytotic pathway in a synchronized wave. By homogenizing the cells at various times after warm-up, populations of time-defined endocytotic vesicles can be isolated (15). The same protein may be used to isolate membrane compartments on the secretory pathway. If cells are infected with VSV, they begin to synthesize large amounts of the viral proteins. VSV G is synthesized on ribosomes bound to the membrane of the rough ER, and then trafficked via the Golgi complex to the plasma membrane. If the cells are incubated at 20 °C, then the transport of the protein is arrested in a compart-ment on the *trans* side of the Golgi complex—the *trans*-Golgi network (TGN). If cells are homogenized following the 20 °C block, antibodies to the cytoplasmic domain of VSV G can be used to isolate 'right-side-out' vesiculated TGN (16).

7.2 Selection of solid support

Several types of solid support have been used successfully for immunoisolation. These include meshworks of cellulose fibres (17), polyacrylamide beads (18), and fixed *Staphylococcus aureus* cells (19). A particularly interesting medium is a magnetic solid support (14), which is commercially available from Dynal A.S., Oslo, Norway. The magnetic beads (Dynabeads® M-500) are 5 μm spheres with smooth surfaces that minimize non-specific binding. The surface of the beads is activated with *p*-toluene sulfonyl chloride, which permits the covalent linkage of antibodies via primary amine or sulfhydryl groups. After incubation of the derivatized beads with a crude source of the organelle of interest, the beads can be attached to the wall of the tube using an external magnet. This allows extensive washing of the captured organelles without the need for pelleting, which greatly reduces structural damage. Furthermore, the organelles may be eluted in any convenient volume.

7.3 Selection of coupling mode

In theory, the specific antibody can be coupled covalently to the solid support either directly or indirectly, via an anti-immunoglobulin second antibody. In practice, indirect coupling has proven much more successful, probably because of the reduction in steric hindrance and additional flexibility conferred by the extra linking immunoglobulin. Ideally, the target epitope in the antigen should be freely exposed on the surface of the organelle, so that it can be bound by the antibody attached to the solid support. If the antigen is in any way 'buried', then it may be necessary to expose the starting material to primary antibody first, and then capture the antigen–antibody complex using a second antibody attached to the beads. The optimal procedure can only be determined empirically, once the organelle, antigen, antibody, and solid support have been chosen.

8 Monitoring purity

The purity of subcellular fractions is most commonly assessed through the use of proteins localized to a particular organelle. Often, these marker proteins are enzymes, and purity is then assessed through the increase in specific activity of the enzyme in the fraction over that in the starting homogenate, and the corresponding decrease in specific activity of marker enzymes for potential contaminating organelles. Of course, for an organelle that occupies a large fraction of the volume of the intact cell (such as the mitochondrion), the degree of enrichment of the marker enzyme that represents purification to near homogeneity will be smaller than the value for a less abundant organelle (such as the Golgi complex). In addition to enzymes, other proteins (usually antigens for which specific antibodies are available) may be used as markers for particular organelles; the enrichment of these antigens may be determined by quantitative immunoblotting. A selection of marker enzymes and antigens for the major organelles is given in Table 1. In addition to marker proteins, electron micro-scopy is also commonly used to assess the purity of organelle fractions. This approach is particularly suitable for organelles that have characteristic struct-ures, such as mitochondria and Golgi stacks.

If the purpose of isolating organelles is to study their function *in vitro*, it is often necessary to reach a compromise between organelle purity and functional

Table 1 Marker enzymes and antigens for major organelles

Organelle	Enzyme	Antigen	Reference
Endoplasmic reticulum (ER)	NADH cytochrome *c* reductase Glucose-6-phosphatase	Calnexin Protein disulfide isomerase	
ER–Golgi intermediate compartment (ERGIC)		p58	20
Golgi complex			
cis	Mannosidase I	GM130	21
medial	*N*-acetylglucosaminyl transferase I		
trans	Galactosyl transferase		
TGN		TGN38	22
Plasma membrane	K$^+$- *p*-nitrophenyl phosphatase 5′-nucleotidase		
Endosome	Horseradish peroxidase (introduced)	Transferrin (introduced)	
Lysosome	β-Hexosaminidase *N*-acetyl-β-glucosaminidase		
Mitochondrion	Cytochrome *c* oxidase Succinate dehydrogenase Monoamine oxidase		

activity; that is, exhaustive and lengthy protocols may produce organelle fractions which, although highly homogeneous, are devoid of activity. In fact, careful design of cell-free assays may remove the requirement for highly purified organelles. Many assays for transport of proteins between organelles, or for organelle fusion, rely on complementation, where two components of a reaction, e.g. enzyme and substrate (7, 23, 24), antigen and antibody (25), or avidin and biotin (26, 27), are included in two separate membrane compartments, and the assay measures the generation of a product that can only be produced as a result of membrane fusion. The reaction components may be introduced into the appropriate compartments either biosynthetically (e.g. for transport steps on the secretory pathway) (7, 23, 24) or by uptake (e.g. for steps on the endocytotic pathway) (25–27). Because the markers are chased into the compartment to be studied before homogenization of the cells, it is common for reactions to be carried out with only partially purified cell extracts, which have been prepared relatively rapidly from crude homogenates, leaving organelles which retain functional activity.

9 Monitoring organelle structural integrity and activity

Functionally competent organelles should be intact, and it is therefore important to establish the structural integrity of a subcellular fraction before studying its behaviour. Electron microscopy will provide a good idea of the condition of the isolated organelle. Another commonly used method involves determining the accessibility of protein epitopes to reagents such as proteases or water soluble biotinylating reagents. For example, if the organelle is predominantly intact, trypsin treatment should remove only proteins present on the outside of the organelles and not internal proteins, unless a detergent is also added.

An early example of the successful reconstitution of the function or an organelle *in vitro* was the processing (i.e. translocation, signal sequence cleavage, and glycosylation) of nascent secretory and membrane-spanning proteins by microsomes, generated by vesiculation of the rough endoplasmic reticulum (28). In this *in vitro* system, translocation is usually assayed through the acquisition of resistance of the nascent proteins to proteases, a method that depends upon the integrity of the microsomes.

Protein transport between two membrane compartments *in vitro* was first demonstrated using the transport of VSV G between the *cis* and *medial* compartments of the Golgi complex (7). The assay used Golgi-enriched fractions from Chinese hamster ovary cells, and transport was measured through the addition of N-acetylglucosamine to VSV G as it arrived in the *medial* Golgi. In this case, the degree of enrichment of the Golgi membrane stacks was assessed through the specific activity of marker enzymes (7), and by electron microscopy (29). It was also shown that the VSV G remained protected from trypsin throughout the transfer between compartments (only the short cytoplasmic tail of the protein being cleaved).

Our own particular interest is in the mechanisms underlying the fusion of secretory vesicles with the plasma membrane. We have developed a model system based upon the fusion of zymogen granules with pancreatic plasma membranes (30). The two membrane compartments are isolated as described in Protocol 4. Examination of the structures of the isolated zymogen granules using

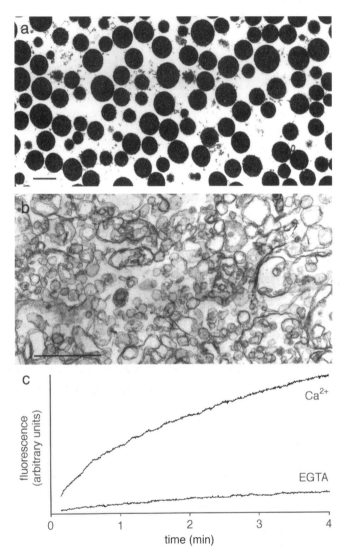

Figure 1 Structure and functional activity of isolated pancreatic zymogen granules and plasma membranes. Electron micrographs of zymogen granule (a) and pancreatic plasma membrane (b) fractions. Bars: 1 μm. (c) Fusion between zymogen granules and plasma membranes measured through the de-quenching of the fluorescence of octadecylrhodamine. The fluorescent probe was loaded into the membranes of the zymogen granules, and a steady baseline fluorescence was obtained. Unlabelled plasma membranes were added at time zero, in the presence of 67 μM free Ca^{2+} or 1 mM EGTA. Fluorescence was monitored for 4 min at 37 °C.

the electron microscope (Figure 1) shows that the granules are highly homogeneous. They are also intact, and still contain their dense cores. Trypsinization of granules prepared in this way causes complete cleavage of the cytoplasmically-oriented membrane protein synaptobrevin 2; in contrast, the luminal enzyme procarboxypeptidase is cleaved only if proteolysis is carried out in the presence of Triton X-100. Note that the granules shown were taken for use after Protocol 4, step 9. Step 10 produces a more pure preparation but the granules are less likely to retain functional activity. A typical plasma membrane preparation consists of membrane vesicles and sheets of varying sizes. Other organelles, such as mitochondria, Golgi complexes, and rough microsomes, are clearly absent. Protocol 5 describes an *in vitro* assay for fusion between these two membrane compartments, to illustrate that the membranes retain functional activity after isolation. Membrane fusion is measured through the relief of the self-quenching of the fluorescent probe octadecylrhodamine (R18). R18 is loaded into the membranes of the zymogen granules to a concentration that causes quenching of its fluorescence. Fusion of the granules with plasma membranes is then detected through the de-quenching that occurs as the probe becomes diluted into the unlabelled target membranes. As shown in Figure 1, addition of plasma membranes to the labelled zymogen granules causes a time-dependent de-quenching signal in the presence but not the absence of Ca^{2+} (at concentrations in the micromolar range). Evidence that this signal is monitoring a genuine membrane fusion event includes the abolition of de-quenching by prior treatment of the membranes with Clostridial neurotoxins, which specifically cleave SNARE proteins that are known to be intimately involved in membrane fusion, and which are present on the two interacting membranes (31).

Protocol 5

Monitoring fusion between pancreatic zymogen granules and plasma membranes

Equipment and reagents

- Fluorescence spectrometer (e.g. an Hitachi F-2000 model)

- Octadecylrhodamine (R18; Molecular Probes)

- Incubation buffer: 280 mM sucrose, 5 mM MES pH 6.0 or 6.5, containing 1 mM EGTA,

1 mM PMSF, 1 μg/ml pepstatin, 1 μg/ml antipain, 1 μg/ml leupeptin, 17 μg/ml benzamidine, 50 μg/ml bacitracin, and 10 μg/ml soybean trypsin inhibitor

Method

1 Prepare zymogen granules and plasma membranes as described in Protocol 4. Prepare the granules fresh each time (Protocol 4, step 9), although the plasma membranes can be stored frozen, and thawed as required.

Protocol 5 continued

2 Store octadecylrhodamine (R18) at −20 °C as a 20 mM stock solution in ethanol. Add 1 μl of the probe solution to 300 μl of a zymogen granule suspension prepared from a single rat pancreas.

3 Incubate the granules with the probe at 37 °C for 5 min, and then collect the labelled granules by centrifugation at 900 g for 10 min. Resuspend the granules in the original volume of sucrose/MES incubation buffer at pH 6.0.

4 Add 10 μl of the labelled granule suspension to 700 μl of incubation buffer (pH 6.5) at 37 °C, in a 1 ml capacity plastic cuvette. Adjust the free Ca^{2+} concentration in the buffer as required. Add any other reagents to be tested (e.g. potential stimulators or inhibitors of fusion) at this stage.

5 Measure the fluorescence continuously at 37 °C in the fluorescence spectrometer at an excitation wavelength of 560 nm and an emission wavelength of 590 nm.

6 After a short (1–2 min) equilibration period, add unlabelled plasma membranes (10–20 μg protein) in a small volume (10–20 μl) and mix with the granule suspension.

7 Monitor the fluorescence for a further 4–5 min.

Acknowledgement

We thank the Wellcome Trust for supporting our work.

References

1. Nagasawa, M., Koide, H., Ohsawa, K., and Hoshi, T. (1992). *Anal. Biochem.*, **201**, 301.
2. Walter, H., Johansson, G., and Brooks, D. E. (1991). *Anal. Biochem.*, **197**, 1.
3. Graham, J. M. (1997). In *Subcellular fractionation: a practical approach* (ed. J. M. Graham and D. Rickwood), p. 1. IRL Press, Oxford.
4. Graham, J. M. and Sandall, J. K. (1970). *Biochem. J.*, **182**, 157.
5. Jacobson, B. S. and Branton, D. (1977). *Science*, **195**, 302.
6. Chapel, J. B. and Perry, S. V. (1954). *Nature*, **173**, 1094.
7. Balch, W. E., Dunphy, W. G., Braell, W. A., and Rothman, J. E. (1984). *Cell*, **39**, 405.
8. Hinton, R. H. and Mullock, B. M. (1997). In *Subcellular fractionation: a practical approach* (ed. J. M. Graham and D. Rickwood), p. 31. IRL Press, Oxford.
9. Spragg, S. P. (1978). In *Centrifugal separations in molecular and cell biology* (ed. G. D. Birnie and D. Rickwood), p. 7. Butterworth, London.
10. Hinton, R. and Dobrota, M. (1976). *Laboratory techniques in biochemistry and molecular biology* (ed. T. S. Work and E. Work). Vol. 6, *Density gradient centrifugation*. Elsevier/North-Holland, Amsterdam.
11. Rickwood, D., Ford, T. C., and Graham, J. M. (1982). *Anal. Biochem.*, **123**, 23.
12. Crawford, N. (1988). In *Progress in clinical and biological research*. Vol. 270, *Cell-free analysis of membrane traffic* (ed. D. J. Morré, K. H. Howell, G. M. W. Cook, and W. H. Evans), p. 51. Alan R. Liss, New York.

13. Pierre, P., Turley, S. J., Gatti, E., Hull, M., Meltzer, J., Mirza, A., *et al.* (1997). *Nature*, **388**, 787.

14. Howell, K. E., Gruenberg, J., Ito, A., and Palade, G. E. (1988). In *Progress in clinical and biological research*. Vol. 270, *Cell-free analysis of membrane traffic* (ed. D. J. Morré, K. H. Howell, G. M. W. Cook, and W. H. Evans), p. 77. Alan R. Liss, New York.

15. Gruenberg, J. and Howell, K. E. (1986). *EMBO J.*, **5**, 3091.

16. de Curtis, I. and Simons, K. (1988). In *Progress in clinical and biological research*. Vol. 270, *Cell-free analysis of membrane traffic* (ed. D. J. Morré, K. H. Howell, G. M. W. Cook, and W. H. Evans), p. 101. Alan R. Liss, New York.

17. Richardson, P. J. and Luzio, J. P. (1986). *Appl. Biochem. Biotech.*, **13**, 133.

18. Ito, A. and Palade, G. E. (1978). *J. Cell Biol.*, **79**, 590.

19. Merisko, E. M., Farquhar, M. G., and Palade, G. E. (1982). *J. Cell Biol.*, **92**, 846.

20. Saraste, J. and Svensson, K. (1991). *J. Cell Sci.*, **100**, 415.

21. Nakamura, N., Rabouille, C., Watson, R., Nilsson, T., Hui, N., Slusarewicz, P., *et al.* (1995). *J. Cell Biol.*, **131**, 1715.

22. Luzio, J. P., Brake, B., Banting, G., Howell, K. E., Braghetta, P., and Stanley, K. K. (1990). *Biochem. J.*, **270**, 97.

23. Patel, S. K., Indig, F. E., Olivieri, N., Levine, N. D., and Latterich, M. (1998). *Cell*, **92**, 611.

24. Urbé, S., Page, L. J., and Tooze, S. A. (1998). *J. Cell Biol.*, **143**, 1831.

25. Woodman, P. G. and Warren, G. (1988). *Eur. J. Biochem.*, **173**, 101.

26. Braell, W. A. (1987). *Proc. Natl. Acad. Sci. USA*, **84**, 1137.

27. Gorvel, J.-P., Chavrier, P., Zerial, M., and Gruenberg, J. (1991). *Cell*, **64**, 915.

28. Blobel, G. and Dobberstein, B. (1975). *J. Cell Biol.*, **67**, 835.

29. Braell, W. A., Balch, W. E., Dobbertin, D. C., and Rothman, J. E. (1984). *Cell*, **39**, 511.

30. MacLean, C. M. and Edwardson, J. M. (1992). *Biochem. J.*, **286**, 747.

31. Hansen, N. J., Antonin, W., and Edwardson, J. M. (1999). *J. Biol. Chem.*, **274**, 22871.

Chapter 5
Plant cell biology

Aldo Ceriotti and Alessandro Vitale
Istituto di Biologia e Biotecnologia Agraria, Via Bassini 15,
20133 Milano, Italy.

Nadine Paris
CNRS UMR 6037, IFRMP 23, Université de Rouen,
76821 Mont Saint Aignan, France.

Lorenzo Frigerio
Department of Biological Sciences, University of Warwick,
Coventry CV4 7AL, UK.

Jean-Marc Neuhaus
Institute de Botanique, Laboratoire de Biochimie,
Université de Neuchâtel, rue Emile-Argand 9, CH-2007 Neuchâtel 7,
Switzerland.

Stefan Hillmer and David G. Robinson
HIP – Zellbiologie, Universitaet Heidelberg, Im Neuenheimer Feld 230,
69120 Heidelberg, Germany.

1 Introduction

The ability to express heterologous proteins in plant cells, and to study the
subcellular localization of heterologous and endogenous polypeptides is central
to many studies in the field of plant cell biology. Heterologous expression and
metabolic labelling are, together, two powerful tools to study intracellular
transport and protein–protein interactions. However, it is often the combination
of biochemical and immunohistochemical approaches that gives the more
comprehensive view of what is really happening within the cell. Accordingly,
the first part of this chapter will present a protocol for transient expression in
tobacco protoplasts, while two other sections will be devoted to immunolabelling for confocal microscopy observation and immunogold electron microscopy, respectively.

In addition, we would like to stress the fact that cell biology had a tremendous
lift with the introduction of the green fluorescent protein (GFP). This polypeptide allows *in vivo* visualization of subcellular compartments by a confocal
line for fluorescein isothiocyanate (FITC) and the signal can still be visible after

fixation. The chapter therefore includes a description of the advantages and possible pitfalls that derive from the use of GFP in plant cells. Some basic rules for optimal use of the confocal microscope will also be described.

2 Expression and analysis of heterologous proteins in plant cells

2.1 Transient or permanent expression?

Most of the studies in plant protein cell biology are performed using proteins transiently expressed in protoplasts or infected tissues or permanently expressed in transgenic plants, often after modification of their amino acid sequence by recombinant DNA techniques. A major goal of the expression of recombinant, modified proteins has to date been the dissection of polypeptide sequences in search of motifs responsible for subcellular sorting, stability, or molecular interactions in general. The identification of protein function through the use of dominant negative mutants is also rapidly expanding. Expression in a heterologous host is also used simply because the level of expression of the protein of interest is too low in the natural tissue to perform certain experiments, or because the tissue of origin is not suited to certain cell biology studies such as, for example, pulse-chase or drug treatments.

The lasting success of transient expression in the era of transgenic plants is mainly due to its rapidity of execution: the time-consuming procedures for selection of transgenic cell cultures or the even slower regeneration of plants are avoided. Another advantage is the possibility to modulate the level of protein expression by simply changing the concentration of plasmid used in the DNA uptake procedure (1, 2). Co-expression of more than one recombinant protein is also very easy (3). Finally, in terms of space and budget, transient expression avoids the need to maintain many different transgenic cell cultures or plants.

Transient expression can be performed in intact tissues or in isolated protoplasts. Protoplasts have some important advantages in cell biology studies: they can be subjected to pulse-chase labelling with radioactive amino acids much more easily, uniformly, and reproducibly than intact tissues and can be treated with drugs that inhibit cellular processes. The major disadvantages are the difficulties in comparing behaviour in different tissues and the unavoidable dilution of the transformed protoplasts in an untransformed population, since transfection is never 100% efficient. Thus, when dominant negative mutants are used, their effect will be only on a proportion of the protoplasts used and therefore in some types of experiments, such as pulse-chase labelling, the effects can be unequivocally detected only when marker proteins are co-expressed. For transient expression in general, the possible saturation of cellular machineries due to the very high level of recombinant protein synthesis that can occur during transient expression should always be taken into account, especially in protein sorting studies. This can actually be both a drawback and an advantage:

transient expression studies established that vacuolar delivery of at least certain proteins is a saturable process, and whose saturation leads to secretion (1, 2).

2.2 Tobacco mesophyll protoplasts

Protoplasts isolated from leaves of tobacco are still the most popular tool for transient expression in cell biology studies. The reasons are in part historical, tobacco being the first plant in which synthesis of a foreign protein could be detected, and in part practical: tobacco plants are propagated very easily in axenic culture (see Protocol 1) and protoplasts can be obtained with high yield from their leaves. These techniques have been improved by the experience of many laboratories and are nowadays well established.

The population of protoplasts that is obtained is not uniform, because it does not result from a single cell type. As such, the protoplasts do not possess a uniform population of vacuoles (4). This should be taken into account when vacuolar protein sorting studies are performed, because the different types of vacuoles have different modes of biogenesis (5).

Protocol 1

Preparation of axenic cultures of *Nicotiana tabacum* plants

Equipment and reagents

- Plant tissue culture containers (V1601 Duchefa)
- *Nicotiana tabacum* seeds (cv. Petit Havana, SR1)
- 70% ethanol
- 10% solution of commercial bleach (sodium hypochlorite)
- ½ MS-agar: half-strength Murashige and Skoog medium (M0221 Duchefa) solidified with 8 g/litre Plant Agar (P1001 Duchefa) and containing 30 g/litre sucrose. Bring pH to 5.6 with 1 N NaOH and sterilize by autoclaving

Method[a]

1. Pour about 100–150 ml of seeds into a 1.5 ml tube.
2. Add 1 ml of 70% ethanol and gently shake the tube for 1 min.
3. Spin for 1 min at 100 g in a microcentrifuge to sediment the seeds, remove the supernatant.
4. Add 1 ml of 10% commercial bleach and gently shake the tube for 10 min.
5. Spin for 1 min at 100 g, remove supernatant.
6. Wash the seeds with 1 ml sterile distilled water for 1 min.
7. Spin as above, repeat the wash two times.
8. Pour the seeds on to a sheet of sterile filter paper and let them dry.
9. Spread the seeds on to a Petri dish containing ½ MS-agar.

Protocol 1 continued

10 Distribute germinated seeds in plant tissue culture containers on ½ MS-agar.

11 Grow the plants at 26 °C in 16 h per day light (2000 lux) subculturing as cuttings every six weeks.

ᵃ Perform all operations in a sterile hood.

2.3 Choice of expression vector for transient expression

To guarantee efficient transcription and translation, several factors should be taken into account when preparing a construct for transient expression. Since transient expression generally requires the use of relatively large quantities of DNA, the plasmid should ideally be small and replicated at high copy number, to guarantee the reproducible isolation of high quantities of high quality DNA. In many cases, the aim is to produce a construct giving the maximal potential transcription/translation efficiency, since the level of expression can then be easily down-regulated by reducing the amount of DNA used in the DNA uptake procedure. Although different promoters can be used, the 35S promoter derived from cauliflower mosaic virus remains the most popular choice. Similarly, the 35S terminator or the nopaline synthase terminators are frequently used to guarantee mRNA processing.

Particular attention should be devoted to the presence of sequences at the 5′ end of the mRNA that can interfere with initiation, which is considered to be a rate limiting step in translation. Some easily recognizable features that have the potential for affecting initiation efficiency are:

(a) Presence of AUG codons upstream of the authentic initiation codon. Some plant mRNAs contain one or more small open reading frames that affect expression of the main coding sequence. Their elimination can enhance translation many-fold. Similarly, the fortuitous insertion of AUG codons upstream of the authentic initiation codon must be avoided while preparing the construct to be used for transient expression.

(b) Base composition of the 5′ untranslated leader. Many plant leaders are AU-rich, reducing the tendency to form stable secondary structures that can interfere with translation initiation (6). The insertion of sequences within the 5′ untranslated leader that are able to form stable secondary structures should be avoided (7). While maintaining the original 5′ untranslated sequence can be a reasonable choice, the coding sequence can also be subcloned into expression vectors which provide 5′ untranslated leaders that have been shown to allow efficient translation of downstream coding sequences (translational enhancers) (6). One of these vectors is pDHA (8), a high copy plasmid that allows 35S promoter-directed gene expression and which contains the 5′ untranslated region of alfalfa mosaic virus coat protein mRNA, followed by a series of cloning sites.

(c) Context of the AUG initiation codon. Consensus contexts for dicots and monocots have been identified (9, 10) and varying the context of the AUG initiation codon has been shown to affect expression in different systems, including tobacco protoplasts (10, 11).

2.4 Preparation of DNA and amount used in transfection

Plasmid DNA suitable for transfection can be purified using commercial anion exchange resin based kits (Quiagen). However, we find that best transformation rates (number of protoplasts expressing the heterologous polypeptide/total number of viable protoplasts) are obtained using DNA which has been purified on a caesium chloride gradient. The concentration of DNA should be >1 mg/ml, to avoid adding an excessive volume of DNA solution to the protoplast. This would negatively affect the transfection procedure.

As a general guideline, 40 μg of average sized plasmid DNA (5 kb, purified on an anion exchange resin) in water can be used in the transfection of 10^6 protoplasts (Protocol 2). The level of expression can be modulated by varying the amount of DNA added to the sample. However, we have noticed that reduction of the amount of DNA often produces a more than proportional reduction in the expression level. If necessary, this effect can be partially counteracted by the addition of compensating amounts of empty vector DNA.

Protocol 2

Protoplast isolation and transfection

Equipment and reagents

- 100 μm mesh nylon filter. Fix filter in a funnel-like shape to the opening of a 50–100 ml beaker (use paper tape to fix the edges of the filter to the walls of the beaker). Cover with foil and autoclave.

- Kova Slide 10 Grids (HYCOR Biomedical 87144)

- K3 medium: 3.78 g/litre Gamborg's B5 basal medium with minimal organics (Sigma G5893), 750 mg/litre $CaCl_2·2H_2O$, 250 mg/litre NH_4NO_3, 136.2 g/litre sucrose, 250 mg/litre xylose, 1 mg/litre 6-benzylaminopurine,[a] 1 mg/litre α-naphthaleneacetic acid.[b] Bring to pH 5.5 with 1 M KOH. Filter sterilize through a 0.22 μm filter and store at −20°C.

- 10 × enzyme mix: 2% Macerozyme R-10 (Duclefa), 4% Cellulase Onozuka R-10 (Duclefa). Dissolve in K3 medium, stir vigorously for 30 min, spin for 15 min at 10 000 g at 4°C to precipitate insoluble materials. Filter sterilize through a 0.22 μm filter, aliquot, and store at −20°C. Avoid repeated freeze–thaw cycles.

- W5 medium: 9 g/litre NaCl, 0.37 g/litre KCl, 18.37 g/litre $CaCl_2·2H_2O$, 0.9 g/litre glucose. Filter sterilize through a 0.22 μm filter and store at −20°C.

- MaCa buffer: 91.08 g/litre mannitol, 2.94 g/litre $CaCl_2·2H_2O$, 1 g/litre 4-morpholine-ethanesulfonic acid (MES). Adjust pH to 5.7 with 1 M KOH, filter sterilize through a 0.22 μm filter, and store at −20°C.

Protocol 2 continued

- 40% PEG 4000: dissolve 4.72 g of $Ca(NO_3)_2 \cdot 4H_2O$ and 14.6 g of mannitol in 120 ml of water. Then add 80 g of polyethylene glycol 4000 (Merck 807490) and dissolve by heating at 40 °C. Adjust pH to 8–10 with 1 M KOH and bring volume to 200 ml with water. Filter sterilize through a 0.22 μm filter and store at −20 °C.

- K3-FDA: just before use, dilute 2 μl of fluorescein diacetate stock (5 mg/ml in acetone, stored at −20 °C) in 1 ml of K3 medium.

Method[c]

1 Dilute the 10 × enzyme mix in K3 medium, keep at room temperature. Add 7 ml of 1 × enzyme mix to each 10 cm Petri dish.

2 Cut green, young (three to five weeks old, 3–6 cm long) tobacco leaves from an axenically grown plant. Using a blade, cut the abaxial surface of the leaf every ~1 mm, taking care not to cut through the whole leaf.[d] Remove the midrib (optional) and float the leaf on the enzyme mix in the Petri dish, abaxial surface down and without wetting the adaxial surface. Try to fill the plates as much as possible with whole leaves and fragments to fill in the gaps. Leave plates overnight in the dark at 25 °C.[e]

3 With a plastic Pasteur pipette, gently remove the digestion mix and discard it.[f] Add 6 ml K3, dropwise, to each plate. Shake the plates gently to release the protoplasts. With the aid of a plastic Pasteur pipette recover the released protoplasts and filter the protoplast suspension through a sterile 100 μm mesh nylon filter, previously wetted with K3 medium.

4 Gently transfer the protoplasts into a 50 ml polypropylene tube.

5 Spin tube for 20 min at 60 *g* in a clinical centrifuge (use swinging buckets throughout the protocol), brake off. Viable protoplasts will float.

6 With a sterile glass Pasteur pipette connected to a peristaltic pump, form a window through the protoplast layer by pushing the cells from the centre to the sides, then activate the pump and carefully suck away the medium and the pelleted material.

7 Add 4 vol. of W5 medium, mix gently. Pellet protoplasts by centrifugation (5 min at 60 *g*) and remove supernatant.

8 Gently resuspend the pellet in the same volume of W5 used in step 7. Pellet protoplast by centrifugation (5 min at 60 *g*) and remove supernatant.

9 Resuspend in W5 medium at an expected concentration of 0.5–1.0 × 10^6 protoplasts/ml and incubate 30 min in the dark at 25 °C.

10 Transfer 50 μl of protoplast suspension (use cut tip) into 450 μl of K3-FDA. Count fluorescent (viable) protoplasts under UV light using a gridded slide.

11 Spin for 5 min at 60 *g*, brake off. Remove the supernatant and slowly resuspend the cells in MaCa buffer, at a concentration of 10^6 protoplasts/ml.

12 Heat shock the protoplasts by placing the tube at 45 °C for 5 min in a water-bath, then let them cool down at room temperature.

13 Prepare 15 ml polypropylene tubes with the DNAs for transfection (usually 40 μg in no more than 100 μl).[g] Transfer 10^6 protoplasts into each tube containing the DNA, mix gently.

14 Tilt the tubes on the rack, remove the caps, and add an equal volume of 40% PEG solution.[h] Mix gently by inverting the tubes several times and incubate at room temperature for 30 min, mixing from time to time.

15 Wash by filling the tubes with W5. Do it very slowly to avoid shock to the cells, taking at least 15 min to get to the final volume (i.e. add 3 ml, mix gently, then add another 3 ml, mix, and so on). Mix gently but thoroughly, trying to dissolve all the protoplast clumps that may have formed.

16 Spin 10 min at 60 g. Remove the supernatant.

17 Resuspend the cells in 1 ml K3.[i] Incubate overnight in the dark at 25 °C.

[a] From a 5 mg/ml stock in 1 N NaOH.

[b] From a 1 mg/ml stock solution (Sigma N1641).

[c] Perform all operations in laminar flow hood. Optimal room temperature: 22–25 °C.

[d] Operate quickly to avoid excessive dehydration of the leaf; if the tip of the leaf is curled and/or wet (due to contact with the jar wall) remove it.

[e] One plate should yield about $1-3 \times 10^6$ protoplasts.

[f] At this point, protoplasts should still be attached to the leaves.

[g] When different plasmids at different concentrations are used, they should all be brought to the same volume with water.

[h] Apply the solution dropwise to the upper part of the tube, and let it slide down the wall so that the impact with the cell suspension will be gentle.

[i] The majority of the protoplasts should float at this point.

2.5 Labelling of protoplasts

Although different labelled substances can be utilized, protein labelling in protoplasts is most commonly performed using mixtures of ^{35}S-labelled methionine and cysteine. Usually 100 μCi of ^{35}S-labelled methionine/cysteine mixture (>1000 Ci/mmole, >10 μCi/μl) are used to label 10^6 protoplasts in a volume of 1 ml. Substantial incorporation of label into newly synthesized proteins can be achieved using a 30 min to 1 h labelling period. Note that some proteins are devoid of methionine and cysteine. ^{3}H-labelled leucine (commercial preparations at >120 Ci/mmole, 5 μCi/μl are available) can be used in these cases.

After separation of protoplasts from the medium and homogenization, standard immunoprecipitation protocols can be used to identify the protein of interest.

Protocol 3

Protoplast labelling and homogenization

Reagents

- [^{35}S]methionine/[^{35}S]cysteine mix (e.g. Pro-mix, Amersham Biosciences Trans^{35}S-Label, ICN)

- Or [^{3}H]leucine (e.g. TRK 683 Amersham Biosciences)

- Unlabelled methionine and cysteine stock (10 ×): 100 mM methionine, 50 mM cysteine dissolved in K3 medium

- Or unlabelled leucine stock (10 ×): 100 mM leucine dissolved in K3 medium[a]

- Protoplast homogenization buffer: 150 mM Tris–HCl pH 7.5, 150 mM NaCl, 1.5 mM EDTA, 1.5% Triton X-100, 1 × complete (Roche 1697498)

Method

1   Label protoplasts[b] by adding 100 μCi of [^{35}S]methionine/[^{35}S]cysteine mixture (or [^{3}H]leucine) to 10^6 protoplasts in K3 medium. Mix gently and incubate at 25 °C in the dark. For the chase, add unlabelled methionine and cysteine from stock to a final concentration of 10 and 5 mM respectively (10 mM leucine if ^{3}H-labelled leucine was used for labelling).

2   At the desired time points, gently mix the tubes and remove an aliquot[c] of the protoplast suspension with a wide-bore tip. Transfer to fresh tube.[d]

3   Add 3 vol. of W5 medium (see Protocol 2), mix gently but thoroughly, and recover protoplast by centrifugation (10 min, 60 g, swinging buckets). Remove the supernatant, and save it for analysis of secreted proteins.

4   Gently resuspend protoplasts in 1 ml of W5 medium, and recover the protoplasts by centrifugation (10 min, 60 g, swinging buckets). Remove and discard the supernatant.

5   Add 200–500 μl homogenization buffer to the protoplast pellet or 2 vol. homogenization buffer to the protoplast medium.

6   Vortex 10 sec.

7   Spin for 2 min at 10 000 g and transfer the supernatant to a fresh tube for immunoprecipitation.

[a] These stocks may be stored at −20 °C, but they should be checked for precipitates after melting. After several freezing and melting the cysteine/methionine solution may become cloudy. If this occurs, discard and prepare a new stock.

[b] See Protocol 2, step 17.

[c] Usually a minimum of 10^5 protoplasts are used for each time point.

[d] Use conical-bottom 1.5 ml or larger polypropylene tubes. From this stage on, perform all operations at 0–4 °C.

2.6 Protoplast subcellular fractionation

An extensive description of the fractionation procedures to which protoplasts can be subjected is outside the scope of this chapter. However, basic fractionation procedures that allow separation of microsomal membranes and vacuoles are described.

2.6.1 Microsomes

Care should be taken in interpreting the results of microsome preparation. We have observed that when a large amount of protein accumulates in the ER, for example, if a recombinant protein with an attached signal for ER localization is expressed, this may be in part lost from the microsomes during the protoplast homogenization procedure and end up in the 'soluble' subcellular fraction. This artefact is possibly due to the formation, inside the ER lumen, of large protein bodies that break the membrane during homogenization. Therefore, recovery of a relevant proportion of a secretory protein in the soluble fraction even after a short pulse should be interpreted with caution.

Protocol 4

Microsome preparation

Reagents

- 12% sucrose buffer: 100 mM Tris–HCl pH 7.5, 10 mM KCl, 1 mM EDTA, 12% (w/w) sucrose
- 17% sucrose buffer: as the 12% buffer, but with 17% (w/w) sucrose

Method[a]

1   Protoplasts (100 000–500 000) are recovered by centrifugation as described in Protocol 3, step 3.

2   The protoplast pellet (less than 50 µl) is homogenized in 400 µl of 12% sucrose buffer, by pipetting 40 times with a Gilson-type micropipette through a 200 µl tip.

3   Remove intact cells and debris by centrifugation for 5 min at 500 g. Discard the pellet.

4   Load the supernatant on top of a 280 µl layer of 17% sucrose buffer, in 5 × 41 mm (0.8 ml, cat. 344090) Beckman Ultra-Clear tubes and centrifuge in a Beckman SW55 Ti rotor at 100 000 g for 30 min. Recover the 12% sucrose layer, carefully aspirate the 17% sucrose pad, and resuspend the pellet in homogenization buffer (see Protocol 3)[b].

[a] Perform all operations at 0–4 °C.

[b] Pellet contains microsomes, the 12% sucrose load contains cytosolic proteins and soluble vacuolar proteins, since large vacuoles break during homogenization.

2.6.2 Vacuoles

The procedure for vacuole isolation results in the recovery of large vacuoles, the classical 'central vacuoles' of mesophyll cells. However, some vacuoles present in protoplasts are in fact rather small (4) and the behaviour of small vacuoles in the isolation procedure detailed below is not known. Therefore, the vacuolar location of a protein cannot be ruled out conclusively solely based upon this procedure.

Recovery of vacuoles should be around 30%, as measured using the activity of the vacuolar marker enzyme α-mannosidase. The contamination by ER should be less than 1%, as measured by protein blot using antiserum prepared against the ER chaperone BiP.

Protocol 5

Vacuole isolation

Equipment and reagents

- Kova Slide 10 Grids (Roche Diagnostics 87144)
- Lysis buffer: 0.2 M sorbitol, 10% (w/v) Ficoll 400, 10 mM HEPES–KOH pH 7.5; sterilize by filtration through a 0.22 μm filter
- 5% Ficoll solution: 0.3 M sorbitol, 0.3 M betaine, 5% (w/v) Ficoll 400, 10 mM HEPES–KOH pH 7.5; sterilize by filtration through a 0.22 μm filter

- Betaine solution: 0.6 M betaine, 10 mM HEPES–KOH pH 7.5; sterilize by filtration through a 0.22 μm filter
- 2,6-dichlorobenzonitrile stock (1000 ×): 5 mg/ml 2,6-dichlorobenzonitrile in ethanol; store at $-20\,°C$
- Neutral red stock solution: 10 mg/ml neutral red in water; store at $+4\,°C$

Method[a]

1. Save about $1.5-2 \times 10^6$ protoplasts for marker enzyme assays and protein pattern control.

2. Pellet $4-10 \times 10^6$ protoplasts[b] at 60 g for 10 min in a Falcon 50 ml polypropylene tube or similar.

3. Discard supernatant and add pre-warmed (42 °C) protoplast lysis buffer to the protoplast pellet (6 ml buffer/5×10^6 protoplasts).

4. Resuspend pellet by gently pipetting the suspension up and down three to four times with a 10 ml pipette.[c] The shearing helps to break the protoplasts and to release the vacuoles. Add 15 μl of the neutral red stock solution per 12 ml lysis buffer, to visualize vacuoles later on.

5. Incubate the suspension for 5 min at room temperature. The first time this protocol is used, analyse an aliquot of the suspension under the microscope after the 5 min incubation, using Kova slides. If release of vacuoles is less than 50%, gently pipette the suspension up and down again two or three times, and re-check.

Protocol 5 continued

6   Place the suspension on ice. Using a wide-bore plastic pipette, transfer the suspension to a 15 ml Corex tube. Use 4–6 ml suspension per tube. To avoid contamination of the suspension with purified vacuoles, which will float upon centrifugation, do not touch the inner sides of the Corex tubes with the pipette during transfer.

7   Overlay the suspension with 3 ml of ice-cold 5% Ficoll solution and then with 300 μl of betaine solution. Work gently and avoid mixing the layers.

8   Centrifuge in a swinging bucket rotor at 5000 g (5500 r.p.m. in a Sorvall HB4 or Beckman JS-13.1) at 10 °C for 30 min.

9   With a glass Pasteur pipette[d] collect the vacuoles from the interface between the 5% Ficoll and the 0.6 M betaine solution layers. Vacuoles will be red. Try to collect them in a small volume (about 600–700 μl).

10   Count vacuoles using a microscope.

11   Freeze in liquid nitrogen and store at −80 °C for subsequent analysis.

12   To break vacuoles for α-mannosidase assay, thaw and freeze them twice. To determine recovery, treat an appropriate number of intact protoplasts in the same way.

13   For protein blot analysis, recover proteins after step 12 by adding 1 vol. of 30% ice-cold trichloroacetic acid, spin for 15 min at 10 000 g, and wash the pellets twice with ice-cold 90% acetone.

14   To prepare the sample for immunoprecipitation, add 2 vol. of homogenization buffer (see Protocol 3) after step 12. Spin for 2 min at 13 000 g and save supernatant for further analysis.

[a] Modified from ref. 12.

[b] In theory, vacuoles should be purified from freshly prepared protoplasts, since protoplasts reconstruct their cell wall leading to a lower recovery of vacuoles. If vacuoles are not purified within a few hours after protoplast preparation (e.g. when protoplasts are subjected to a pulse-chase protocol), 5 μg/ml 2,6-dichlorobenzonitrile, an inhibitor of cellulose synthesis, should be included in the medium in which protoplasts are maintained.

[c] The bore of the pipette should not be narrow. A Costar 10 ml plastic sterile pipette is adequate.

[d] At this stage vacuoles tend to stick to plastic pipettes or tips.

3 Immunolabelling, GFP, and confocal microscopy

3.1 Sample preparation for fluorescence immunostaining

Here we describe the preparation of individual fixed cells or protoplasts for immunolabelling. In plant tissue, chlorophyll is a strongly fluorescent material that will restrict the use of the confocal line to FITC. For this reason, it is recommended to use tissues that are devoid of chloroplasts, such as a material originating from root tips. As a source of protoplasts, we either use an *Arabidopsis thaliana* cell suspension culture (13) that has mostly no chlorophyll-linked fluorescence or tobacco (*Nicotiana tabacum*) leaves. Transient transformation of these protoplasts with GFP constructs can be performed using a classical PEG protocol (see Protocol 2).

Protocol 6

Fixation of pea and barley root tips and release of individual cells[a]

Equipment and reagents

- Ventilated hood to handle the fixative
- Dried pea or barley seeds
- Paraformaldehyde fixative: to prepare 20 ml of 3.7% paraformaldehyde fixative, weigh 0.74 g of paraformaldehyde (Merck 818715) into a small Erlenmeyer (work under a hood and use gloves), add 16 ml of water, and gently warm while mixing in a boiling water-bath. When the paraformaldehyde is almost in solution

(characterized by its slightly slurry aspect) stop warming and add a few drops of 0.1 N NaOH in order to complete the dissolution. Adjust to 20 ml and to the final composition of Na-phosphate buffer. The fixative can be stored for three weeks in the dark at 4 °C.

- Cellulysin cellulase (Calbiochem 219466)
- Na-phosphate buffer: 50 mM Na-phosphate pH 7, 5 mM EGTA, and 0.02% azide

Method

1   Sterilize dry peas or barley seeds with 1% sodium hypochlorite for 15 min. Rinse thoroughly with autoclaved deionized water, three times for 15 min each. Grow in the dark on sterile and humidified papers or gauze for five days (peas) or three days (barley).

2   Cut off the root tips (1–2 mm) from the germinated seeds with a razor blade, and immediately immerse in fixative at room temperature. Keep in the fixative for an hour at room temperature, gently mix occasionally, and transfer to 4 °C overnight.[b]

3   Remove some root tips from the fixative (one pea tip or two to three barley tips are sufficient for one immunostaining), rinse them three times for a few minutes, and then for an hour in Na-phosphate buffer.

4   Quickly drain the tips on a paper towel and transfer them to a solution of 1% cellulase in Na-phosphate buffer. Incubate for 20 min at room temperature mixing continuously.

5   Rinse the tips with Na-phosphate buffer (two times for 5 min) and quickly dry them on a paper towel. Still on the paper towel, gently and partially flatten the tips to remove some of the moisture, and transfer them into the bottom of a 1.5 ml conical tube cut to a volume of 0.5 ml. Using the outside of a 0.5 ml tube as a pillar, squeeze the tips onto the side of the 1.5 ml tube in order to release individual cells (see Figure 1).

Protocol 6 continued

6   Add 200 μl of Na-phosphate buffer to rinse the outside of the 0.5 ml microcentrifuge tube, pipette the single cells out of the large debris using a cut tip, and transfer into a new 0.5 ml microcentrifuge tube. To recover the cells at the bottom of the tube, use a short spin at approx. 150 g in a swing-out rotor. Rinse the cells three times with Na-phosphate buffer.

[a] Adapted from refs 14 and 15.

[b] The root tips can be stored in fixative for up to a month (the antigenicity is mainly preserved but should be checked individually for each primary antibody).

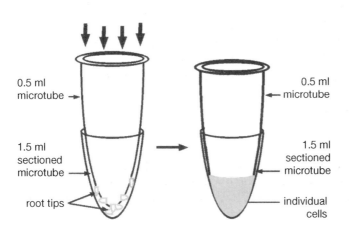

Figure 1 Schematic representation of the procedure used to release individual cells from root tips. The tips are squeezed between the outside of a 0.5 ml microcentrifuge tube and a sectioned 1.5 ml conical microcentrifuge tube, similar to a pillar and mortar (respectively).

Protocol 7

Fixation of tobacco or *Arabidopsis* protoplasts

Equipment and reagents

- Ventilated hood to handle the fixative
- Protoplasts either untransfected or transiently transfected with the construct of choice (see Protocol 2)
- 3.7% paraformaldehyde fixative: dissolve 0.74 g of powder (Merck 818715) in 16 ml

water (see Protocol 6); adjust to 20 ml to obtain the final composition of mannitol buffer

- Mannitol buffer: 0.5 M mannitol, 50 mM HEPES pH 5.8

Protocol 7 continued

Method

1   Recover transiently transfected protoplasts (from Protocol 2, step 17) by adding 3 vol. of W5 medium and spinning for 10 min at 60 g in a clinical centrifuge.[a]

2   Add the fixative at room temperature to the pelleted protoplasts. Keep in fixative for an hour at room temperature, gently mix occasionally, and transfer to 4 °C overnight.

3   Rinse the fixed protoplasts three times for a few minutes and then for an hour in fresh mannitol buffer before using them for the immunolabelling step.

[a] Approx. 0.2×10^6 protoplasts are necessary per labelling.

3.2 Fluorescence immunolabelling

This section describes possible artefacts in double immunolabelling that are due to the technical approach and proposes some controls to ensure the fidelity of the double staining obtained.

Secondary antibodies, in addition to reacting with the species against which they are raised may also cross-react with immunoglobulins from other species. For double labelling experiments, therefore, it is recommended that affinity purified preparations from which cross-reacting antibodies have been removed should be purchased.

Protocol 8

Single or double fluorescence immunolabelling

Equipment and reagents

- Fluorescence or confocal microscope
- Immunolabelling buffer (IL buffer): 1 × PBS, 0.25% BSA, 0.25% gelatin, 0.05% Nonidet P-40, and 0.02% azide; filter sterilize through a 0.45 μm filter

- Antibodies raised against immunoglobulins from the hosts from which the primary antibodies were obtained, coupled to a fluorochrome[a]

Method

1   Incubate the cells (see Protocol 6) or the protoplasts (see Protocol 7) for 5 min exactly in 0.5% Triton X-100 at room temperature to perforate the membranes.

2   Wash the cells quickly with IL buffer and wash twice more for 5 min. Block the cells with IL buffer plus 1% BSA for 15 min at room temperature. Aliquot the cell suspension into individual tubes, one for each primary antibody.

3   Spin down the cells, remove as much supernatant as possible, and add a small volume (50 μl) of the primary antibody diluted in IL buffer.[b] Incubate for 1 h at room temperature, mixing from time to time.

Protocol 8 continued

4 Rinse the cells quickly with 100 μl of IL buffer, and repeat three times for 10 min each. Remove as much supernatant as possible, and add a small volume (50 μl) of the secondary antibody diluted in IL buffer. Incubate for 1 h at room temperature, mixing from time to time.

5 Rinse the cells quickly with 100 μl of IL buffer, and repeat three times for 10 min each with 100 μl of fresh IL buffer.

6 For double immunostaining with two primary antibodies from different species, repeat steps 3–5 with the second set of primary and secondary antibodies.

7 Remove most of the supernatant and leave a few microlitres in the bottom of the tube with the labelled cells. Pipette out with a cut yellow tip a maximum of 5 μl to mount between a slide and coverslip for microscopic observation.

[a] For double labelling, ideally use serum devoid of any cross-affinity for different species (Jackson Immunoresearch).

[b] The dilution should be determined independently for each antibody but an average of 1/100 is recommended for crude polyclonal antisera.

If the two primary antibodies were raised in the same species, a different immunolabelling procedure (developed from ref. 15) must be followed. Follow Protocol 8 as for a normal double labelling except that the first secondary antibody (step 4), usually a whole IgG, should be replaced by a large excess of a Fab equivalent. Two Fab fragments are produced from a whole IgG molecule by papain digestion, and each of them presents only a single antigen recognition site. Figure 2 represents the combination of antibodies that can be found after a

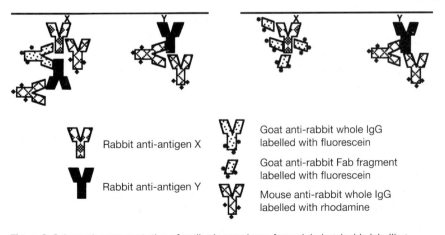

Rabbit anti-antigen X

Rabbit anti-antigen Y

Goat anti-rabbit whole IgG labelled with fluorescein

Goat anti-rabbit Fab fragment labelled with fluorescein

Mouse anti-rabbit whole IgG labelled with rhodamine

Figure 2 Schematic representation of antibody complexes formed during double labelling experiments against X antigen in first and Y antigen in second. Left: when only whole IgG are used non-specific labelling may occur on antigen X location. Right: when the first secondary IgG is replaced by a large excess of its Fab equivalent the labelling appears specific.

double labelling, first against antigen X and second against antigen Y. Figure 2 (left), shows the antibody complexes that are formed at the end of the usual double labelling protocol. Figure 2 (right), shows the complexes obtained when using Fab fragments in excess as a first secondary antibody.

In more detail, when a classical procedure is used two possible labelling artefacts may occur. Remaining free rabbit sites from the anti-X antibody may be recognized by the second anti-rabbit antibody. Additionally, the whole anti-rabbit IgG, bound to the anti-X antibody, will provide a free site that can bind to the anti-Y rabbit antibody added in the second step. These two phenomenon will lead to an apparent double labelling pattern (star plus diamond in Figure 2) that will be mis-interpreted as a co-localization of both antigens. To partially over-come these problems, as shown in Figure 2 (right), anti-rabbit Fab fragments are used in excess instead of a whole IgG molecule during the first step of labelling. These Fabs:

(a) Will block all the available rabbit sites from anti-X antibody.

(b) Will not provide any binding site for the anti-Y antibody added later to the sample.

In addition, a post-fixation step with 3.7% paraformaldehyde for 30 min can be added to maintain the X-linked complex of antibodies (after Fab incubation and before adding the second primary anti-Y antibody).

3.3 Confocal microscopy

Several points should be underlined when using the confocal in order to avoid excessive and therefore potentially misleading conclusions. The definition limit of a confocal is approximately 100 nm; that is much larger that the average of plant vesicles which can therefore not be visualized using this technique. That definition limit is therefore restrictive in term of defining between 'proximity' and co-localization of two labelled structures. Another important point to keep in mind is that some extremely bright structures may emit light that is detected by the microscope in optical sections beyond the actual physical size of the structure. This may lead to an over-estimation of size of the structure of interest and to an apparent co-localization of two structures that are in fact separated.

Although the definition limit can not yet be overcome, the other phenom-enon can be limited by a few working tricks. On most confocal microscopes, the signal emitted by a labelled structure can be visualized in a 'glowover mode' which basically attributes a colour value in correspondence to the intensity of the detected signal. For example, the absence of signal corresponds to black while the linear signals are shown in a gradient of colours from red, orange, yellow, to end with white. Each signal over this linear range is shown with a blue colour. The goal is to remain in a linear mode as much as possible. Using this glowover imaging, the amount of incident light delivered to the sample by the laser should be empirically adjusted in order to remain in a linear range as wide as possible, which ensures the visualization of the maximum of labelled struct-ures. This can be achieved either by lowering the total amount of light coming from the laser beam or by decreasing the pinhole size. In general, a smaller aperture is preferred since it limits the optical section thickness and is more likely to correspond to a real confocal image.

Confocal microscopy is also used to address the question of the co-localization of two antigens. There are few artefacts possible when double immunolabelling is performed with two primary antibodies coming from two different species. When the two fluorescent signals obtained are very different in intensity, a higher laser power is required to detect the faintest signal which sometimes leads to the recovery of a bleed-through from the strongest signal. When two antibodies from the same species are used, even the use of Fab does not fully avoid artefactual double labelling. The only indirect way to control the double labelled sample is to perform first single labelling, individually with each of the primary antibodies. The single labelled sample should be analysed on a wide range of cells to define in detail the various patterns that are found. Three main points should be taken into consideration: the frequency of each pattern found within the cell population, shape of the labelled structures, and relative intensity from one cell to another as well as within one cell. Each pattern represents the combination of a given antigen, the antibodies, and the sample used. In double labelling experiments, each cell that does not fit with the typical pattern obtained in single labelled control should be rejected since it is an indication that some artefact may have occurred.

3.4 GFP marker

The identification of signals sufficient for organelle-specific retention or sorting, combined with the use of GFP, allowed visualization of a subclass of compartments in living tissue. This confirmed the dynamic feature of these structures, and highlighted a control of organelle movements through the cell. For example, fusion of the sequence coding for a mammalian sialyl transferase with the GFP allowed the movement of the dictyosomes along the ER to be followed, suggesting a vacuum cleaner-like function of the Golgi that may directly pick up the material from the ER (16). The GFP reporter commonly serves as two principal tools:

(a) To visualize a given compartment and then address dynamic questions *in vivo*.

(b) To determine the subcellular location of a new protein or to study a potential trafficking motif by fusing them to the GFP coding sequence.

So far, several locations have been labelled using GFP in plant, including the Golgi complex (16), the nucleus (17), the ER (18), neutral and acidic vaculoles (19, 20), chloroplasts (21), and mitochondria (22). With a little experience, it is therefore now feasible to identify these structures simply on the basis of the GFP pattern, as illustrated in Figure 3.

Nevertheless, what is true for the large and easily defined locations remains to be established for smaller structures. In the terminology 'small' we include every 1–3 μm structure that can be found all over the cytoplasm and vary in

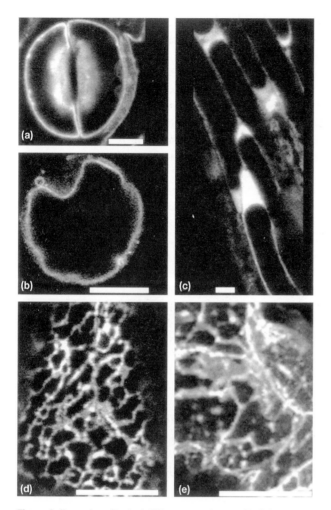

Figure 3 Examples of typical GFP patterns observed in living plant tissue. (a) Membrane-bound GFP accumulated in the plasma membrane in tobacco guard cells. (b) Membrane-bound GFP accumulated in the tonoplast of a tobacco protoplast. (c) GFP accumulated in the apoplast of *Arabidopsis* stem cells. (d) Soluble GFP in the ER of *Arabidopsis* trichome. (e) Soluble GFP in the cytoplasm of *Arabidopsis* trichome. We thank Dr Klaus Palme for providing the tonoplast construct of GFP. Scale bar: 10 μm.

shape from punctuated to elongated patterns. Dichtyosomes fall in this category but, in opposition to the other GFP structures, they can easily be identified under electron microscopy (EM) by their typical and unique organization. At the confocal level a strict identification of Golgi labelling has to be confirmed using a Golgi reference such as ERD2 fused to colour variants of GFP (23). Beside dichtyosomes, almost none of the small compartments involved in the secretory pathway in plants have been similarly characterized either at EM or GFP pattern levels. In addition, membrane-linked GFPs, when overexpressed in a continuous way, may disturb the morphology of the destination compartmenty by a higher membrane requirement than normal. In conclusion, the exclusive use of fluorescence pattern for the identification of small GFP compartments is not yet possible and will require more descriptive data.

Both fixation procedures described in Protocol 6 and Protocol 7 can be used on samples already expressing the GFP. The fluorescence emitted by the marker will still be visible on fixed samples especially when an enhanced variant of GFP is used, namely the F64L/S65T mutant (24). This feature allows a given GFP labelled subcellular location to be studied *in vivo* as well as by immuno-fluorescence on the same tissue. Extensive studies combining *in vivo* fluorescence with immunolabelling data should lead to useful GFP patterns being defined in the future.

4 Immunogold electron microscopy

4.1 General remarks

As defined by Griffiths (25), immunocytochemistry is the cellular location of biochemically-defined antigens. At the level of the electron microscope antigen–antibody coupling is most frequently visualized by immunogold labelling of ultrathin plastic or cryosections, and the exposure of the sections to primary and secondary antibody solutions follows the same rules as for the detection of antigens in Western blots. Ideally, the antigen under consideration is highly stable towards fixatives and embedding media, so that it can be detected under conditions where the morphological preservation of cell organelles is optimal. Normally, a compromise has to be reached between the retention of antigenicity on the one hand and the maintenance of structural integrity on the other.

There is no 'golden rule' in immunogold labelling: every antigen must be treated separately. We have encountered stable antigens, e.g. storage globulins, as well as others, whose detection can only occur under 'homeopathic' conditions.

Unfortunately, although antigen preservation through cryo-methods is becoming increasingly the method-of-choice, the necessary equipment is expensive to acquire and operate. As a consequence, only few plant EM laboratories are 'state-of-the-art' equipped and experienced in these cryo-techniques.

With only few possible exceptions, virtually every method in immunoelectron microscopy is applicable to any type of eukaryotic cell. Figure 4 presents an overview of these methods, and indicates how to select the most appropriate strategy for plant cells. Numerous reviews and books are available on general immunoelectron microscopy; in our opinion the best reference works are Griffiths (25), and Newman and Hobot (26).

4.2 Assessing antigenic stability

A simple method by which the sensitivity of an antigen towards chemical fixatives can be ascertained is the dot blot screening procedure of Riederer (27).

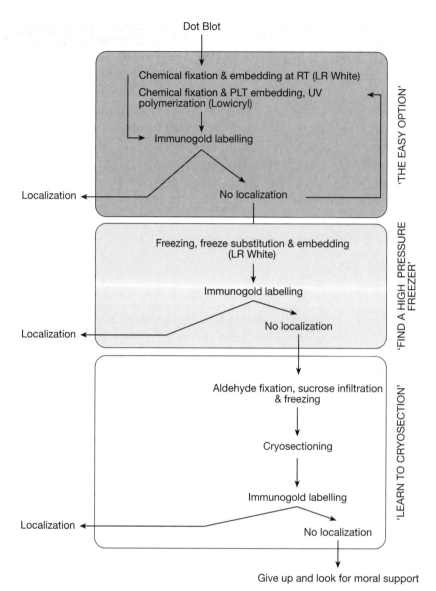

Figure 4 Flow chart presenting the different strategies for EM immunogold localization.

Since this is an *in vitro* assay, and the cellular concentrations of individual fixatives may be different to those presented to the organism, it can only deliver recommendations of an empirical nature. Pure preparations of antigen are not required, a subcellular fraction (membranes or cytosol) known to contain the antigen under investigation will suffice. If the antigen is membrane-bound it is advisable to resuspend the membrane pellet in a detergent (e.g. 0.5% Triton X-100)-containing buffer.

Protocol 9

Determining antigen stability by dot blot screening

Equipment and reagents

- Dot blot apparatus (commercially available, or self-made)
- Nitrocellulose (NC) filters (e.g. Sartorius)
- ECL-Kit (Amersham Biosciences)
- Tris-buffered saline (TBS): 0.9% (w/v) NaCl, 50 mM Tris–HCl pH 7.5
- TBS-T: 0.5% (v/v) Tween 20 in TBS
- Blocking solution: 5% (w/v) powdered skimmed milk, 1% fatty acid-free BSA in TBS-T
- Primary antibody solution (diluted as for a Western blot in TBS containing 1% BSA)
- Secondary antibody solution (diluted 1:15 000 in TBS)
- Fixatives (glutaraldehyde, paraformaldehyde, osmium tetroxide, uranyl acetate)

Method[a]

1 Add 2 μl antigen solution (protein concentration not exceeding 0.25 μg/μl) to a series of 2 μl fixative solutions (end concentrations of fixative ranging from 0.01–2%). Mix well, incubate for 30 min.

2 Transfer to NC filter, allow to dry. The NC filter will have upwards of 20 dots on it.

3 Immerse NC filter in blocking solution for 30 min.

4 Wash NC filter 3 × 10 min in TBS.

5 Immerse NC filter in primary antibody solution for 1 h.

6 Wash NC filter 3 × 10 min in TBS-T.

7 Immerse NC filter in secondary antibody solution for 1 h.

8 Wash NC filter 3 × 10 min in TBS-T.

9 Subject NC filter to ECL procedure according to the manufacturer's instructions. Try different exposure times.[b]

[a] All operations are to be performed at room temperature.

[b] Antigen–antibody coupling is recognized as a black spot on the negative film. The intensity of the spot will decrease according to the sensitivity of the antigen towards a particular concentration of fixative.

4.3 Chemical fixation for immunocytochemical studies

The composition of the fixative should be dependent on the results of the dot blot screening described above. As a starting point for antigens which are relatively insensitive to fixatives we commonly use Protocol 10 for plant specimens. This fixation can be used in combination with Protocol 11 for ethanol dehydration and embedding in LR White. If a PLT scheme for Lowicryl embedding is planned (see below), one should avoid the secondary osmium tetroxide fixation and greatly reduce, or possibly also avoid, uranyl acetate for *en bloc* fixation and contrast enhancement.

Protocol 10

Scheme for a conventional chemical fixation

Reagents

- 100 mM potassium phosphate buffer pH 7
- 25% stock solution glutaraldehyde (EM grade, Sigma)
- 10% (w/v) stock solution formaldehyde, freshly prepared from paraformaldehyde powder (Sigma)

- 1% uranyl acetate in water (Merck)
- 2% (w/v) osmium tetroxide in water (Merck)
- Primary fixative: 1.5% formaldehyde, 0.2% glutaraldehyde, 50 mM phosphate buffer
- Secondary fixative: 0.5% osmium tetroxide in 50 mM phosphate buffer

Method

1 Immerse a small sample (max. 1 mm^3) from the specimen of choice in primary fixative overnight at 4 °C.

2 Remove the fixative and wash with 50 mM phosphate buffer for 3 × 15 min at 4 °C.

3 Replace buffer with secondary fixative for 2–6 h at 4 °C.

4 Wash 4 × 15 min (twice with buffer, twice with water) at room temperature. Replace water with 0.5% aqueous uranyl acetate and leave overnight at room temperature.

4.4 Dehydration

Since water is the main component of all biological tissues, its removal causes changes that can affect cellular constituents. To minimize the deleterious effects of dehydration, new embedding media have been developed which are to some extent water miscible and can even tolerate small amounts of water during polymerization. This is important, because the water that is tightly bound to biological molecules and forms their hydration shells can thus be retained. Therefore, conformational changes, aggregation, precipitation, and extraction will be reduced, and the immunodetection of the components in question improved. There are at least two possible strategies for an incomplete dehydration in combination with chemical fixation. The first and easiest to achieve is carried out at room temperature, most often in combination with water miscible acrylics (Protocol 11). The time schedule may be adjusted to the specimen in question, but it should be kept in mind that it must be as short as possible to avoid negative effects on antigenicity.

Protocol 11

Room temperature dehydration

Reagents

- London Resin White (LR White, hard grade; Plano)
- Ethanol solutions

Method

1   Immerse the sample in 30% ethanol for 10 min at room temperature.

2   Immerse the sample in 50% ethanol for 10 min at room temperature.

3   Immerse the sample in 70% ethanol for 10 min at room temperature.

4   Immerse the sample in 2 parts 100% LR White and 1 part 70% ethanol (mix well before adding to the specimen) for 45 min.

5   Immerse the sample in 100% LR White for 45 min.

The second way to achieve incomplete dehydration is known as progressive lowering of temperature (PLT). The principle of this technique is a reduction of the temperature during dehydration, and utilizes the fact that the freezing point of a solvent mixture decreases with higher solvent concentrations. Important here is that at a concentration of 70% solvent the temperature should be $-35\,°C$, where the deleterious effects of dehydration are less. The time schedule must be adapted to the specimen, e.g. the sample has to be in equilibrium with the surrounding solvent, otherwise it may be damaged by ice crystal formation during the next step at a lower temperature. The disadvantage of the PLT method is procuring the necessary equipment to achieve and maintain constant subzero temperatures (e.g. $-20\,°C$ and $-35\,°C$). In addition, since resin infiltration also takes place at these temperatures it is recommended that this procedure be carried out in a fume cupboard.

Protocol 12

Suggested scheme for progressive lowering of temperature

Equipment and reagents

- $-20\,°C$ freezer
- Ethanol cooling bath (Thermomix UB, Braun)
- Lowicryl HM20 (Plano)

Method

1   Immerse the fixed and washed sample in 30% ethanol at $0\,°C$ (ice-bath) for 30 min.

2   Transfer to 50% ethanol at $-20\,°C$ (freezer) for 60 min.

Protocol 12 continued

3 Transfer to 70% ethanol at −35 °C (cooling bath) for 60 min.

4 Transfer to 100% ethanol at −35 °C (cooling bath) for 60 min.

5 Transfer to fresh 100% ethanol at −35 °C (cooling bath) for 60 min.

6 Transfer to 2 parts 100% ethanol and 1 part Lowicryl at −35 °C (cooling bath) for 60 min.

7 Transfer to 1 part 100% ethanol and 1 part Lowicryl at −35 °C (cooling bath) for 60 min.

8 Transfer to 1 part 100% ethanol and 2 parts Lowicryl at −35 °C (cooling bath) for 60 min.

9 Transfer to pure Lowicryl at −35 °C (cooling bath) for 60 min.

10 Transfer to fresh pure Lowicryl at −35 °C (cooling bath) overnight.

11 Transfer to fresh pure Lowicryl at −35 °C (cooling bath) for 60 min.

4.5 Polymerization

4.5.1 London Resin White

Polymerization of LR White can be done either with heat (60 °C for 16–24 h) or chemical accelerators (at room temperature or on ice inside a microwave oven). Care has to be taken that polymerization occurs in an oxygen-free atmosphere (for this a vacuum oven is an ideal solution, but flushing the samples with nitrogen gas prior and during polymerization is also a possibility). An excellent overview on the use of LR White is given by Newman (28).

4.5.2 Lowicryl HM20

Specimens immersed in pure Lowicryl HM20 must be transferred to gelatin cups which have been pre-filled with fresh resin. The tips of these cups should be submerged in 100% ethanol during polymerization at −35 °C with indirect UV light for at least 24 h, to ensure adequate removal of heat during the exothermic polymerization reaction. This will improve the sectioning quality of the blocks. Polymerization should then be continued at room temperature for another 24 h to improve further the sectioning quality of the cured blocks. For an excellent review on the Lowicryl resins the reader is referred to Villinger (29).

4.6 Rapid freezing and freeze substitution

Physical fixation through rapid freezing represents an important alternative to chemical fixation (30), especially if aldehydes must be avoided. Rapid freezing is often followed by freeze substitution, which can be performed in the presence of OsO_4, providing the antigen is insensitive to the latter fixative.

Protocol 13

High pressure freezing and freeze substitution

Equipment and reagents

- High pressure freezer (e.g. Bal-Tec)
- Freeze substitution apparatus (e.g. AFS, Leica)
- Beem capsules for embedding the samples (Plano)
- Hexadecene (Sigma)
- Osmium tetroxide (Merck)
- Dry acetone
- Dry ethanol
- LR White (hard grade, Plano)

Method

1 Remove a small piece of tissue from a specimen.

2 Immerse in hexadecene.

3 Freeze the sample with a high pressure freezer (follow the manufacturer's operating instructions).

4 Transfer samples to liquid nitrogen.

5 Transfer samples to freeze substitution apparatus.

6 Freeze substitute with 1% OsO_4 in dry acetone for two days at $-85\,°C$.

7 Slowly ($<4\,°C/h$) warm the sample to $-35\,°C$.

8 Wash with dry pre-chilled acetone once, then replace acetone with ethanol, and wash 3 × with ethanol at $-35\,°C$.

9 Warm the sample to room temperature ($<4\,°C/h$).

10 Immerse the sample in 2 parts 100% LR White and 1 part 100% ethanol for 45 min.

11 Immerse the sample in 100% LR White overnight.

12 Replace LR White with fresh resin and transfer sample to Beem capsules. After 4 h polymerize in a vacuum oven at $60\,°C$ for 24 h.

4.7 Immunogold labelling of acrylic sections

Optimal gold labelling is specific, i.e. background labelling is negligible, and the only gold particles to be seen lie over the antigenic sites. Non-specific immuno-label can have many causes, the main ones being: insufficient blocking, non-specific IgGs, too high a concentration of primary antibody, and poor washing of the (nickel) grids. Visualization of the bound primary antibody is normally done by application of a secondary anti-primary IgG coupled to colloidal gold or protein A gold (as a rule 5 or 10 nm gold particles). Density of labelling is less important, but can be increased, albeit at the danger of loss of specificity and fine structural detail, by revealing additional antigenic binding sites. This can be done, in the case of acrylic sections, by pre-treating the sections with saturated NaOH in ethanol (31), in the case of cryosections, by pre-treating with sodium dodecyl sulfate (32). A less 'invasive' method is to increase the concentration and exposure time to the secondary antibody solution, while at the same time reducing the primary antibody concentration. A typical immunogold labelling procedure for acrylic sections is given in Protocol 14.

Protocol 14

Gold labelling of antigenic sites on acrylic sections

Reagents

- Tris-buffered saline (TBS, see Protocol 9)
- BSA (fatty acid-free)
- BSA-C (acetylated BSA, AurIon)
- Gold-coupled secondary antibodies (British BioCell)
- Blocking solution: 1% (w/v) BSA, 0.1% (w/v) BSA-C in TBS

Method

1 Float grids on drops of blocking solution for 30 min.

2 Transfer to primary antibody solution in TBS for 1 h.[a]

3 Transfer to blocking solution (4 × 10 min).

4 Transfer to secondary gold-conjugated antibody solution (either in TBS or blocking solution) for 1 h.

5 Transfer to blocking solution (3 × 5 min).

6 Transfer to double-distilled water (2 × 5 min).

7 Double stain with lead and uranyl salts as for conventional electron microscopy.

[a] As a guide, the concentration of primary antibody should be roughly 10-fold higher than for Westerns.

4.8 Preparation and immunogold labelling of cryosections

The validity of dot blot screening assays is hampered by the fact that this technique can not predict the influence of a particular dehydration/embedding medium on a given antigen. Thus a complete lack of specific immunolabelling may result, although fixation conditions were correctly chosen. The reason for this disappointment is either that the antigen is negatively influenced by dehydration or masked by the embedding resin after polymerization, and is therefore not accessible to the antibodies. Such problems are often encountered and have led to the increasing popularity of the Tokujasu method (33). This method avoids totally dehydration and embedding of the sections prior to the labelling procedure. To achieve this samples are gently pre-fixed with aldehydes, then slowly infiltrated with high concentrations of osmotica (mostly sucrose) and frozen. Ultrathin sections are cut from frozen samples which then are thawed and subsequently immunolabelled. Afterwards, the labelled sections are embedded on grid with a mixture of methylcellulose and uranyl acetate to allow inspection of the samples in an electron microscope. For aldehyde-insensitive antigens this procedure often produces ten times higher labelling densities as compared to labelling which can be observed on identically fixed samples subsequently processed by dehydration and plastic embedding.

Protocol 15

Specimen preparation for cryosectioning

Equipment and reagents

- A rotator
- Specimen stubs (Leica)
- Liquid nitrogen
- 8.0% glutaraldehyde stock (EM-grade, Sigma G-7526)

- Formaldehyde stock: 10% (w/v), freshly prepared from paraformaldehyde powder
- Fixative: 1.5% formaldehyde, 0.2% glutaraldehyde, 50 mM phosphate buffer (pH 7.0)

Additional equipment and reagents for cryosectioning

- Ultramicrotome and Cryochamber: Leica UCT-FCS (Leica)
- A knife-maker to prepare glass knives for trimming and sectioning (Leica)
- A diamond knife suitable for cryosectioning without knife boat (Diatome)

- An anti-static device (Diatome)
- Carbon coated formvar filmed grids
- A dewar for storage of the frozen samples
- A binocular (Leica)
- Methylcellulose (Sigma M-6385)
- Uranyl acetate (Merck)

Method

1   Cut a small sample from the specimen of choice and immerse it in fixative overnight at 4 °C.

2   Replace fixative with 50 mM phosphate buffer (three washes, 15 min each).

3   Replace with 0.8 M sucrose in 50 mM phosphate buffer. Keep this for 3 h on a rotator at 4 °C.

4   Replace the solution with 1.6 M sucrose in 50 mM phosphate buffer. Rotate for 3 h at 4 °C.

5   Replace the solution with 2.3 M sucrose in 50 mM phosphate buffer. Rotate overnight at 4 °C.

6   Replace the solution with fresh 2.3 M sucrose. Rotate for 1 h at 4 °C.

Protocol 15 continued

7 Place the sample on a clean specimen stub, remove as much sucrose as possible with a piece of tissue paper, and immediately freeze the sample by throwing it into a beaker filled with liquid nitrogen.

8 Afterwards the specimen stubs can be transferred into the cryo chamber of the ultramicrotome for trimming and cryosectioning. Since the cryosectioning process itself is fairly difficult to master the reader is referred to the excellent description in Griffiths (25) for details.

Protocol 16

Immunogold labelling of cryosections

Reagents

- Phosphate-buffered saline (PBS): 0.9% (w/v) NaCl in 50 mM potassium phosphate pH 7.5
- 20 mM glycine in PBS (aldehyde blocking)
- 1% (w/v) BSA in PBS
- Uranyl oxalate: 2% (w/v) uranyl acetate in 150 mM aqueous oxalic acid, adjust to pH 7 with 25% (w/v) ammonium hydroxide
- 1.8% (w/v) methylcellulose, 0.4% aqueous uranyl acetate pH 4

Method

1 Wash grids 3 × for 3 min on drops of PBS–glycine at 37 °C.

2 Block for 5 min with 1% (w/v) BSA in PBS at room temperature.

3 Transfer grids to 30 µl drops of primary antibody diluted in PBS for 1 h.

4 Wash grids 3 × for 5 min on drops of 0.1% (w/v) BSA in PBS.

5 Transfer grids to gold-coupled secondary antibodies for 1 h.

6 Wash grids for 5 min on drops of 0.1% (w/v) BSA in PBS.

7 Wash grids 3 × 5 min on drops of PBS.

8 Wash grids 5 × for 2 min on double-distilled water.

9 Contrast for 5 min on a drop of uranyl oxalate.

10 Wash grids 2 × for 30 sec on double-distilled water.

11 Transfer grids to a drop of methylcellulose/uranyl acetate for 2 min.

12 Transfer grids to a fresh drop of methylcellulose/uranyl acetate for 2 min.

13 Transfer grids to a fresh drop of methylcellulose/uranyl acetate for 5 min on ice.

14 Take up the grids with a wire loop (Ø 3.5 mm) and gently remove the embedding medium by moving the loop over a moist filter paper.

15 Dry the grids for 15 min in the loop.

16 Remove grids from the loop with the tip of a fine forceps.

References

1. Frigerio, L., de Virgilio, M., Prada, A., Faoro, F., and Vitale, A. (1998). *Plant Cell*, **10**, 1031.
2. Neuhaus, J.-M., Pietrzak, M., and Boller, T. (1994). *Plant J.*, **5**, 45.
3. Leborgne-Castel, N., Jelitto-Van Dooren, E. P. W. M., Crofts, A. J., and Denecke, J. (1999). *Plant Cell*, **11**, 459.
4. Di Sansebastiano, G.-P., Paris, N., Marc-Martin, S., and Neuhaus, J.-M. (1998). *Plant J.*, **15**, 449.
5. Vitale, A. and Raikhel, N. V. (1999). *Trends Plant Sci.*, **4**, 149.
6. Gallie, D. (1996). *Plant Mol. Biol.*, **32**, 145.
7. Kozak, M. (1986). *Proc. Natl. Acad. Sci. USA*, **83**, 2850.
8. Tabe, L. M., Wardley-Richardson, T., Ceriotti, A., Aryan, A., McNabb, W., Moore, A., *et al.* (1995). *J. Anim. Sci.*, **73**, 2752.
9. Joshi, C. P., Zhou, H., Huang, X., and Chiang, V. L. (1997). *Plant Mol. Biol.*, **35**, 993.
10. Lukaszewicz, M., Feuermann, M., Jérouville, B., Stas, A., and Boutry, M. (2000). *Plant Sci.*, **154**, 89.
11. Gallie, D., Sleat, D., Watts, J., Turner, P., and Wilson, T. (1987). *Nucleic Acids Res.*, **15**, 8693.
12. Dombrowsky, J. E., Gomez, L., Chrispeels, M. J., and Raikhel, N. V. (1994). In *Plant molecular biology manual* (ed. S. B. Gelvin and R. A. Schilperoort), pp. J3, 1–29. Kluwer Academic Publishers.
13. Axelos, M., Curie, C., Mazzolini, L., Bardet, C., and Lescure, B. (1992). *Plant Physiol. Biochem.*, **1**, 123.
14. Wick, S. M., Muto, S., and Duniec, J. (1985). *Protoplasma*, **126**, 198.
15. Paris, N., Stanley, C. M., Jones, R. L., and Rogers, J. C. (1996). *Cell*, **85**, 563.
16. Boevink, P., Oparka, K., Cruz, S., Martin, B., Betteridge, A., and Hawes, C. (1998). *Plant J.*, **15**, 441.
17. Haasen, D. C., Köler, C., Neuhaus, G., and Merkle, T. (1999). *Plant J.*, **20**, 695.
18. Haseloff, J., Siemering, K. R., Prasher, D. C., and Hodge, S. (1997). *Proc. Natl. Acad. Sci. USA*, **94**, 2122.
19. Di Sansebastiano, G.-P., Paris, N., Marc-Martin, S., and Neuhaus, J.-M. (1998). *Plant J.*, **15**, 449.
20. Di Sansebastiano, G. P., Paris N., Marc-Martin, S., and Neuhaus, J.-M. (2001). *Plant Physiol.*, **126**, 78.
21. Hibberd, J. M., Linley, P. J., Khan, M. S., and Gray, J. C. (1998). *Plant J.*, **16**, 627.
22. Köler, R., Zipfel, W. R., Webb, W., and Hanson, M. (1997). *Plant J.*, **11**, 613.
23. Brandizzi, F., Frange, N., Marc-Martin, S., Hawes, C., Neuhaus, J.-M., and Paris, N. (2002). *Plant Cell*, **14**, 1077.
24. Cormack, B. P., Valdivia, R. H., and Falkow, S. (1996). *Gene*, **173**, 33.
25. Griffiths, G. (1993). *Fine structure immunocytochemistry*. Springer–Verlag, Berlin.
26. Newman, G. R. and Hobot, J. A. (1987). *J. Histochem. Cytochem.*, **35**, 971.
27. Riederer, B. M. (1989). *J. Histochem. Cytochem.*, **37**, 675.
28. Newman, G. R. (1987). *Histochem. J.*, **19**, 118.
29. Villinger, W. (1991). In *Colloidal gold* (ed. M. A. Hayat), Vol. 3. Academic Press, San Diego.
30. McDonald, K. (1999). *Methods Mol. Biol.*, **117**, 77.
31. Matsubara, A., Laake, J. H., Davanger, S., Usami, S., and Ottrsen, O. P. (1996). *J. Neurosci.*, **16**, 4457.
32. Braun, D., Lyndon, J., McLaughlin, M., Stuart-Tilley, A., Tyszkowski, R., and Alper, S. L. (1996). *Histochem. Cell Biol.*, **104**, 261.
33. Tokuyasu, K. T. (1980). *Histochem. J.*, **12**, 381.

Chapter 6
Protein purification

Susan E. Slade and Howard Dalton
Department of Biological Sciences, University of Warwick,
Coventry CV4 7AL, UK.

1 Introduction

This chapter outlines the main features and requirements for the successful purification of a target protein from either its native or recombinant host. For each of the commonly used separation techniques an experimental protocol is described.

2 General principles of protein purification

The aim of purification is to obtain pure target protein in a structurally or functionally active form from a crude lysate. In many cases, pure protein is required only so its gene can be isolated using amino acid sequence information, in which case only a small quantity of protein may be required. Therefore, it is important to establish the purity and quantity of protein required from the procedure before embarking on any purification method. Often this decision is determined by the availability of raw material and the concentration of target protein within the host cell or culture supernatant. The increasing use of mass spectrometry to determine molecular mass, subunit composition, amino acid sequence, and establish the nature of protein–ligand interactions, requires a high level of target protein purity

Purification will generally involve a number of steps, each based on the molecular characteristics of the protein of interest—molecular weight and shape, net charge, hydrophobicity, and biological properties. Therefore, a selection of each type of separation media will be needed to optimize the purification procedure. Important consideration must be given to the buffer used at each stage to ensure that its constituents do not interfere with the analysis following purification.

The purification procedure can be divided into four stages. The initial stage involves cell disintegration with the subsequent release of the cell contents. Many techniques are available, from ultrasonication, French press and homogenization, to less aggressive methods using enzymes or chemicals to disrupt the cell wall. Once the target protein has been removed from its protective environment within the cell it is susceptible to a loss of biological activity through the action of proteases, denaturants (pH, temperature, and chemical solutes), and inactivation of the active site/s. Stabilizers can be added to the homogenization buffer to ensure maximum activity in the extract (e.g. protease inhibitors, EDTA, DTT, metal ions, glycerol, etc.). A series of small scale trials should be undertaken to investigate target protein stability under the various conditions encountered during purification (pH, ionic strength, temperature, etc.) and establish optimum storage conditions.

The second stage aims to remove as many contaminating proteins from the target as possible. The sample may be crude and viscous due to the presence of nucleic acids, carbohydrates, lipids, lipoproteins, and particulate matter or alternatively may be very dilute. Separation of the target protein from contaminants using an adsorptive technique (ion exchange, hydrophobic interaction, or affinity chromatography) is generally best employed at this stage.

For the next stage of the purification process, it is preferable to use techniques that exploit different properties of the target protein than those used in the previous step. A number of different chromatographic steps may be required at this stage to improve target protein purity.

The final finishing or 'polishing' step aims to resolve the target protein from any minor contaminants, molecular variants, or isoforms. A method should be employed which utilizes an attribute of the target protein not previously used in the preparation. This may include gel filtration, chromatofocusing, reversed phase or affinity chromatography.

Following each step of the purification process, the total amount of protein present is calculated and the specific activity of the target protein is determined. An increase in specific activity indicates the loss of contaminating proteins. A purification table summarizes the information on total protein, total activity, specific activity, yield, and purification factor for each step. It helps to determine which steps of the purification provide good recovery of target protein in terms of yield and activity whilst highlighting any that may need to be removed or further optimized.

3 Determination of protein concentrations

In order to determine the recovery of the target protein at each stage of the purification procedure a rapid and accurate method of determining the enzyme activity or bioactivity is required. This figure is expressed relative to the total protein content of the sample. Unfortunately, all methods for the measurement of total protein are subject to problems with interference and are only accurate if the method has been calibrated against a standard of pure target protein. A number of protein concentration assay kits are available commercially and the manufacturer's specifications should be consulted to determine the most suitable for a particular application.

Total protein concentration is also used to ensure that the protein-binding capacity for a particular column is never reached in any purification run.

3.1 Spectroscopic methods—UV absorption

Protein concentration is determined by absorption at a particular wavelength. Proteins absorb UV light at 280 nm due to the presence of the aromatic residues tryptophan and tyrosine, whereas at 205 nm the peptide bond is responsible. The method described in Protocol 1 requires little sample, is non-destructive, and analysis is rapid, although interference from buffer constituents can cause problems, especially at 205 nm. The concentration of the protein sample is calculated either by comparison of the measured absorbance with the published extinction coefficient, or a calibration curve prepared with a protein standard, usually BSA. The absorbance of a 1 mg/ml protein solution will generally lie in the region of 0.4–1.5 at 280 nm and approximately 31 at 205 nm.

Protocol 1

Protein concentration determination using absorption at 280 nm (A_{280})

Equipment and reagents

- Spectrophotometer with UV lamp
- Quartz cuvette 1 cm path length
- 1 mg/ml standard protein solution for calibration

Method

1. Prepare a concentration series of the standard protein solution in the same buffer as the sample protein.

2. Zero the spectrophotometer at 280 nm using a buffer blank.

3. Measure the absorbance of each of the standard solutions and the sample protein. Any samples having an absorbance >1.0 should be diluted using buffer and measured again.

4. If the extinction coefficient for the sample protein is known this can be compared against the measured value and its concentration determined.

5. A calibration curve for the standard protein is prepared, plotting A_{280} versus protein concentration. The concentration of the sample protein is calculated using its absorbance from the calibration curve.

3.2 Colorimetric dye-binding assays

3.2.1 Biuret method

The Biuret assay (1) described in Protocol 2 is based on the formation of a purple colour (with a λ_{max} of 540 nm) when the alkaline copper reagent reacts with the peptide chain of a protein. Although the least sensitive of the colorimetric methods, it has the advantage of being less susceptible to chemical interference, although it is not recommended for use with ammonium sulfate fractions. The absorption of each sample is measured at 540 nm and compared with a calibration plot for BSA or other suitable protein standard.

Protocol 2

Biuret assay for protein concentration

Equipment and reagents

- Spectrophotometer
- Cuvettes 1 cm path length
- Standard protein solution of known concentration

- Biuret reagent: prepare in 500 ml pre-boiled, cooled H_2O 1.5 g $CuSO_4 \cdot 5H_2O$ and 6.0 g sodium potassium tartrate; add 300 ml of 10% NaOH and dilute to 1 litre

Method

1 Prepare a concentration series of the standard protein solution in the same buffer as the sample protein.

2 To separate test-tubes, add 0.5 ml of sample protein, a dilution of the protein standard, or sample buffer as a blank.

3 To each tube add 2.5 ml Biuret reagent and incubate at room temperature for 20–30 min.

4 Zero the spectrophotometer at 540 nm (A_{540}) using the buffer blank.

5 Measure the A_{540} for the sample protein and each of the protein standard solutions. Any samples having an absorbance >1.0 should be discarded and a more dilute sample used in step 2.

6 A calibration curve for the standard protein is prepared of the A_{540} versus concentration. The concentration of the sample protein is calculated using its absorbance from the calibration curve.

3.2.2 Hartree–Lowry method

The Hartree–Lowry colorimetric assay (2) described in Protocol 3 involves the reduction of Folin–Ciocalteu reagent by a copper complex under alkaline conditions with the formation of a strong, dark blue colour. It is more sensitive than the Biuret method and maintains a linear response over a greater concentration range, although the method is subject to interference from a number of components used in the purification process. The absorption of each sample is measured at 650 nm and compared to a calibration plot for BSA or other suitable protein standard.

Protocol 3

Hartree–Lowry assay for protein concentration

Equipment and reagents

- 50 °C water-bath
- Spectrophotometer
- Cuvettes 1 cm path length
- Standard protein solution of known concentration
- Reagent A: 7 mM sodium potassium tartrate·4H$_2$O, 0.81 M Na$_2$CO$_3$, 0.5 M NaOH

- Reagent B: 70 mM sodium potassium tartrate·4H$_2$O, 40 mM CuSO$_4$·5H$_2$O, 0.1 M NaOH
- Reagent C: 1 ml Folin–Ciocalteu reagent diluted with 15 ml H$_2$O

Method

1. Prepare a concentration series of the standard protein solution (50–150 μg/ml) in the same buffer as the sample protein.

2. To separate test-tubes, add 1 ml of sample protein, a dilution of the protein standard, or sample buffer as a blank.

3. To each tube add 0.9 ml reagent A and incubate at 50 °C for 10 min.

4. Cool the tubes to room temperature.

5. Add 0.1 ml of reagent B to each tube and incubate at room temperature for 10 min.

6. Add 3 ml of reagent C to each tube and mix thoroughly before incubation at 50 °C for 10 min. Cool tubes to room temperature prior to measurement.

7. Zero the spectrophotometer at 650 nm (A$_{650}$) using the buffer blank.

8. Measure the A$_{650}$ for the sample protein and each of the protein standard solutions. Any samples having an absorbance >1.0 should be discarded and a more dilute sample used in step 2.

9. A calibration curve for the standard protein is prepared of the A$_{650}$ versus concentration. The concentration of the sample protein is calculated using its absorption from the calibration curve.

3.2.3 Bicinchoninic acid (BCA) method

The Lowry method, improved by Smith (3), involved replacing the Folin–Ciocalteu reagent with bicinchoninic acid, resulting in a method less sensitive to interference. The purple coloration is left to develop for 30 min and the absorption is measured at 562 nm and compared with a calibration plot of BSA. This method is more prone to variation due to incubation temperatures but is simple and very sensitive. This method is described in Protocol 4.

Protocol 4

Bicinchonic acid assay for protein concentration

Equipment and reagents

- 37 °C water-bath
- Spectrophotometer
- Cuvettes 1 cm path length
- Standard protein solution of known concentration

- Reagent A: 26 mM 4,4'-dicarboxy-2-2'-biquinoline, disodium salt, 0.16 M Na_2CO_3, 7 mM disodium sodium tartrate, 0.1 M NaOH, 0.11 M $NaHCO_3$ pH adjusted to 11.3 with solid NaOH or $NaHCO_3$
- Reagent B: 16 mM $CuSO_4 \cdot 5H_2O$

Method

1   Prepare a concentration series of the standard protein solution (0.2–1 mg/ml) in the same buffer as the sample protein.

2   To separate test-tubes, add 100 μl of sample protein, a dilution of the protein standard, or sample buffer as a blank.

3   Mix 50 ml BCA reagent A with 1 ml BCA reagent B.

4   To each tube add 2 ml of the BCA reagent mix, prepared in step 3 and incubate at 37 °C for 30 min.

5   Cool the tubes to room temperature.

6   Zero the spectrophotometer at 562 nm (A_{562}) using the buffer blank.

7   Measure the A_{562} for the sample protein and each of the protein standard solutions. Any samples having an absorbance >1.0 should be discarded and a more dilute sample used in step 2.

8   A calibration curve for the standard protein is prepared of the A_{562} versus concentration. The concentration of the sample protein is calculated using its absorbance from the calibration curve.

3.2.4 Coomassie dye binding (Bradford) method

Bradford used the binding of dyes to proteins to estimate protein concentration (4). It is the most widely used method due to its simplicity, high sensitivity towards most proteins, and rapid colour development. The method is described in Protocol 5. Coomassie dye (Brilliant Blue G-250) binds to protein molecules at acidic pH, producing a colour change from red–brown to blue measured at 595 nm. BSA is commonly used to prepare a calibration curve but has a greater dye-binding capacity than most proteins, hence the reference standard should ideally be related to the target protein. Interfering chemical compounds include the detergents SDS and Triton X-100, and the blue colour can adsorb to glassware and cuvettes.

Protocol 5

Bradford assay for protein concentration

Equipment and reagents

- Spectrophotometer
- Cuvettes 1 cm path length
- Standard protein solution of known concentration
- Coomassie dye reagent: 100 mg Coomassie Brilliant Blue G-250 dissolved in 50 ml of

95% ethanol, then mixed with 100 ml of 85% phosphoric acid, and diluted to 1 litre with H_2O. Filter solution through Whatman No. 2 filter paper.

Method

1 Prepare a concentration series of the standard protein solution (0.1–0.75 mg/ml) in the same buffer as the sample protein.

2 To separate test-tubes, add 20 μl of sample protein, a dilution of the protein standard, or sample buffer as a blank.

3 To each tube add 1 ml Coomassie dye reagent and mix.

4 Incubate at room temperature for 2–30 min.

5 Zero the spectrophotometer at 595 nm (A_{595}) using the buffer blank.

6 Measure the A_{595} for the sample protein and each of the protein standard solutions. Any samples having an absorbance >1.0 should be discarded and a more dilute sample used in step 2.

7 A calibration curve for the standard protein is prepared of the A_{595} versus concentration. The concentration of the sample protein is calculated using its absorption from the calibration curve.

4 Fractionation of proteins using adsorptive techniques

4.1 Fractionation by charge—ion exchange chromatography

Ion exchange chromatography separates molecules according to their surface charge. Separation depends on the reversible adsorption of the accessible charged groups on the protein to the oppositely charged groups on the ion exchange matrix. The net charge on a protein is pH dependent, so when the pH is greater than the isoelectric point (pI) the protein will have a net negative charge and will have a positive charge if the pH drops below the pI.

These properties are exploited in ion exchange chromatography through the choice of separation media. By increasing the buffer pH above the target protein's pI value, the negatively charged protein molecules will bind to the positively charged groups on an anion exchange matrix, but will bind to negatively charged cation exchange matrices when the pH is dropped.

Choice of ion exchange media and buffering system requires knowledge of the p*I*, pH stability, and solubility of the target protein. The buffering ion must have the same charge as the ion exchange matrix to prevent binding, as must any detergents or additives in the buffers (see Table 1 for a selection of ion exchange buffers). However, if the p*I* is unknown it is assumed to be below pH 7 (for most proteins or enzymes) and therefore the use of anion exchange media at pH 8.5 is a common starting point. Peptides and many regulatory proteins have a basic p*I* and therefore a cation exchange matrix may be more suitable at low pH. Using small scale trials the operating conditions can be optimized for the target protein.

Binding of the target protein to the matrix occurs whilst unbound and weakly bound proteins are washed away. Salt molecules selectively displace column-bound proteins as the ionic strength of the buffer is increased using a linear or stepwise gradient. Elution of proteins can also be achieved by decreasing the

Table 1 Buffers suitable for use in ion exchange chromatography[a]

Name	pK_a (25 °C)
Lactic acid	3.86
Acetic acid	4.76
N-Methylpiperazine	4.8
Piperazine	5.7
Histidine	6.0
N-Morpholinoethanesulfonic acid (MES)	6.15
Bis-(2-hydroxyethyl)imino-tris-(hydroxymethyl)methane (Bis-Tris)	6.5
Bis-Tris propane	6.8
Phosphate	7.2
N-Morpholinopropanesulfonic acid (MOPS)	7.2
N-Tris(hydroxymethyl)methyl-2-aminoethane sulfonic acid (Tes)	7.5
Triethanolamine	7.75
Tris(hydroxymethyl)aminomethane (Tris)	8.06
N-Methyldiethanolamine	8.5
Diethanolamine	8.9
Ethanolamine	9.5
Piperazine	9.7
Glycine	9.8
1-Aminopropan-3-ol	9.95
Carbonate	10.3

[a] The ionic strength of the buffer should be kept low to ensure sample binding to the ion exchanger. Ideally, buffer pH should be within 0.3 pH units of its pK_a and at least one pH unit above the target protein's p*I*.

buffer pH for anion exchange matrices or increasing the buffer pH for cation exchange matrices in a stepwise manner, giving improved reproducibility. The addition of 20% (v/v) water miscible organic solvents (methanol, glycerol, or acetonitrile) to the buffer system reduces hydrophobic interactions between protein molecules and the column matrix and can improve peak tailing and the poor recovery of target protein.

Following elution, the ion exchange media is regenerated prior to further use either by washing with a buffer of high ionic strength (2 M) or a large change in pH. The column is then equilibrated with sufficient binding buffer to ensure that the eluate is of the same pH and ionic strength as the binding buffer.

It is important to consider the conditions under which the target protein is eluted from the ion exchange column. Further ion exchange or chromato-focusing steps may require the protein to be desalted prior to column application. Hydrophobic interaction chromatography may require the salt concentration be increased to allow binding to the matrix. A typical example of anion exchange chromatography is described in Protocol 6.

Protocol 6

Sample fractionation by anion exchange chromatography using a linear salt gradient

Equipment and reagents

- Anion exchange column packed with a suitable matrix
- Distilled or deionized water[a]
- Binding buffer: 25 mM HEPES pH 8.5[a]
- Eluent: 25 mM HEPES pH 8.5 containing 2 M NaCl[a]

Method

1 Wash the column with distilled water to remove the bacteriostatic storage buffer.

2 Wash the column with 5 column volumes of elution buffer.

3 Equilibrate the column with 10 column volumes of binding buffer.

4 Dissolve the sample in binding buffer.

5 Filter or centrifuge the sample to remove particulates.

6 Apply the sample to the column.

7 Maintain 100% binding buffer for 5 column volumes.

8 Elute protein using a linear gradient of 0–50% elution buffer over 20 column volumes.

9 Maintain 50% elution buffer for 5 column volumes.

10 Monitor the detector response and collect fractions.

Protocol 6 continued

11 Assay fractions and pool those with related activity.

12 Regenerate column by washing with 5 column volumes of 100% elution buffer.

13 Equilibrate column as described in step 3.

^a Filter all solutions through a 0.45 μm filter and degas prior to use.

4.2 Fractionation by hydrophobicity

4.2.1 Hydrophobic interaction chromatography (HIC)

Hydrophobic interaction chromatography is based on the reversible adsorption of protein molecules to a hydrophobic matrix, through the hydrophobic groups on the exterior of the protein. The sample is prepared in a solution containing a high concentration of ammonium sulfate and is applied to the matrix under the same high salt conditions. The more hydrophobic proteins will bind preferentially to the surface of the matrix if the column is more hydrophobic than the solution itself. Bound proteins are selectively eluted using a decreasing stepwise or linear salt gradient, see Protocol 7, or through the addition of non-ionic detergents or organic solvents.

HIC is frequently used as the initial step in a purification procedure following ammonium sulfate precipitation, as the protein is already present in a suitable buffer. Ideally, the protein should be in starting buffer at the highest concentration of salt that will not cause it to precipitate. The most frequently used salts are $(NH_4)_2SO_4$, NaCl, and Na_2SO_4.

The target protein elutes at low salt concentrations so ion exchange is frequently used as the next purification step. Highly hydrophobic proteins (e.g. membrane proteins) may remain bound to the matrix even under low salt conditions. An increase in concentration of a water miscible solvent (ethanol, methanol, or acetonitrile), chaotropic agents (ethylene glycol, urea, sodium or magnesium chloride), or non-ionic detergent (Triton X-100) may be required to elute the target protein. A decrease in temperature or increase in pH will also reduce the hydrophobic interactions between the matrix and protein and result in elution from the column.

An ideal HIC matrix will bind the target protein with the majority of the contaminating proteins being eluted during the initial stages of the run. Failure of the target protein to bind to the matrix or early elution in the gradient may require the use of a more hydrophobic bonded phase or increased salt concentrations. If the target protein binds strongly, a less hydrophobic matrix may be more suitable.

Protocol 7

Sample fractionation by hydrophobic interaction chromatography using a decreasing salt gradient

Equipment and reagents

- HIC column packed with a suitable matrix
- Distilled or deionized water[a]
- Binding buffer: 50 mM sodium phosphate buffer pH 7.0 containing 1.7 M $(NH_4)_2SO_4$[a]
- Eluent: 50 mM sodium phosphate buffer pH 7.0[a]

Method

1 Wash the column with distilled water to remove the bacteriostatic storage buffer.

2 Equilibrate the column with 10 column volumes of binding buffer.

3 Dissolve the sample in binding buffer.

4 Filter or centrifuge the sample to remove particulates and precipitated protein.

5 Apply the sample to the column and wash with binding buffer whilst monitoring the detector response.

6 When the baseline stabilizes, elute protein using a linear gradient, reducing the ammonium sulfate concentration to zero over 10 column volumes.

7 Monitor the detector response and collect fractions.

8 Assay fractions and pool those with related activity.

9 Regenerate column by washing with 5 column volumes of distilled water.

10 Equilibrate column as described in step 2.

[a] Filter all solutions through a 0.45 μm filter and degas prior to use.

4.2.2 Reverse phase chromatography (RPC)

Reverse phase chromatography is related to HIC as it involves interactions between the non-polar groups on the exterior of a protein and hydrophobic ligands on the matrix.

Under aqueous conditions, adsorption of the protein to the matrix occurs in the presence of an ion-pairing agent, such as trifluoroacetic acid (TFA), formic, phosphoric, or acetic acid. A more hydrophobic ion-pairing agent such as heptafluorobutyric acid (HFBA) can improve the selectivity of a RPC method.

Proteins are selectively eluted from the column by reducing the polarity of the mobile phase, with the addition of a water miscible organic solvent in a linear concentration gradient. Acetonitrile, methanol, and isopropanol are suitable solvents for RPC, and a suitable method is described in Protocol 8.

The environment required for protein separation in RPC is more denaturing than HIC, with the target protein often eluting in an irreversibly non-native state. Hence, RPC is most frequently used in the purification of small proteins (<30 kDa) and peptides for amino acid sequencing, or the isolation of peptides from an enzymic digestion (e.g. trypsin).

Protocol 8

Sample fractionation by reverse phase chromatography

Equipment and reagents

- RPC column packed with a suitable matrix
- Binding buffer: distilled or deionized water[a] containing 0.1% TFA
- System filters for use with organic solvents
- Eluent: 100% acetonitrile containing 0.1% TFA[b]

Method

1 The equipment requires priming with organic solvents prior to use. Check manufacturer's recommendations for this procedure.

2 Wash the column with distilled water to remove the methanol storage solvent.

3 Wash the column with 10 column volumes of elution buffer.

4 Equilibrate the column with 10 column volumes of binding buffer.

5 Dissolve the sample in binding buffer and filter or centrifuge to remove particulates.

6 Apply the sample to the column.

7 Maintain 100% binding buffer for 5 column volumes.

8 When the baseline stabilizes, elute protein using a linear gradient of 0–100% elution buffer over 10 column volumes.

9 Monitor the detector response and collect fractions.

10 Maintain 100% elution buffer for 5 column volumes.

11 Equilibrate column as described in step 4.

[a] Filter solution through a 0.45 μm filter and degas prior to use.

[b] HPLC grade acetonitrile degassed in a sonicating water-bath for 30 min.

4.3 Hydroxyapatite chromatography

Hydroxyapatite is the inorganic adsorbent calcium phosphate. Protein binding to the matrix is dependent on interactions between the crystalline particles and the terminal protein groups.

Hydroxyapatite can be used to remove nucleic acids from a crude lysate, allowing further purification of the protein solution. It can also remove contaminants to leave the target protein in solution, and has been used to separate proteins that are highly homologous (e.g. truncated proteins).

The binding characteristics of a protein are loosely associated to its pI value, although it is usually necessary to establish the suitability of the technique in each case. Basic proteins bind via electrostatic interactions between their amino groups and the phosphate groups on the matrix. Elution is achieved by increasing the concentration of the buffer (see Protocol 9), or by the addition of chloride salts or divalent cations such as Ca^{2+} or Mg^{2+}. Neutral and acidic proteins are thought to bind via their carboxyl groups to the Ca^{2+} sites on the matrix and are eluted by increasing the concentration of the buffer.

Protocol 9

Protein separation with hydroxyapatite using a linear gradient

Equipment and reagents

- Hydroxyapatite column packed with a suitable matrix
- Distilled or deionized water[a]
- Binding buffer: 10 mM sodium phosphate pH 6.8[a]
- Eluent: 0.4 M sodium phosphate pH 6.8[a]

Method

1 Wash the column with distilled water to remove the bacteriostatic storage buffer.

2 Equilibrate the column with 10 column volumes of binding buffer.

3 Dissolve the sample in binding buffer.

4 Filter or centrifuge the sample to remove particulates.

5 Apply the sample to the column.

6 Maintain 100% binding buffer for 5 column volumes.

7 Elute protein using a linear gradient (0–100%) over 10 column volumes with elution buffer.

8 Monitor the detector response and collect fractions.

9 Assay fractions and pool those with related activity.

10 Regenerate the column by washing with 5 column volumes of water.

11 Repeat equilibration step 2.

[a] Filter solution through a 0.45 μm filter and degas prior to use.

4.4 Immobilized metal affinity chromatography (IMAC)

Immobilized metal affinity chromatography has been steadily growing in popularity since its introduction in 1975 by Porath (5). Transition metal ions immobilized on a column via a chelating group will bind reversibly certain amino acids, primarily histidine but also tryptophan, tyrosine, phenylalanine, and cysteine if exposed on the surface of protein molecules.

A number of metal chelating groups are available, with imino diacetate (IDA), tris(carboxymethyl) ethylene diamine (TED), and nitrilotriacetic acid (NTA) being used most frequently.

A number of phosphoproteins have been successfully purified using IMAC with immobilized Fe^{3+} ions, although it has found greatest use in the selective purification of recombinant fusion proteins engineered to contain six consecutive histidine amino acids at either terminus of the protein (the 6-His tag). Immobilized Cu^{2+} and Ni^{2+} ions are predominantly used, although Zn^{2+}, Co^{2+}, and Al^{2+} ions are also available.

The crude lysate is applied to the column at a high ionic strength (0.5–1 M NaCl) to reduce ion exchange effects. The 6-His tag of the target protein binds to the immobilized Ni^{2+} ions as the contaminating proteins pass through. The target protein is eluted using a stronger complexing agent (EDTA or imidazole), or by protonation of the histidine residues by reducing the pH to below 4.5, or a combination of the two. A method for the purification of a recombinant 6-His tag fusion protein using IMAC is described in Protocol 10.

Protocol 10

Purification of a recombinant 6-His tag fusion protein using IMAC

Equipment and reagents

- IMAC column charged with Ni^{2+} ions
- Distilled or deionized water[a]
- Binding buffer: 50 mM phosphate pH 7.0 containing 0.5 M NaCl and 0.5 mM imidazole[a]
- 50 mM EDTA containing 1 M NaCl

- Eluent: 50 mM phosphate pH 7.0 containing 0.5 M NaCl and 100 mM imidazole[a]
- 1–5 mg/ml of appropriate metal salt in water

Method

1 Wash the column with distilled water to remove the bacteriostatic storage buffer.

2 Wash the column with 10 column volumes of elution buffer.

3 Equilibrate the column with 5 column volumes of binding buffer.

4 Dissolve the sample in binding buffer.

5 Filter or centrifuge the sample to remove particulates.

6 Apply the sample to the column.

7 Elute protein using a linear gradient (0–100%) over 15 column volumes with elution buffer.

8 Monitor the detector response and collect fractions.

9 Assay fractions and pool those with related activity.

Protocol 10 continued

10 Maintain 100% elution buffer over 5 column volumes.

11 Strip the column using 2–3 column volumes EDTA solution.

12 Recharge column with sufficient metal salt solution to ensure that the eluate contains metal ions.

13 Repeat step 2 to equilibrate column.

[a] Filter solution through a 0.45 μm filter and degas prior to use.

5 Affinity purification

5.1 Immobilized natural ligands

Affinity chromatography is based on the biospecific binding between a ligand chemically bound to a support matrix and the target protein. Interactions include the binding of antigen to antibody, glycoproteins to lectins, and a substrate, cofactor, or inhibitor to an enzyme. Due to the high specificity of such interactions, a high level of purification can be achieved in a single step.

One of the ligands is covalently coupled to the chromatographic matrix in an orientation that allows the binding site to be fully exposed. A linker or spacer arm places extra distance between the matrix and the immobilized ligand, giving a higher degree of accessibility. When the sample mixture is applied to the column, depending on the nature of the immobilized ligand, only the target molecules will bind to the stationary bonded phase. Contaminating proteins will pass through the column. The target molecule is then selectively eluted from the column in a purified form.

Due to the potentially large number of ligands that can be coupled to a matrix, most manufacturers supply activated affinity media, allowing the user to generate a purpose-made stationary phase. The researcher makes a series of choices regarding each stage of the preparation process.

First, the ligand to be attached is chosen by its ability to bind specifically and reversibly to the target protein. Following coupling to the matrix, the ligand must still bind to the target protein.

The next stage is to choose an activated matrix from those commercially available. Ideally, the matrix should be chemically stable to allow repeated use and regeneration of the column, and should demonstrate low, non-specific adsorption with a high binding capacity for the ligand. Most matrices are based on agarose, cellulose, or crosslinked dextrans. Many of the activation methods involve hazardous reagents. Cyanogen bromide-activated matrices are most frequently used, but others are activated by epoxy, carboxyldiimidazole, N-hydroxysuccinimidyl, epichlorohydrin, bisoxirane, and tosyl/tresyl chloride. Choice of matrix will depend on those groups available on the ligand that are not required for target binding. Coupling of ligands to the matrix may be achieved through their amino, carboxyl, thiol, or hydroxyl groups.

Spacer arms are frequently used between the matrix and ligand to reduce steric hindrance, and many activated matrices are supplied with spacer atoms already introduced. These arms can be up to 12 atoms long, although in some cases they can cause unwanted non-specific hydrophobic binding. An additional spacer can be attached to either the matrix or the ligand prior to coupling.

The coupling procedure will vary according to the nature of the activated matrix and the manufacturer's instructions should be followed. A typical procedure is described briefly below.

The activated matrix (e.g. CNBr-activated agarose) is swollen and washed in bicarbonate or borate buffer (buffers containing amino groups such as Tris are unsuitable as they couple preferentially to the matrix). An alkaline coupling buffer maintains a partially unprotonated ligand state without causing inactivation or denaturation and may contain up to 0.5 M NaCl to reduce protein–protein adsorption. With the addition of ligand the coupling procedure takes place overnight at 4 °C in a sealed bottle, rotating end-over-end. Any remaining active groups on the matrix are blocked by soaking the matrix in Tris–HCl buffer pH 8.0 for 2 h or by the addition of a primary amine, e.g. ethanolamine. Finally, any uncoupled ligand is removed using a washing step, alternating between high and low pH buffers. The matrix is equilibrated in a suitable buffer and packed into a column.

Elution of the target protein can be achieved by a number of methods and a series of small scale trials to establish a suitable protocol is advisable. Once the sample has been applied to the column, buffers are preferably applied in a reverse direction to prevent components bound at the inlet passing through the column during the elution step. The ideal elution buffer contains either free ligand or a substance that competitively binds to the ligand. Cost can be prohibitive, especially if high concentrations of ligand are required to displace the target protein. Other elution strategies include increasing the ionic strength of the buffer up to 1 M NaCl, a decrease in buffer pH, reducing the polarity of the buffer with the addition of ethylene glycol, or the addition of guanidine or urea at subdenaturing levels. Another effective strategy for tightly-bound proteins involves halting the flow of buffer through the column for a period of up to 2 h before continuing the elution. It is important to clean and regenerate the column immediately after use to prevent denaturation of the bound ligand under the harsh conditions required for target protein elution.

A desalt step may be required for the target protein after elution to remove interfering buffer constituents or prevent denaturation.

5.2 Group-specific adsorbents

A selection of commercially available group-specific affinity adsorbents is discussed in the following section. These affinity techniques will purify biomolecules with related structure or function. Although these techniques lack the high specificity described in the previous section, a high degree of purity is achievable in a single step.

There is a large number of commercially available media, and the reader is directed to the recommended texts for more detailed discussions of these and other related techniques.

5.2.1 Lectin affinity chromatography

Lectin affinity chromatography utilizes the binding of glycoproteins to an immobilized lectin through exposed sugar side chains. During secretion from the cell, carbohydrate moieties are added by post-translational modification to plant and animal proteins. Lectin affinity chromatography lacks specificity, as the crude sample may contain a number of different proteins with similar side chains that bind with varying affinities to the immobilized lectin.

Most glycoproteins contain mannose residues and, as described in Protocol 11, can be purified using the lectin concanavalin A (Con A) which also binds proteins containing glucose. Bound proteins are eluted using buffers containing α-D-methylmannoside or α-D-methylglucoside. Elution can also be achieved with a buffer gradient of increasing ionic strength. Con A is particularly suitable for the purification of solubilized surface membrane glycoproteins.

Commercially available media include wheat germ lectin (elution achieved with N-acetyl-D-glucosamine) and a lectin from the snail *Helix pomatia* with specificity for N-acetyl-α-D-galactosaminyl residues.

Protocol 11

Purification of a glycoprotein using Con A Sepharose

Equipment and reagents

- Affinity chromatography column packed with Con A Sepharose
- Distilled or deionized water[a]
- Binding buffer: 20 mM Tris–HCl pH 7.4 containing 0.5 M NaCl, 1 mM CaCl$_2$, 1 mM MgCl$_2$, and 1 mM MnCl$_2$[a]

- Elution buffer: 20 mM Tris–HCl pH 7.4 containing 0.5 M NaCl, 1 mM CaCl$_2$, 1 mM MgCl$_2$, 1 mM MnCl$_2$, and 0.5 M α-D-methylmannoside[a]

Method

1 Equilibrate the column with at least 10 column volumes of binding buffer.

2 Dissolve the sample in binding buffer.

3 Filter or centrifuge the sample to remove particulates.

4 Apply the sample to the column.

5 Maintain 100% binding buffer for 5 column volumes.

6 Elute protein using either a stepwise or linear gradient (0–100%) elution buffer over 5 column volumes.

7 Monitor the detector response and collect fractions.

Protocol 11 continued

8 Assay fractions and pool those with related activity.

9 Maintain 100% elution buffer for 5 column volumes.

10 Equilibrate column as described in step 1.

[a] Filter solution through a 0.45 μm filter and degas prior to use.

5.2.2 Dye affinity chromatography

Dye affinity chromatography is based on the high affinity that many proteins have for immobilized dyes. The dye ligands have some structural similarities to nucleotides and the cofactors $NAD^+/NADP^+$, although binding of many proteins to the ligand is via electrostatic and/or hydrophobic interaction.

A large number of dyes are available commercially and a series of small scale screenings, as described by Scopes (6), should be used to establish the most suitable dye for purification of the target protein. Elution of the target protein is achieved by the addition of NaCl to the buffer, an increasing pH gradient, or the addition of a suitable ligand (e.g. a cofactor) to the buffer.

5.2.3 Immobilized heparin chromatography

Immobilized heparin will bind a variety of biomolecules, including restriction endonucleases, nucleic acid binding proteins, serine protease inhibitors, growth factors, lipoproteins, hormone receptors, and coagulation proteins. Binding occurs through a combination of affinity chromatography and/or cation exchange. Elution is achieved with the addition of heparin to the buffer or an increase in ionic strength (2 M NaCl) in a linear or stepwise gradient.

5.2.4 Immobilized small ligands

Examples of small ligands used to isolate specific groups of biomolecules include benzamidine (trypsin-like proteases), lysine (plasminogen, plasminogen activator), arginine (serine proteases), streptavidin (biotinylated molecules), nucleotides, and DNA. This group has also found use in the purification of fusion proteins engineered to contain a sequence tag, allowing purification in a single step with ligands such as glutathione, IgG, and amylose.

Recombinant fusion proteins bearing a glutathione S-transferase (GST) 'tag' at one of the termini bind to the glutathione ligand on the affinity matrix (see Protocol 12). Elution is achieved by one of two methods. The recombinant protein (minus the fusion tag) is eluted if a site-specific protease (thrombin or factor Xa) is added to the buffer. Alternatively, the fusion protein is eluted by the addition of imidazole or glutathione to the buffer.

Protocol 12

Purification of a GST fusion protein

Equipment and reagents

- Glutathione affinity column
- Binding buffer: 25 mM Tris–HCl pH 8.0
- Elution buffer: 25 mM Tris–HCl pH 8.0 containing 10 mM glutathione

Method

1 Equilibrate the column with at least 10 column volumes of binding buffer.

2 Dissolve the sample in binding buffer.

3 Apply the sample to the column.

4 Maintain 100% binding buffer for 10 column volumes.

5 Elute protein using 100% elution buffer.

6 Monitor the detector response and collect fractions.

7 Assay fractions and pool those with related activity.

8 Maintain 100% elution buffer for 5 column volumes.

9 Equilibrate column as described in step 1.

5.2.5 Immobilized protein A or protein G

The immobilized ligands used in protein A or protein G chromatography were originally isolated from the surface of the Gram-negative bacteria *Staphylococcus* and *Streptococcus* respectively. They selectively bind a range of immunoglobulins with varying affinities. Protein A is preferred because of its lower cost, higher stability under the regeneration/cleaning, and higher yields of the target antibody. Protein G has high affinity for sheep, goat, and horse immunoglobulins and binds to all human IgG subclasses. A method for the purification of an antibody using protein G affinity chromatography is described in Protocol 13.

Binding of the target antibody occurs at pH > 6.0 and the addition of salt to the buffer reduces non-specific binding. Mouse IgG_1 can bind to protein A under high pH conditions in the presence of 3 M NaCl.

Elution is achieved by reducing the buffer pH to 3.0 or below. Under these conditions, the eluate should be returned rapidly to pH 7 to prevent denaturation, either by the addition of 1 M Tris–HCl pH 9.0 or by buffer exchange. Antibodies bound to protein G may require harsher elution conditions, e.g. 1 M acetic acid, 3 M potassium isothiocyanate, the addition of denaturing agents (guanidine, urea), or the addition of chaotrophic agents.

Other purification techniques used with success for immunoglobulins include cation exchange chromatography (pH ~ 6.0) and HIC. Where possible, the respective antigen should be coupled to a suitable activated matrix and the target antibody purified from a polyclonal mixture.

Protocol 13

Purification of an antibody using protein G affinity chromatography

Equipment and reagents

- Affinity protein G chromatography column
- Binding buffer: 50 mM sodium phosphate pH 7.0 containing 0.15 M NaCl[a]
- Elution buffer: 100 mM glycine pH 3.0 containing 0.15 M NaCl[a]
- Neutralizing buffer: 1 M Tris–HCl pH 9.0

Method

1 Equilibrate the column with 10 column volumes of binding buffer.

2 Centrifuge sample to remove intact cells, debris, etc.

3 Filter sample.

4 Apply the sample to the column.

5 Maintain 100% binding buffer to wash unbound proteins through the column.

6 Monitor the baseline until stable.

7 Place enough neutralizing solution in each fraction tube to ensure a pH of 7 on addition of column eluate.

8 Elute target antibody with 100% elution buffer.

9 Maintain 100% elution buffer for 10 column volumes to regenerate the column.

10 Equilibrate the column as described in step 1.

[a] Filter solution through a 0.45 μm filter and degas prior to use.

---

6 Fractionation by size—gel filtration chromatography

Gel filtration chromatography (also known as size exclusion or gel permeation chromatography) separates proteins according to their molecular size using a porous matrix. Small molecules diffuse into the pores whereas larger molecules may be excluded and confined to the surrounding liquid. Passing eluent through the column results in elution of the molecules in decreasing order of size.

Gel filtration is generally used as a final finishing step in a purification scheme. Sample volume is limited to 0.1–1% of the bed volume to maintain good resolution of the sample peaks, so a concentration step is usually required prior to separation. The lower the flow rate through the column, the better the resolution obtained. The three main applications of gel filtration are protein fractionation over a defined molecular size range (see Protocol 14), molecular size estimation, and desalting.

Choice of a column for protein fractionation is dependent on the sizes of the proteins to be separated. Supplier's technical data should be used to determine the most suitable matrix. The target protein should elute between one-third and two-thirds through the separation range with the majority of the contaminants eluting later. Non-specific interactions between the protein components or with the chromatographic media can be reduced by addition of 0.15–1.5 M NaCl to the filtration buffer.

Protocol 14

Fractionation by gel filtration

Equipment and reagents

- Gel filtration column packed with a suitable matrix
- Distilled or deionized water[a]
- Gel filtration buffer: usually 50 mM Tris–HCl[a]

Method

1. Equilibrate the column with distilled water to remove the bacteriostatic storage buffer.

2. Equilibrate the column with a minimum of 2–3 column volumes of gel filtration buffer.

3. Dissolve the sample in gel filtration buffer, typically 0.1–1% column bed volume.

4. Filter or centrifuge the sample to remove particulates.

5. Apply the sample to the column using an appropriate flow rate, see supplier's technical data.

6. Monitor the detector response and collect fractions.

7. Assay fractions and pool those with related activity.

[a] Filter solution through a 0.45 μm filter and degas prior to use.

Estimation of protein size by gel filtration is achieved by calibrating the column with reference samples of known molecular size. The unknown protein is then eluted under identical conditions (flow rate and sample volume) and the molecular size determined from the calibration curve of elution volume. The tertiary structure of biological molecules means that calculation of the molecular mass from the elution volume is not recommended. The references and unknown sample may be denatured and separated in 6 M guanidine hydrochloride to reduce the effects of molecular shape.

A desalt step will remove low molecular mass contaminants from the target protein, which elutes first. Selection of a matrix for desalting is dependent on the exclusion limit of the gel. A desalt step is most frequently used to decrease

Table 2 Common detergents

Non-ionic

        Digitonin

     Polyoxyethylene detergents

        Genapol X-080

        Brij 35

        Triton X-100

        Triton X-114

        LuBrol PX

        Tween 20

        Tween 80

     Glycoside detergents

        Decylglucoside

        Octylglucoside

        Dodecyl-β-D-maltoside

        n-Decylsucrose

        n-Dodecylsucrose

   BigCHAP

Zwitterionic

        CHAPS

        CHAPSO

        Lysolecithin

        Zwittergents

        LDAO

Anionic

        Cholate

        Deoxycholate

        SDS

Cationic

     Trimethylammonium bromides

        CTAB

        DTAB

        MTAB

---

solubilization of the protein, this level may need to be exceeded, although as the purification proceeds it is preferable to maintain the minimum level of detergent required for protein solubility and stability.

The first stage of membrane protein purification usually involves isolating the membrane of interest from the rest of the cell. The cell disintegration process should not disrupt the membrane of interest to the extent that biological activity is lost. Protease enzymes can be activated under certain conditions, so the inclusion of inhibitors during the purification process should be considered. Differential centrifugation can be used to isolate the required membrane fraction, which is then resuspended in a suitable buffer and washed to remove contaminants. This step may be repeated several times to ensure that all soluble proteins are removed.

The next step is to release the protein from the membrane in an active form. Unfortunately, it is not possible to predict the most effective method of solubilizing and stabilizing the target protein and this stage may involve small scale trials using various detergents at different concentrations (0.01–5%, w/v) to establish a suitable protocol. Contaminating membrane proteins can be removed by washing the membranes with buffers and sequentially increasing the detergent concentration in the buffer until the target protein is released. A high-speed centrifugation step (100 000 g spin for 1 h) will produce a supernatant containing not only solubilized proteins and lipids but the detergent in single molecule or micellar form.

Following solubilization, the protein can be treated for purification, but with the following factors being taken into consideration:

(a) All buffers must contain detergent to ensure that the protein does not aggregate or lose activity. The presence of detergent can interfere with the accuracy of some protein concentration measurements. The concentration of ionic detergents in the buffer will contribute towards its ionic strength.

(b) Stabilizers may be required in buffers to preserve activity during the purification process, e.g. 10–50% glycerol, up to 250 mM sucrose, or the addition of natural lipids.

(c) In ion exchange chromatography, the detergent should not adsorb to the resin, and the concentration should maintain solubility and activity but not interfere with the interaction between the protein and the matrix. Under such conditions most proteins will be eluted at concentrations of <1 mg/ml.

(d) The hydrophobic nature of detergent micelles reduces the effectiveness of HIC and ammonium sulfate precipitation, making these techniques of little use.

(e) Immunoaffinity techniques and lectin affinity chromatography for the purification of glycosylated membrane proteins have proved to be highly successful. However, it is necessary to ensure that any detergents containing carbohydrate moieties do not bind to the lectin active site.

(f) The detergent–protein complex will have a greater molecular weight than the protein alone, which will have implications during gel filtration chromatography.

(g) Above the critical micelle concentration (CMC) the detergent molecules can form micelles that reduce the effectiveness of gel filtration chromatography.

8.2 Recombinant proteins

The expression and purification of recombinant proteins has increased in the last 15 years, primarily due to the ease with which these proteins can be produced in higher quantities than in the native organism. Recombinant proteins can also be engineered to be more (or less) stable than the wild-type protein and can be used to investigate the importance of specific amino acid/s on protein function. They may be directed to a specific location within the cell (cytoplasmic or periplasmic) or secreted into the culture medium. The addition of an N or C terminal peptide 'tag' allows the resulting fusion protein to be purified using affinity chromatography. The tag can then be removed from the protein using proteolytic or chemical cleavage releasing the protein of interest.

A number of host vectors are available for use in *Escherichia coli*, *Saccharomyces cerevisiae*, and *Bacillus subtilis* as well as fungal, mammalian, and cultured insect cell lines. Depending on the vector used, not all recombinant proteins are expressed in a soluble form. Expression of some recombinant proteins at high levels and incorrect protein folding or disulfide bond formation can result in the aggregation of the recombinant protein into inclusion bodies.

Soluble recombinant proteins lacking a purification 'tag' can be purified using the techniques described previously. The purification protocol used when the target protein is in the native host may be unsuitable for the recombinant protein, due to the host's different protein complement, thus requiring a new protocol to be established. Recombinant proteins secreted into the medium may require a concentration step prior to purification, although an adsorptive or affinity technique may be suitable. Proteins released into the periplasm can be extracted using osmotic shock to disrupt the cell wall whilst maintaining the intact cell membrane, effectively removing all cytoplasmic proteins in a single step.

Most fusion proteins are purified by an affinity chromatography step. In most cases, only the recombinant protein binds to the affinity column with the host proteins passing straight through. After a washing step to remove any loosely bound proteins, the recombinant protein is eluted using a suitable buffer. In some cases, the recombinant protein may have full bioactivity in the presence of the tag and cleavage of the fusion peptide may not be necessary. Alternatively, following a suitable incubation period with a specific protease, the target protein is purified using either a gel filtration step or a second passage through the affinity column. The target protein now passes through the column whilst the peptide tag binds to the column and is eluted separately.

Insoluble recombinant proteins expressed as inclusion bodies are present in a non-native conformation and lack bioactivity. After cell disintegration, differential centrifugation is used to separate the inclusion bodies from the remaining cell components. This yields protein of high purity but a 'naturation' process is required to allow the protein to regain its native conformation. This is achieved by solubilizing the protein in a denaturant (6–8 M guanidine or urea) and allowing it to refold in a reduced environment as the concentration of denaturant is gradually reduced by dialysis. The addition of molecular chaperones, such as *E. coli* GroEL and GroES, or the enzyme protein disulfide isomerase, can help reduce aggregation of the folding intermediates. A purification step is then required to remove unfolded, degraded, or incorrectly folded protein. Alternatively, the denatured protein can be purified using gel filtration, ion exchange with a buffer containing urea, or IMAC chromatography prior to the refolding step.

9 Example purification

Methanotrophs are aerobic bacteria capable of utilizing methane as their sole source of carbon and energy. The first reaction in the methane oxidation pathway is the hydroxylation of methane to methanol by the enzyme methane monooxygenase (MMO). MMO is also capable of co-oxidizing a wide range of alkanes, alkenes, ethers, cyclic alkanes, aromatic compounds, styrene, pyridine, and chlorinated organic compounds. MMO exists as two forms, a membrane-bound, particulate complex (pMMO) found in almost all methanotrophs and a soluble cytoplasmic form (sMMO) found only in certain strains. *Methylosinus trichosporium* OB3b produces both forms of the enzyme depending on the growth conditions, with sMMO expressed under low copper ion concentration / biomass ratios. The sMMO enzyme complex consists of three components, a hydroxylase, a regulatory Protein B, and a reductase.

The hydroxylase component consists of three subunits in a $\alpha_2\beta_2\gamma_2$ arrangement and is the site of methane and oxygen activation. Although the three proteins of OB3b sMMO had previously been purified by Fox (7), hydroxylase in a highly purified form was required for amino acid sequencing using mass spectrometry. The hydroxylase component was purified using a combination of ion exchange and gel filtration chromatography. During analysis by mass spectrometry, the β and γ subunits of the hydroxylase are easily and preferentially ionized using an electrospray source, but the α subunit produces peaks of low intensity. To improve the mass spectrum of the α subunit, the purified hydroxylase was separated into its component subunits using a reverse phase chromatography step and the α subunit analysed independently.

Fractions containing the hydroxylase protein were assayed for enzyme activity by their ability to oxidize propylene in the presence of three-fold molar excess Protein B and reductase with NADH as electron donor (8). Propylene is used as a substrate in the reaction as the product, propylene oxide (PO), is not metabolized by the organism. PO is detected by gas chromatography using a flame ionization detector.

Cells are grown in chemostat culture as described by Colby (9) and following harvest are drop frozen in liquid nitrogen and stored at $-70\,^{\circ}C$. All purification operations are performed at $4\,^{\circ}C$.

9.1 Preparation of cell-free extract and ion exchange chromatography to fraction sMMO proteins

The first stage of the purification involves cell disintegration followed by anion exchange chromatography to separate the three protein components of sMMO. These methods are described in Protocol 16

Protocol 16

Cell disintegration and ion exchange chromatography

Equipment and reagents

- Cell disrupter (Constant Systems, UK)
- Glass column 3 cm × 12.5 cm
- DEAE Sepharose (Amersham Pharmacia) ion exchange matrix (80 ml settled bed volume) resuspended in binding buffer
- Stirred ultrafiltration cell (Millipore)
- Binding buffer: 25 mM MOPS pH 7.0 containing 1 mM benzamidine and 5 mM DTT[a]
- Elution buffer: binding buffer and 1 M NaCl[a]
- DNase I

Method

1 Thaw the cell paste (350 g wet weight) and dilute with 1 litre of binding buffer.

2 Centrifuge at 17 000 g for 15 min at 4 °C.

3 Resuspend the pellet in 350 ml binding buffer and add sufficient DNase I to bring final concentration to 20 μg/ml.

4 Break the cells by two passes through the cell disrupter at 137 Mpa cooled to 4 °C.

5 Centrifuge at 48 000 g for 1 h at 4 °C.

6 Add the cell-free extract (supernatent) to the ion exchange matrix and stir gently on ice for 10 min.

7 Centrifuge at 2750 g for 10 min at 4 °C.

8 Pour off the supernatant and resuspend the matrix in 350 ml binding buffer.

9 Repeat steps 7 and 8 until the supernatant is colourless.

10 Resuspend the matrix up to a final volume of 120 ml with binding buffer and pour into the glass column.

11 Elute any remaining unbound proteins with binding buffer at a linear velocity of 90 cm/h until the baseline is stable.

12 Elute the three component proteins using a linear gradient of 0% to 40% (0.4 M) elution buffer over 7 column volumes, collecting 5 ml fractions. The elution order for the proteins is:

   (a) Hydroxylase at 0.08 M NaCl.

   (b) Protein B at 0.21 M NaCl.

   (c) Reductase at 0.32 M NaCl.

13 Assay hydroxylase fractions and pool those with activity.

14 Concentrate hydroxylase fractions to a final volume of 2 ml using a stirred ultrafiltration cell.

15 Drop freeze the protein in liquid nitrogen and store at −70 °C.

[a] Filter solution through a 0.45 μm filter and degas prior to use.

9.2 Gel filtration chromatography of hydroxylase protein

The hydroxylase component is further purified using gel filtration chromatography (see Protocol 17). The elution profile is shown in Figure 1, with fractions of >150 nmol/min/mg specific activity eluting between the marked arrows.

Protocol 17

Gel filtration chromatography

Equipment and reagents

- Superdex 200 (Amersham Pharmacia) gel filtration column 2.6 cm × 60 cm
- Stirred ultrafiltration cell (Millipore)
- Buffer: 25 mM MOPS pH 7.0[a]

Method

1 Equilibrate the column with a minimum of 2–3 column volumes of buffer.

2 Filter the sample to remove particulates.

3 Apply the hydroxylase to the column and elute at a linear velocity of 23 cm/h.

4 Collect and assay hydroxylase fractions and calculate the specific activity for each.

5 Fractions having a specific activity >150 nmol/min/mg were pooled and concentrated using a stirred ultrafiltration cell.

[a] Filter solution through a 0.45 μm filter and degas prior to use.

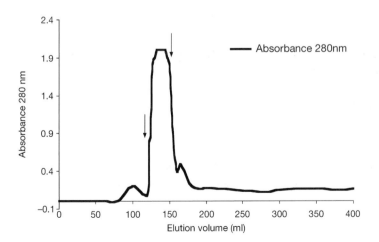

Figure 1 Elution profile of hydroxylase purification on Superdex 200. Fractions collected between arrows have specific activity of >150 nmol/min/mg.

9.3 High resolution ion exchange purification of the hydroxylase

The final purification step involves high resolution anion exchange chromatography, described in Protocol 18, to ensure that only fractions containing high purity hydroxylase protein are used for MS analysis.

The results of a typical purification are shown in Table 3. The elution profile is shown in Figure 2, with fractions of >220 nmol/min/mg specific activity eluting between the marked arrows.

Protocol 18

High resolution anion exchange chromatography

Equipment and reagents

- Mono Q HR16/10 (Amersham Pharmacia) anion exchange column
- Binding buffer: 25 mM MOPS pH 7.0[a]
- Elution buffer: 25 mM MOPS pH 7.0 containing 1 M NaCl[a]

Method

1 Equilibrate the column with 10 column volumes of binding buffer.

2 Filter or centrifuge the hydroxylase to remove particulates.

3 Apply the sample to the column.

4 Elute unbound proteins with 5 column volumes of binding buffer, monitor the baseline until stable.

5 Elute protein using a linear gradient of 5–25% elution buffer (0.05–0.25 M NaCl) over 10 column volumes.

6 Monitor the detector response and collect fractions.

7 Elute remaining proteins using 100% elution buffer (1 M NaCl) for 2 column volumes.

8 Collect and assay hydroxylase fractions and calculate the specific activity for each.

9 Analyse fractions having a specific activity >220 nmol/min/mg by SDS–PAGE (12% acrylamide).

10 Pool and concentrate fractions meeting both specific activity and SDS–PAGE purity requirements, using a stirred ultrafiltration cell.

[a] Filter solution through a 0.45 μm filter and degas prior to use.

9.4 Separation of the hydroxylase subunits by reverse phase chromatography

Resolution of the hydroxylase subunits is achieved with a reverse phase chromatography column using a linear acetonitrile gradient, as described in Protocol 19. The elution profile is shown in Figure 3.

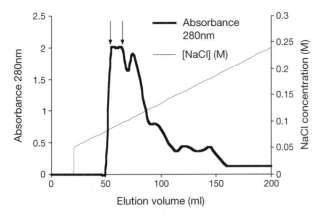

Figure 2 Elution profile of hydroxylase purification on Mono Q. Fractions collected between arrows have specific activity of >220 nmol/min/mg.

Table 3 Purification table for the hydroxylase protein of the soluble methane monooxygenase enzyme of *Methylosinus trichosporium* OB3b[a]

Purification step	Total protein (mg)	Specific activity (nmol/min/mg)	Total activity (nmol/min)	Yield (%)	Purification (fold)
Cell-free extract	14 040	81.2	1 140 668	100	100
DEAE	775	10	7750	0.7	0.1
Superdex 200	225	179	40 275	3.5	2.2
Mono Q	70	260	18 200	1.6	3.2

[a] The method is described in Section 9. Activity was determined using propylene as a substrate in the presence of three-fold molar excess Protein B and reductase.

Protocol 19

High resolution reverse phase chromatography

Equipment and reagents

- ProRPC (Amersham Pharmacia) reverse phase column
- Distilled or deionized water containing 0.1% trifluoroacetic acid[a]
- Elution buffer: 90% acetonitrile containing 9.9% distilled or deionized water and 0.1% trifluoroacetic acid[b]

Method

1 Equilibrate the column with 10 column volumes of binding buffer.

2 Filter or centrifuge hydroxylase to remove particulates.

3 Apply the sample to the column.

Protocol 19 continued

4 Maintain 100% binding buffer for 1.5 column volumes.

5 Increase the concentration of elution buffer to 40% over 1 column volume.

6 Elute hydroxylase subunits using a linear gradient of 40–65% elution buffer over 15 column volumes.

7 Monitor the detector response and collect fractions.

8 Maintain 100% elution buffer for 5 column volumes.

9 Analyse fractions by SDS–PAGE (12% acrylamide) and ESI-MS.

[a] Filter solution through a 0.45 μm filter and degas prior to use.

[b] HPLC grade acetonitrile solution degassed in a sonicating water-bath for 30 min.

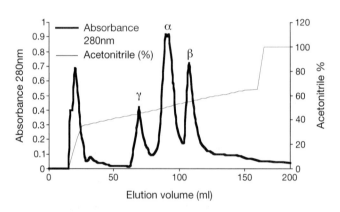

Figure 3 Elution profile of hydroxylase protein resolved into its component subunits.

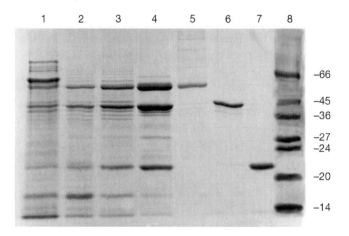

Figure 4 SDS–PAGE analysis of hydroxylase purification. Lane 1, cell-free extract; lane 2, DEAE Sepharose; lane 3, Superdex 200; lane 4, Mono Q; lane 5, α-subunit; lane 6, β-subunit; lane 7, γ-subunit; lane 8, markers (kDa).

Analysis by SDS–PAGE (Figure 4) demonstrates that the purification yields hydroxylase protein that can be further resolved into its component subunits for amino acid sequencing by MS.

References

1. Itzhaki, R. F. and Gill, D. M. (1964). *Anal. Biochem.*, **9**, 401.
2. Hartree, E. F. (1972). *Anal. Biochem.*, **48**, 422.
3. Smith, P. K., Krohn, R. I., Hermanson, G. T., Mallia, A. K., Gartner, F. H., Provenzano, M. D., *et al.* (1985). *Anal. Biochem.*, **150**, 76.
4. Bradford, M. M. (1976). *Anal. Biochem.*, **72**, 248.
5. Porath, J., Carlsson, J., Olsson, I., and Belfrage, G. (1975). *Nature*, **258**, 598.
6. Scopes, R. K. (1993). In *Protein purification principles and practice*, 3rd edn, p. 219. Springer–Verlag New York, Inc.
7. Fox, B. G., Froland, W. A., Dege, J. E., and Lipscomb, J. D. (1989). *J. Biol. Chem.*, **264**, 10023.
8. Pilkington, S. J. and Dalton, H. (1990). In *Methods in enzymology* (ed. M. E. Lidstrom), Vol. 188, p. 181. Academic Press, New York.
9. Colby, J. and Dalton, H. (1978). *Biochem. J.*, **171**, 461.

Recommended reading

Current Protocols on CD-ROM – Protein Science. (1999). John Wiley & Sons Inc.

Bollag, D. M., Edelstein, S. J., and Rozycki, M. D. (1996). *Protein methods*. Wiley-Liss.

Dean, P. D. G., Johnson, W. S., and Middle, F. A. (ed.) (1985). *Affinity chromatography: a practical approach*. IRL Press, Oxford.

Deutscher, M. P. (ed.) (1990). *Methods in enzymology. Guide to protein purification*. Vol. 182. Academic Press, New York.

Goding, J. W. (1986). *Monoclonal antibodies: principles and practice*, 2nd edn. Academic Press, London.

Harris, E. L. V. and Angal, S. (ed.) (1990). *Protein purification applications: a practical approach*. IRL Press, Oxford.

Harris, E. L. V. and Angal, S. (ed.) (1989). *Protein purification methods: a practical approach*. IRL Press, Oxford.

Janson, J.-C. and Ryden, L. (1998). *Protein purification*. John Wiley & Sons Inc.

Mohr, P. and Pommerening, K. (1985). *Affinity chromatography: practical and theoretical aspects*. Dekker, New York.

Scopes, R. K. (1998). *Protein purification principles and practice*, 4th edn. Springer–Verlag New York, Inc.

Turkova, J. (1978). *Affinity chromatography*. Elsevier Scientific, Amsterdam.

Von Jagow, G. and Schaegger, H. (ed.) (1994). *A practical guide to membrane protein purification*. Academic Press, London.

Chapter 7
Gel electrophoresis of proteins

David E. Garfin

Life Science Group, Bio-Rad Laboratories, 2000 Alfred Nobel Drive, Hercules, CA 94547, USA.

1 Introduction

At one time or another during the course of protein analysis or purification, researchers are likely to make use of gel electrophoresis. All laboratories working with proteins have some capability for carrying out gel electrophoresis and all researchers have at least rudimentary knowledge of the technique.

Gel electrophoresis can provide information about the molecular weights and charges of proteins, the subunit structures of proteins, and the purity of a particular protein preparation. It is relatively simple to use and it is highly reproducible. The most common use of gel electrophoresis is the qualitative analysis of complex mixtures of proteins. Microanalytical methods and sensitive, linear image analysis systems make gel electrophoresis popular for quantitative and preparative purposes as well. The technique provides the highest resolution of all methods available for separating proteins. Polypeptides differing in molecular weight by as little as a few hundreds of daltons and proteins differing by less than 0.1 pH unit in their isoelectric points are routinely resolved in gels.

Gel electrophoresis is a broad subject encompassing many different techniques. Sodium dodecyl sulfate–polyacrylamide gel electrophoresis (SDS–PAGE) is the most commonly practiced gel electrophoresis technique used for proteins. The method provides an easy way to estimate the number of polypeptides in a sample and thus assess the complexity of the sample or the purity of a preparation. SDS–PAGE is particularly useful for monitoring the fractions obtained during chromatographic or other purification procedures. It also allows samples from different sources to be compared for protein content. One of the more important features of SDS–PAGE is that it is a simple, reliable method with which to estimate the molecular weights of proteins. SDS–PAGE requires that proteins be denatured to their constituent polypeptide chains, so that it is limited in the information it can provide. In those situations where it is desirable to maintain biological activity or antigenicity, non-denaturing electrophoresis systems must be employed. However, the gel patterns from non-denaturing gels are more difficult to interpret than are those from SDS–PAGE. Non-denaturing systems also give information about the charge isomers of proteins, but this information is best obtained by isoelectric focusing (IEF). An IEF run will often show heterogeneity due to structural modifications that is not apparent in other types of electrophoresis. Proteins thought to be a single species by SDS–PAGE analysis are sometimes found by IEF to consist of multiple species. A true determination of the purity of a protein preparation is obtained with two-dimensional polyacrylamide gel electrophoresis (2D PAGE) that combines IEF with SDS–PAGE. Since 2D PAGE is capable of resolving over 2000 proteins in a single gel it is important as the primary tool of proteomics research where multiple proteins must be separated for parallel analysis. Proteins can be definitively identified by immunoblotting, which combines antibody specificity with the high resolution of gel electrophoresis. Finally, gel electrophoresis lends itself to protein purification for which purpose various devices have been developed.

Although methods have been refined since the introduction of gel electrophoresis as an analytical technique, the basic principles and protocols have not changed appreciably. The topic has been covered in numerous, readily accessible texts, methods articles, and reviews (1–10).

2 Gel electrophoresis

The subject of electrophoresis deals with the controlled motion of charged particles in electric fields. Since proteins are charged molecules, they migrate under the influence of electric fields. From the point of view of electrophoresis, the two most important physical properties of proteins are their electrophoretic mobilities and their isoelectric points. The electrophoretic mobility of a protein depends on its charge, size, and shape, whereas its isoelectric point depends only on its net overall charge. Various electrophoresis systems have been developed to exploit the differences between proteins in these two fundamental properties.

The rate of migration of a protein per unit of field strength is called its 'electrophoretic mobility'. The units of electrophoretic mobility are those of velocity (cm/sec) divided by the units of electric field (V/cm), or cm^2/V-sec. Separations between proteins result from differences in their electrophoretic mobilities. It is relatively easy to show that in free solution the electrophoretic mobility of a particular protein is a function of the ratio of its charge to its frictional coefficient (shape) (1, 5, 11). Both quantities are established by the composition of the protein and by the makeup of the surrounding medium. Electrophoretic mobilities are influenced by factors such as pH and the amounts and types of counter ions and denaturants that are present in the medium.

Proteins are amphoteric molecules. As such, they can carry positive, negative, or zero net charge depending on the pH of their local environment. For every protein there is a specific pH at which its net charge is zero. This pH is called the 'isoelectric point', or pI, of the protein. A protein is positively charged in solutions at pH values below its pI and negatively charged when the pH is above its p*I*. This pH dependence on charge obviously affects the mobilities of proteins in terms of both magnitude and the direction of migration. It is exploited in gel electrophoresis, especially in the technique of isoelectric focusing.

The electrophoretic mobilities of proteins are very different in gels than in free solution. Gels can act as molecular sieves for molecules the size of proteins. They consist of three-dimensional networks of solid material and pores. During electrophoresis in gels, the polymeric material acts as a barrier to the motion of proteins, forcing them to move between the buffer-filled pores of the gels.

Equipment and reagents for gel electrophoresis are readily available and familiar to laboratory workers. Particularly noteworthy is the steady increase in the popularity of precast polyacrylamide gels since their introduction in the early 1990s. Precast gels provide researchers with 'off-the-shelf' convenience and help to make gel electrophoresis an 'everyday' laboratory procedure.

2.1 Gels

The key element in a gel electrophoresis system is, obviously, the gel itself. It determines the migration rates of proteins and holds proteins in place at the end of the run until they can be stained for visualization. Polyacrylamide gel is the principal medium for protein electrophoresis. Agarose is used in some applications such as for the separation of proteins larger than about 500 kDa and for immunoelectrophoresis (6, 12). However, agarose gels are not used much in protein work and they are not discussed in this chapter.

Polyacrylamide gels are well suited for protein electrophoresis.

(a) The gels can be cast in a range of pore sizes suitable for sieving proteins.

(b) The polymerization reaction is easy and reproducible and gels can be cast in a variety of shapes.

(c) Pore size is determined by the conditions of polymerization and can be easily altered by changing the monomer concentration.

(d) Polyacrylamide gels are hydrophilic and electrically neutral at the time they are cast.

(e) They are transparent to light at wavelengths above about 250 nm and do not bind protein stains.

The gel-forming reaction is shown in Figure 1. Acrylamide and *N,N'*-methylenebisacrylamide (bis) are mixed, then co-polymerized by means of a vinyl addition reaction initiated by free radicals. Gels are formed as acrylamide monomer polymerizes into long chains that are linked together by bis molecules. The resultant structure has both solid and liquid components. It can be thought of as a mass of relatively rigid fibres that create a network of spaces, all immersed in liquid. The liquid (buffer) in the gel maintains the gel's three-dimensional shape. Without the liquid, the gel will dry to a thin film. At the same time, the polymer fibres prevent the liquid from flowing away.

For protein electrophoresis, the pores of the gel are the important structures. During electrophoresis, proteins move through the pores of a gel. Nevertheless,

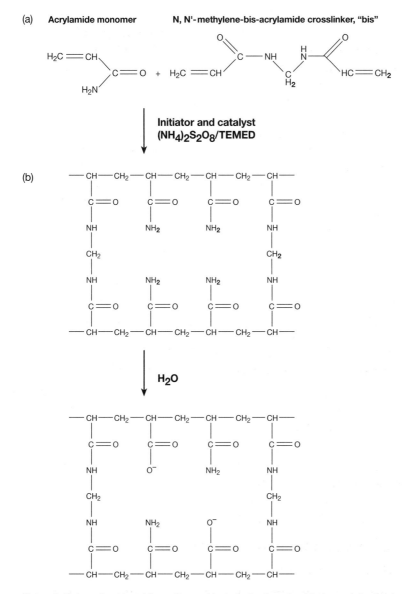

Figure 1 Polyacrylamide gel formation and hydrolysis of acrylamide to acrylate. Acrylamide and bis are co-polymerized in a reaction catalysed by ammonium persulfate and TEMED. This is shown in the upper portion of the figure. The lower portion of the figure shows how pendant, neutral carboxamide groups can become hydrolysed to charged carboxyls.

the pore size of a gel is difficult to measure directly. It is operationally defined by the size limit of proteins that can be forced through the gel.

From a macroscopic point of view, migrating proteins segregate into discrete regions, or zones, corresponding to their individual gel-mediated mobilities. When the electric field is turned off, the proteins stop moving. The gel matrix constrains the proteins at their final positions long enough for them to be stained to make them visible. An example of a one-dimensional separation of proteins is shown in Figure 2. In this configuration, the protein pattern is one of multiple bands with each band containing one protein or a limited number of proteins with similar molecular weights.

By convention, polyacrylamide gels are characterized by a pair of values, %T and %C (13). In this convention, %T is the weight percentage of total monomer (acrylamide + bis) in g/100 ml and %C is the proportion of bis as a percentage of total monomer. The effective pore size of a polyacrylamide gel is an inverse function of the total monomer concentration (%T) and a biphasic function of %C. When %T is increased at a fixed %C, the number of chains increases and the pore size decreases. On the other hand, when %T is held constant and %C is increased from low values, pore size decreases to a minimum at about 5%C. With further increases in %C from the minimum, pore size increases, presumably because of the formation of shorter, thicker bundles of polymer chains. Gels with low %T (e.g. 7.5%T) are used to separate large proteins, while gels with high %T (e.g. 15%T) are used with small proteins.

An example of the effect of pore size on the separation of a set of native proteins is shown in Figure 3. The 4%T, 2.67%C gel is essentially non-sieving. Proteins migrate in it more-or-less on the basis of their free mobility. The 8%T, 2.67%C gel acts as a sieve for the proteins shown, demonstrating the combined effects of charge and size on protein separation. The relative positions of some proteins are shifted in the sieving gel as compared to the non-sieving one.

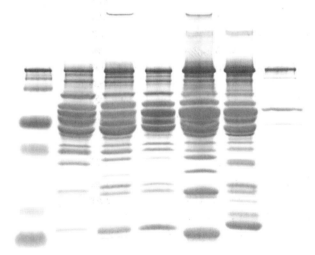

Figure 2 A typical analytical SDS–PAGE gel. Extracts of muscle proteins from five different fish varieties were separated by SDS–PAGE in a precast mini-gel and stained with colloidal CBB G-250 as described in the text. The lanes contain proteins from the following sources (left to right): pre-stained marker proteins, shark, salmon, trout, catfish, sturgeon, and rabbit (actin and myosin). The salmon and trout patterns (lanes 3 and 4) are very similar, as expected given the close evolutionary relationship between the two species. The gel image also indicates that all of the muscle extracts contain myosin and actin.

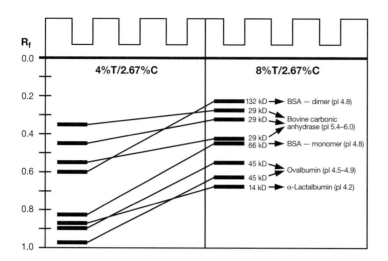

Figure 3 Effect of pore size on migration. The diagram shows the migration patterns of a set of proteins in the Ornstein–Davis, native, discontinuous system. The diagram on the left was obtained from 4%T, 2.67%C gels, while the pattern on the right was obtained from gels with 8%T, 2.67%C. The slanted lines connect bands representing the same proteins in the two gels. Note the large relative mobility shifts of BSA dimer and α-lactalbumin between the two gels types.

2.1.1 Polymerization reactions

In the formation of polyacrylamide gels, the chemical system most often used to form the free radicals needed for polymerization consists of ammonium persulfate (APS) and N,N,N′,N′-tetramethylethylenediamine (TEMED). TEMED accelerates the decomposition of persulfate molecules into sulfate free radicals and these in turn initiate the polymerization. The free base of TEMED is required for this reaction, so that polymerization is most efficient at alkaline pH. Polymerization efficiency falls rapidly at pH values below about pH 6 (14). Photopolymerization with riboflavin and TEMED is used for low pH gels (below pH 7).

The rate of polymerization is dependent on:

(a) The net concentration of monomers and initiators.

(b) The temperature.

(c) The purity of the reagents.

All three should be controlled for reproducibility. Reagents should be electrophoresis grade and water should be thoroughly deionized or distilled. For highest quality results, dissolved oxygen should be removed from the monomer mixtures by degassing them, since oxygen decreases the rate of polymerization (13). Nevertheless, completely acceptable gels can be obtained without removing oxygen, so that degassing can be omitted if it is inconvenient.

Monomers are commonly made as concentrated stock solutions containing 30%T or 40%T in water. For proteins, the usual crosslinker concentration is

2.7%C. These stock solutions can be purchased commercially. Other gel casting reagents include buffers and initiators. Monomer stock and buffer are combined at the desired concentrations and deaerated under moderate vacuum for about 15 min. Initiators are then added and the solution is poured into the casting apparatus.

Initiator concentrations are determined empirically to give visible polymerization in 15–20 min after addition. Under these conditions, gelation is essentially complete in 90 min (15). Final ammonium persulfate and TEMED concentrations of 0.05% each are usually sufficient for polymerization of resolving gels (see Section 2.2.2). When stacking gels are used (see Section 2.2.2), all that is required of them is that they have large pores. Because of this, stacking gels can be set to polymerize rapidly, in 8–10 min, with final concentrations of 0.05% ammonium persulfate and 0.1% TEMED.

Polyacrylamide gels are inherently unstable (16). At basic pH, the pendant neutral carboxamide groups $(-CO-NH_2)$ of acrylamide monomers hydrolyse to ionized carboxyl groups $(-COO^-)$ which can interact with some proteins (Figure 1). In addition, counter ions from the buffer neutralize the carboxyl groups. The waters of hydration associated with the counter ions disrupt the integrity of the pores. Over extended periods of storage, band sharpness and resolution slowly deteriorate. This becomes noticeable after storage at 4°C for three to four months depending on the gel type and the buffer. The ageing of polyacrylamide gels has commercial significance because it limits the shelf-lives of precast gels (see Section 2.10).

2.2 Buffers

The electrical current in an electrophoresis cell is carried largely by the ions supplied by buffer compounds—proteins constitute only a small portion of the current carrying ions in an electrophoresis cell. Buffers supply current carrying ions, maintain desired pH, and provide a medium for heat dissipation. In native systems, electrophoresis buffers also maintain the pH environment needed for protein activity.

Many useful buffer systems have been devised, but only a few are in widespread use (1–7, 13). Buffer systems for electrophoresis are classified as either continuous or discontinuous, depending on whether one or more buffers are used. Both types of buffer system are useful. Dilute samples require discontinuous systems for best results. With high concentration protein samples, above about 1 mg/ml, continuous buffer systems provide adequate resolution (17).

2.2.1 Continuous buffer systems

Continuous systems use the same buffer, at constant pH, in the gel, sample, and electrode reservoirs. With continuous systems, the sample is loaded directly on the gel in which separation will occur. It is helpful to dilute the sample buffer to at least half-strength. The decrease in conductivity achieved by dilution causes a localized voltage drop across the sample that helps drive proteins into the gel. As proteins migrate through the pores of the gel they are separated on the basis of (gel-mediated) mobility differences. Bandwidths are highly dependent on the height of the applied sample volume, which should be kept as small as possible, thus restricting continuous systems to high concentration samples for best results.

Almost any buffer can be used for continuous buffer electrophoresis. Solutions of relatively low ionic strength are best suited for electrophoresis, because these keep heat generation at a minimum. On the other hand, protein aggregation may occur if the ionic strength is too low. The choice of buffer will depend on the proteins being studied, but in general, the concentrations of electrophoresis buffers are in the range of from 0.01 to 0.1 M (but see Protocol 3).

2.2.2 Discontinuous buffer systems

Discontinuous buffer systems (often called multiphasic buffer systems) employ different ions in the gel and electrode solutions. These systems are designed to sharpen starting zones for high resolution separations, even with dilute samples. The sharpening of sample starting zones is called 'stacking'. It is an electro-chemical phenomenon based on mobility differences between proteins and carefully chosen leading and trailing buffer ions. Samples are diluted in gel buffer and sandwiched between the gel and the electrode buffer. When the electric field is applied, *leading ions* from the gel move ahead of the sample proteins while *trailing ions* from the electrode buffer migrate behind the proteins. The proteins in the sample become aligned between the leading and trailing ion fronts in the order of decreasing mobility. Proteins are said to be *stacked* between the two buffer ion fronts. The width of the *stack* is no more than a few hundred micrometres with protein concentrations there approaching 100 mg/ml (18). Electrophoretic stacking concentrates proteins into regions narrower than can be achieved by mechanical means. This has the effect of minimizing overall bandwidths during a run.

In order to allow the stack to develop, the gels used with discontinuous systems are usually divided into two distinct segments. The smaller, upper portion is called the *stacking gel*. It is cast with appreciably larger pores than the lower *resolving gel* (or separating gel) and serves mainly as an anticonvective medium during the stacking process. Separation takes place in the resolving gel, which has pores of roughly the same size as the proteins of interest. Once proteins enter the resolving gel their migration rates are slowed by the sieving effect of the small pores. In the resolving gel, the trailing ions pass the proteins and electrophoresis continues in the environment supplied by the electrode buffer. The proteins are said to become 'unstacked' in the resolving gel. They separate there on the basis of size and charge.

Runs are monitored and timed by means of the buffer front. The migration of the buffer front as it moves through the gel can be followed by the change in the index of refraction between the regions containing the leading and trailing ions. Tracking dye that moves with the buffer front aids in visualization of its motion.

For detailed descriptions of the electrochemical processes that operate with discontinuous buffer systems, consult refs 1–7, 13, and 19. Mathematically inclined readers might want to follow the development of multiphasic buffer theory as presented in refs 20–22.

2.2.3 Native systems

The choice of electrophoresis system depends on the particular proteins of interest. There is no universal buffer system ideal for the electrophoresis of all native proteins. Both protein stability and resolution are important considerations in buffer selection. Recommend choices are the Ornstein–Davis discontinuous system (20, 23) and McLellan's continuous buffers (24).

The set of buffers compiled by McLellan provide the simplest way to carry out the electrophoresis of proteins in their native state (24). McLellan's buffers range from pH 3.8 to pH 10.2, all with relatively low conductivity (Table 1). By using different buffers from the set it is possible to compare the effect of pH changes on protein mobility while maintaining similar electrical conditions. This is demonstrated in Figure 4. The illustration is a line drawing representation, drawn to scale, of the relative positions of two haemoglobin variants, A and C,

Table 1 Continuous buffers for electrophoresis of native proteins[a]

Buffer pH[b]	Basic component	Amount for 5 × solution	Acidic component	Amount for 5 × solution
3.8	β-Alanine $1 \times = 30$ mM	13.36 g/litre	Lactic acid $1 \times = 20$ mM	7.45 ml/litre[c]
4.4	β-Alanine $1 \times = 80$ mM	35.64 g/litre	Acetic acid $1 \times = 40$ mM	11.5 ml/litre
4.8	Gaba $1 \times = 80$ mM	41.24 g/litre	Acetic acid $1 \times = 20$ mM	5.75 ml/litre
6.1	Histidine $1 \times = 30$ mM	23.28 g/litre	MES $1 \times = 30$ mM	29.28 g/litre
6.6	Histidine $1 \times = 25$ mM	19.4 g/litre	MOPS $1 \times = 30$ mM	31.40 g/litre
7.4	Imidazole $1 \times = 43$ mM	14.64 g/litre	HEPES $1 \times = 35$ mM	41.71 g/litre
8.1	Tris $1 \times = 32$ mM	19.38 g/litre	EPPS $1 \times = 30$ mM	37.85 g/litre
8.7	Tris $1 \times = 50$ mM	30.29 g/litre	Boric acid $1 \times = 25$ mM	7.73 g/litre
9.4	Tris $1 \times = 60$ mM	36.34 g/litre	CAPS $1 \times = 40$ mM	44.26 g/litre
10.2	Ammonia $1 \times = 37$ mM	12.5 ml/litre	CAPS $1 \times = 20$ mM	22.13 g/litre

[a] Adapted from ref. 24.

[b] Listed buffer pH is ± 0.1 unit. Do not adjust the pH with acid or base. Remake buffers outside the given range.

[c] Lactic acid from an 85% solution.

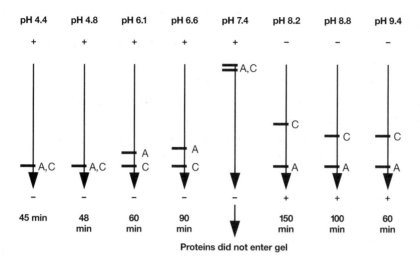

| pH 4.4 | pH 4.8 | pH 6.1 | pH 6.6 | pH 7.4 | pH 8.2 | pH 8.8 | pH 9.4 |

Figure 4 Effect of pH on mobility. Haemoglobin A (p*I* 7.1) and haemoglobin C (p*I* 7.4) were run in native, continuous systems using seven of the McLellan buffers (Table 1). Bands marked 'A' and 'C' show the positions of the two haemoglobin variants in each gel representation. The polarity of the voltages applied to the electrophoresis cell is indicated by + and − signs above and below the vertical arrows. Note the polarity change between the gel at pH 7.4 and the one at pH 8.2. This reflects the p*I*s of the two proteins (and was accomplished by reversing the leads of the electrophoresis cell at the power supply).

run under comparable electrical conditions in different McLellan buffers. HbA has a lower isoelectric point (p*I* 7.1) than HbC (p*I* 7.4). At pH 7.4 neither protein carries enough charge to move into the gel. At the acidic pHs tested, HbC is more highly charged and moves further through the gel than HbA. The situation is reversed at the basic pHs tested. Note the differences in polarity and run times of the various runs.

Protocol 1

Casting gels for the McLellan native, continuous buffer systems[a]

Equipment and reagents

- Gel cassettes and casting apparatus for the electrophoresis cell being used (e.g. the Bio-Rad Mini-PROTEAN 3 system)

- Acrylamide concentrate: 30%T, 2.7%C[b]

- 5 × gel buffer (see Table 1)
- 10% ammonium persulfate (APS)[c]
- N,N,N',N'-tetramethylethylenediamine (TEMED)[d]

Method

This method is for casting two gels in the Bio-Rad Mini-PROTEAN 3 system.

1   Prepare the desired 5 × buffer (see Table 1).

Protocol 1 continued

2 Thoroughly clean the glass plates, spacers, and combs with detergent and rinse them well with water.

3 Assemble the cassettes and place them in the casting apparatus.

4 Determine the gel volume from the manufacturer's specifications, by calculation, or by measuring the amount of water needed to fill a cassette.

5 Insert the well-forming combs between the gel plates of the cassettes and tilt them at a slight angle to provide a way for air bubbles to escape.

6 Mix:

- 12 ml of water
- 4 ml of 5 × buffer
- 4 ml of stock 30%T, 2.7%C acrylamide solution
- 100 μl of 10% APS
- 20 μl of TEMED

Swirl the solution gently and transfer it to the cassettes using a pipette and bulb.

7 Align the comb to its proper position being careful to not trap bubbles under the teeth. The gel should be ready to use in about 90 min.

Note: The APS–TEMED pair is a less efficient polymerization initiator below pH 6 than it is above pH 6. To compensate for this loss of efficiency in buffers below pH 6, increase the concentration of TEMED five-fold. Alternatively, add 100 μg of riboflavin 5'-monophosphate (100 μl of a 0.1% solution) to the solution prepared as in step 6 and initiate photopolymerization as is done with IEF gels (see Section 3.2).

[a] See ref. 24.

[b] To prepare 30%T, 2.7%C acrylamide stock solution, dissolve 29.2 g of acrylamide and 0.8 g of bisacrylamide in 72.5 ml of water. The final volume will be 100 ml. Both this solution and pre-mixed acrylamide–bisacrylamide powder can be purchased from Bio-Rad.

[c] To prepare 10% APS, dissolve 100 mg of APS in 1 ml of water. APS solutions should be made fresh daily, but may perform satisfactorily for up to a week. Store at room temperature.

[d] TEMED is used undiluted as supplied.

Caution: Acrylamide monomer is toxic. Avoid breathing acrylamide dust, do not pipette acrylamide solutions by mouth, and wear gloves when handling acrylamide powder or solutions containing it. Allow unused monomer solutions to polymerize and discard the resultant gels.

Other buffers that have been used for continuous, native electrophoresis are Tris–glycine (pH range 8.3–9.5) (17), Tris–borate (pH range 8.3–9.3) (25), and Tris–acetate (pH range 7.2–8.5) (26). Borate ions can form complexes with some sugars and can therefore influence resolution of some glycoproteins. Very basic proteins, such as histones and ribosomal proteins, are separated in acetic acid–urea gels (27). These buffer systems are not presented in protocols here in the interest of space.

Ornstein (20) and Davis (23) developed the first high resolution PAGE system for *native* (non-denatured) proteins. Their popular system is still in widespread use. It was designed for the analysis of serum proteins, but works well for a broad range of protein types. The Ornstein–Davis buffers should be the first discontinuous system tried when working with a new, native sample.

Gels for the Ornstein–Davis method are cast in two sections. A large-pore stacking gel (4%T, 2.7%C) is cast on top of a small-pore resolving gel (from 5–30%T depending on the proteins being studied). The two gel sections also contain different buffers. The stacking gel contains 0.125 M Tris–HCl pH 6.8, and the resolving gel contains 0.375 M Tris–HCl pH 8.8. Sample is diluted in 0.0625 M Tris–HCl pH 6.8. The electrode (or running) buffer is 0.025 M Tris, 0.192 M glycine pH 8.3. The pH discontinuity between the two sections of the gel was designed to regulate the effective mobility of glycinate ions from the cathode chamber. The concentrations of all four buffers were derived from electro-chemical considerations based on the properties of serum proteins. The porosity of the resolving gel must be empirically determined to match the mobilities of the proteins in the sample. There is no reliable way to predict the correct gel concentration of an untested protein mixture without analysing the proteins in gels. The choice is made such that the proteins of interest in the sample mixture are resolved in the gel. It is common to begin with a 7.5%T gel for the initial electrophoresis of a sample of unknown mobilities, then to try higher concentration gels (and sometimes lower concentration gels, such as 5%T).

For basic proteins, the low pH alanine–acetate system of Reisfeld *et al.* (28) is often used. Allen and Budowle suggest Tris–sulfate/Tris–borate, Tris–formate/Tris–borate, and Tris–citrate/Tris–borate (4) (not presented as protocols).

2.2.4 Denaturing systems

Because it is not yet possible to calculate the physical properties of proteins from mobility data, researchers have taken to denaturing SDS–PAGE in order to estimate protein molecular weights. Sample treatment for SDS–PAGE breaks all inter- and intramolecular bonds, both covalent and non-covalent, and leaves the polypeptide subunits of proteins in forms that can be separated on the basis of their molecular weights. Moreover, SDS solubilizes most proteins, so SDS–PAGE is applicable to a wide range of sample types. The electrophoretic band patterns obtained by SDS–PAGE are appreciably easier to interpret than those from native PAGE.

The most popular electrophoresis system is the discontinuous buffer system devised by Laemmli (29). Laemmli added SDS to the standard Ornstein–Davis buffers (Tables 2 and 3) and developed a simple denaturing treatment.

Sample preparation for SDS–PAGE is quite easy. Proteins are simply brought to near boiling in dilute gel buffer containing 5% (v/v) 2-mercaptoethanol, a thiol reducing agent, and 2% (w/v) SDS. The treatment simultaneously breaks disulfide bonds and dissociates proteins into their constituent polypeptide subunits. SDS monomer binds to the polypeptides and causes a change in their conformations. For most proteins, 1.4 g of SDS binds per gram of polypeptide (approx. one

Table 2 Compositions of electrode buffers

McLellan native, continuous buffer systems

5 × buffer stocks

    See Table 1

Store at room temperature.

To prepare 1 × electrode buffer, dilute 200 ml of 5 × buffer stock with 800 ml of water (see Table 1).

....

Ornstein–Davis native, discontinuous buffer

10 × electrode buffer stock

Tris base	30.3 g
Glycine	144.0 g

Dissolve in water to a total volume of 1 litre. Do not adjust the pH. Store at room temperature.

To prepare 1 × electrode buffer (0.025 M Tris, 0.192 M glycine pH 8.3), dilute 100 ml of 10 × buffer stock with 900 ml of water.

....

Laemmli SDS–PAGE buffer

10.× electrode buffer stock

Tris base	30.3 g
Glycine	144.0 g
SDS	10.0 g

Dissolve in water to a total volume of 1 litre. Do not adjust the pH. Store at room temperature.

To prepare 1 × electrode buffer (0.025 M Tris, 0.192 M glycine, 0.1% SDS pH 8.3), dilute 100 ml of 10 × buffer stock with 900 ml of water.

....

Tricine SDS–PAGE buffer

10 × electrode buffer stock

Tris base	12.1 g
Tricine	17.9 g
SDS	10.0 g

Dissolve in water to a total volume of 1 litre. Do not adjust the pH. Store at room temperature.

To prepare 1 × electrode buffer (0.1 M Tris, 0.1 M Tricine, 0.1% SDS pH 8.3), dilute 100 ml of 10 × buffer stock with 900 ml of water.

SDS molecule per two amino acids) (30). The properties of the detergent overwhelm the properties of the polypeptides. In particular, the charge densities of SDS-polypeptides are independent of pH in the range from 7–10 (4, 31). Most significantly, the SDS-polypeptides all assume the same hydrodynamic shape (30, 32). This means that their electrophoretic mobilities are nearly identical. If electrophoresis were done on a mixture of SDS-polypeptides in free solution, they would all migrate together. In gels, SDS-polypeptides separate by sieving on the basis of size (molecular weight of the polypeptide).

Laemmli's buffers, as usually described, are more elaborate than strictly necessary. Most presentations of this method utilize the two different gel buffers of the Ornstein–Davis system with SDS added to them. Because SDS so dominates the electrophoresis system, the buffer in the stacking gel can be the same as the buffer in the resolving gel. Results are the same whether the stacking gel is cast at pH 6.8 or at pH 8.8. Also, gels do not need to be cast with SDS in them (31). The SDS in the sample buffer is sufficient to saturate the proteins with the

Table 3 Compositions of sample buffers

McLellan native, continuous buffer systems

0.5 × gel buffer, 10% (w/v) glycerol, 0.01% bromophenol blue

5 × gel buffer[a]	1.0 ml
50% glycerol[b]	2.0 ml
0.5% bromophenol blue[c]	0.2 ml
Water	6.8 ml

Store at room temperature.

Ornstein–Davis native, discontinuous buffer

0.0625 M Tris–HCl pH 6.8, 10% glycerol, 0.01% bromophenol blue

Stacking gel 4 × buffer stock[d]	1.25 ml
50% (w/v) glycerol[b]	2.0 ml
0.5% bromophenol blue[c]	0.2 ml
Water	6.55 ml

Store at room temperature.

Laemmli SDS–PAGE buffer

0.075 M Tris–HCl pH 8.8, 2% SDS, 5% 2-mercaptoethanol, 10% glycerol, 0.01% bromophenol blue

Stock sample buffer

Resolving gel 4 × buffer stock[e]	0.5 ml
10% SDS[f]	2.0 ml
50% (w/v) glycerol[b]	2.0 ml
0.5% bromophenol blue[c]	0.2 ml
Water	4.8 ml

Store at room temperature. Prior to use, add 50 µl of 2-mercaptoethanol to 950 µl of sample buffer stock. An *alternative sample buffer* that is often used is obtained by adjusting the Ornstein–Davis sample buffer to 2% SDS and 5% 2-mercaptoethanol.

Tricine SDS–PAGE sample buffer

0.1 M Tris–HCl pH 8.45, 1% SDS, 2% 2-mercaptoethanol, 20% glycerol, 0.04% CBB G-250

Stock sample buffer

Tricine gel 3 × buffer stock[g]	0.33 ml
10% SDS[f]	1 ml
50% (w/v) glycerol[b]	4 ml
0.5% CBB G-250[h]	0.4 ml
Water	3.77 ml

Store at room temperature. Prior to use, add 20 µl of 2-mercaptoethanol to 980 µl of buffer.

---

[a] 5 × McLellan gel buffer (see Table 1).
[b] 50% (w/v) Glycerol (see Protocol 3).
[c] 0.5% Bromophenol blue. Dissolve 50 mg of bromophenol blue in 10 ml of water.
[d] Stacking gel 4 × buffer stock (see Protocol 2).
[e] Resolving gel 4 × buffer stock (see Protocol 2).
[f] 10% SDS. Dissolve 10 g of SDS in ≈70 ml of water, then adjust the volume to 100 ml.
[g] Tricine gel 3 × buffer stock (see Protocol 3).
[h] 0.5% Coomassie Brilliant Blue G-250. Dissolve 50 mg of CBB G-250 in 10 ml of water.

detergent. SDS in the cathode buffer overtakes the proteins in the sample and at 0.1% is sufficient for maintaining protein saturation during electrophoresis. This distinction is important for the commercial manufacturing of gels for SDS–PAGE (see Section 2.10).

Protocol 2

Casting gels for the Ornstein–Davis native buffer system[a] and the Laemmli SDS–PAGE system[b]

Equipment and reagents

- Gel cassettes and casting apparatus
- Acrylamide concentrated stock solution: 30%T, 2.7%C (see Protocol 1)
- APS and TEMED (see Protocol 1)
- 2-Butanol (water saturated)[c]
- Resolving gel 4 × buffer stock[d]
- Stacking gel 4 × buffer stock for native PAGE[e]
- Stacking gel 4 × buffer stock for SDS–PAGE[f]

Method

This method is for preparing two gels in the Bio-Rad Mini-PROTEAN 3 system.

1   Thoroughly clean the glass plates, spacers, and combs with detergent and rinse them well with water. Assemble the casting apparatus according to the manufacturer's instructions.

2   Mark the glass at the height of the resolving gels, which should be 0.5–1 cm below the bottoms of the wells (comb teeth).

3   Calculate the amount of 30%T acrylamide stock solution is needed to make 20 ml of monomer solution at the desired %T for the *resolving gel*.

$V_{resolver} = (2/3) (\%T_{resolver})$, in ml

4   Calculate the amount of water to be used for the resolving gel.

$V_{water} = (15 - V_{resolver})$, in ml

5   Combine the components of the *resolving gel* monomer mixture:

- 30%T acrylamide stock solution $V_{resolver}$ ml
- Water V_{water} ml
- Resolving gel 4 × stock solution   5 ml
- 10% APS   100 μl
- TEMED   10 μl

Swirl the solution gently and transfer it immediately to the gel cassettes using a pipette and bulb. Add the monomer mixture only to the mark showing gel height.

6   Overlay the resolving gel solution with water-saturated 2-butanol to exclude air from the top of the gel. Allow the gel to polymerize for about 1 h. Polymerization is evidenced by the appearance of a sharp interface beneath the overlay, which should start to become evident in about 15 min. Polymerization is essentially complete in about 90 min.

7   Once the resolving gels have hardened, rinse the tops of them thoroughly with water and dry the areas above them with filter paper. Place well-forming combs between the plates and tilt them at a slight angle to provide a way for bubbles to escape.

Protocol 2 continued

8 Combine the components of the *stacking gel* monomer mixture (10 ml):

- Water   6.2 ml
- 30%T acrylamide stock solution   1.3 ml
- Stacking gel 4 × stock solution   2.5 ml
- 10% APS   50 μl
- TEMED   10 μl

Swirl the solution gently and transfer it to the cassettes on top of the resolving gels using a pipette and bulb. Align the comb to its proper position being careful to not trap bubbles under the teeth. No overlay is required because the comb excludes air from the wells. Allow the gel to polymerize for about an hour.

[a] See refs 20 and 23.

[b] See ref. 29.

[c] To prepare 2-butanol, saturate a quantity (≤100 ml) of 2-butanol with water.

[d] To prepare resolving gel 4 × stock (1.5 M Tris–HCl pH 8.8.), dissolve 18.2 g of Tris base in ≈80 ml of water, adjust the solution to pH 8.8 with HCl, and add water to a final volume of 100 ml. Store this solution at room temperature. This solution can be purchased from Bio-Rad.

[e] To prepare 4 × stacking gel buffer stock for native PAGE (0.5 M Tris–HCl pH 6.8) dissolve 6.1 g of Tris base in ≈80 ml of water, adjust the solution to pH 6.8, and add water to 100 ml. Store this solution at room temperature. This solution can be purchased from Bio-Rad.

[f] 4 × stacking gel buffer stock for SDS–PAGE is the same as resolving gel 4 × stock buffer.

Caution: Acrylamide monomer is toxic. Avoid breathing acrylamide dust, do not pipette acrylamide solutions by mouth, and wear gloves when handling acrylamide powder or solutions containing it. Allow unused acrylamide monomer solutions to polymerize and discard the resultant gels.

Other systems for SDS–PAGE have been developed (not presented in protocols). Weber and Osborn's continuous, denaturing SDS–PAGE system uses pH 7 sodium phosphate buffer (33). This system helped establish the utility of SDS in electrophoresis as a means for estimating the molecular weights of proteins. The Weber–Osborn system is a popular one, but the lack of stacking limits its use to high concentration samples for best resolution. A protocol for the Weber–Osborn system can be found in ref. 34.

Neville (35) adapted a Tris–sulfate/Tris–borate buffer system to fractionate SDS-saturated proteins in the 2–300 kDa range. The Neville system produces very sharp bands. Wykoff *et al.* (31) replaced Tris in the Laemmli SDS–PAGE system with its analogue ammediol (2-amino-2-methyl-1,3-propanediol). The ammediol system resolves better in the 1–10 kDa size range than either the Laemmli or Neville systems, but the bands are less sharp.

The cationic detergent cetyltrimethylammonium bromide (CTAB) has been used as an alternative to SDS for gel electrophoresis of proteins. Akins *et al.* (36) devised the most successful application of CTAB. The Akins method employs a discontinuous buffer system with sodium (from NaOH) as the leading ion and arginine as the trailing ion with Tricine as the counter ion and buffer. This method uses no reducing agent in the sample buffer and protein solutions are not boiled prior to electrophoresis. As a result, many enzymes retain their activities. Nonetheless, CTAB coats proteins thoroughly enough that it can be used for molecular weight determinations in analogy with SDS–peptides. See ref. 34 for a protocol for CTAB-PAGE.

A system based on the use of Tricine instead of glycine in the electrode buffer provides excellent separation of small polypeptides (37). Peptides as small as 1 kDa are resolvable in Tricine–SDS gels. In particular, 16.5%T, 3%C separating gels are used for separations in the range from 1–70 kDa. Stacking gels in this system are 4%T, 3%C. Resolution is sometimes enhanced by inclusion of a 10%T, 3%C spacer gel between the resolving and stacking gels. Tricine–SDS resolving gels contain 1 M Tris–HCl pH 8.45, and 13% glycerol. It is not necessary to include SDS in the gel buffer, but the glycerol is important to impart a viscosity that seems necessary for resolving small peptides. Electrode buffer is 0.1 M Tris, 0.1 M Tricine, 0.1% SDS pH 8.25. Sample buffer is 0.1 M Tris–HCl pH 8.45, 1% (w/v) SDS, 2% (v/v) 2-mercaptoethanol, 20% (w/v) glycerol, 0.04% Coomassie Brilliant Blue G-250. Sample buffer should contain no more than 1% SDS for best resolution of small polypeptides (1–5 kDa). Proteins of very low molecular mass are not completely fixed and may diffuse from the gels during staining. This system is quite popular for polypeptide analysis (Protocol 3).

Protocol 3

Casting gels for Tricine SDS–PAGE[a]

Equipment and reagents

- See Protocol 2
- 40%T, 3%C acrylamide stock solution[b]
- Gel buffer 3 × stock solution[c]

Method

This method is for casting two 16.5%T gels for the Bio-Rad Mini-PROTEAN 3 system.

1 Assemble the gel cassettes and casting apparatus as in Protocol 2.

2 Combine the components of the *resolving gels* (20 ml):

• Water	3.0 ml
• 40%T acrylamide stock solution	8.2 ml
• Gel buffer 3 × stock solution	6.7 ml
• Glycerol	2.6 g
• 10% APS	100 μl
• TEMED	10 μl

Protocol 3 continued

Mix until all the components dissolve and transfer the solution to the gel cassettes using a pipette and bulb. Leave space in the cassettes for the stacking gels. Allow the gels to polymerize for about 1 h.

3   Prepare the cassettes for the stacking gels by rinsing the tops of the resolving gels with water, drying the spaces above the resolving gels with filter paper, and inserting combs (at a slight angle).

4   Combine the components of the *stacking gels* (10 ml):

- Water   5.7 ml
- 40%T acrylamide stock solution   1.0 ml
- Gel buffer 3 × stock solution   3.3 ml
- 10% APS   50 μl
- TEMED   10 μl

Swirl the solution gently to mix it and transfer it to the cassettes using a pipette and bulb.

5   Allow the gels to polymerize for about an hour before using them.

[a] See ref. 37.

[b] To prepare 40%T, 3%C acrylamide solution, dissolve 38.8 g of acrylamide and 1.2 g of bisacrylamide in 63.6 ml of water. The final volume will be 100 ml. This solution is available from Bio-Rad.

[c] To prepare 3 × gel buffer (3 M Tris–HCl pH 8.45), dissolve 36.3 g Tris base in ≈80 ml of water, adjust the solution to pH 8.45 with HCl, and add water to a final volume of 100 ml. Store the solution at room temperature.

Caution: see note in Protocol 2.

Compositions of electrode and sample buffers can be found in Tables 2 and 3, respectively.

2.3 Pore-size gradient gels

Acrylamide allows for the possibility of casting pore-gradient gels. Pore-gradient gels are used in SDS–PAGE applications where they have some advantages over homogeneous gels. The most common gradient gels are usually cast with acrylamide concentrations that increase linearly from top to bottom so that the pores get smaller with the distance into the gels. As proteins move through gradient gels from regions of relatively large pores to regions of relatively small pores, their migration rates slow down. Small proteins remain in gradient gels much longer than they do in single-percentage gels, so that they allow both large and small molecules to be run in the same gel. This makes gradient gels popular for analyses of complex mixtures spanning wide molecular mass ranges. A gradient gel, however, cannot match the resolution obtainable with a properly chosen single concentration of acrylamide. A good approach is to use gradient gels for estimates of the complexities of mixtures. This may be sufficient for some purposes. However, best resolution requires the appropriate single-concentration gel.

Various devices are commercially available for producing polyacrylamide gradients of almost any desired shape. These devices range in complexity from programmable pumps to simple cylinder pairs that form gradients hydrostatically. It is much simpler to purchase commercially available precast gradient gels (Section 2.10). A protocol for casting gradient gels is available in ref. 34.

Discontinuous buffer systems give best resolution in gradient gels: 4–15%T gradient mini-gels, based on the Laemmli buffers, resolve SDS–polypeptides in the 40–200 kDa size range, 4–20%T gels separate 10–100 kDa SDS–proteins, and 10–20%T gradients are useful in the 10–100 kDa range.

2.4 Apparatus

Apparatus for gel electrophoresis are relatively simple (Figure 5). Electrophoresis cells are essentially plastic boxes with anode and cathode buffer compartments, electrodes (usually platinum wire), and jacks for making electrical contact with the electrodes. Gels are held vertically between the electrode chambers during the run. Gel cassettes have open tops and bottoms. The bottom is sealed with a gasket during gel formation and the top is open to receive monomer solution. The top and bottom ends are open and in contact with buffer for electrophoresis. High voltage direct current supplies provide electrical power for electrophoresis. Micropipettes, test-tubes, and heating blocks are sample handling necessities. Many suitable devices are available from a number of suppliers.

Gels are cast as rectangular slabs in glass or plastic cassettes. The slab format provides uniformity, so that different samples can be compared in the same gel. In a small number of applications, gels are cast in cylindrical glass tubes. For comparative purposes, slab gels are far superior to tube gels. Conventional gels are of the order of 20 cm long and 20 cm wide. Gel thickness is varied by means of spacers inserted into the cassettes prior to gel formation. A thickness of 0.75 mm or 1 mm allows for adequate loads for good sensitivity of detection while at the same time allowing relatively high voltages to be applied to the gel without excessive heating. Runs take from 2.5–6 h to complete.

Some cells provide means for cooling, but this is not usually necessary. Cooling is often required to maintain the activity of proteins in the native state. It is not necessary with denaturing conditions in which protein conformations are intentionally destroyed. In some cases, cooling is undesirable. SDS will begin to crystallize out of solution at temperatures below about 10 °C, so excessive cooling should be avoided in SDS–PAGE. When cooling is distributed unevenly across a gel cassette the apparent bandwidths can be distorted. For example, with some cell designs, cooling is applied to only one face of the gel cassette. Proteins near the cooled face of the gel migrate slightly slower than the proteins on the other face. This causes the bands to slant downward from the cooled to

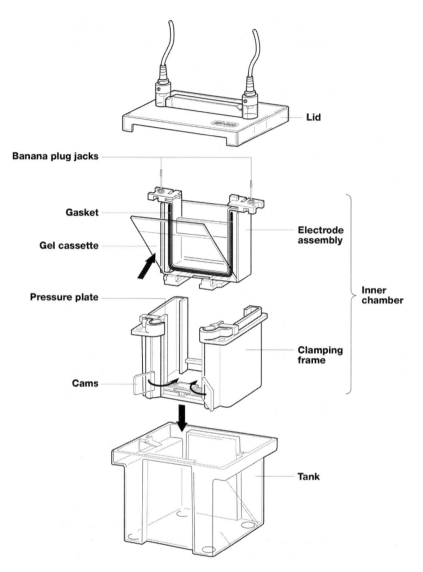

Figure 5 Exploded view of an electrophoresis cell. The components of the Bio-Rad Mini-PROTEAN 3 cell are shown. The inner chamber assembly can hold two gels. It contains an electrode assembly and a clamping frame. The interior of the inner assembly constitutes the upper buffer compartment (usually the cathode). The chamber is placed in the tank, which is filled with buffer and constitutes the lower (anode) buffer compartment. Electrical contact is made through the lid.

uncooled face across the thickness of the gel. Although bandwidths may be the same as without cooling, the protein bands can appear broadened when viewed face on. The solution to this problem is to not apply cooling to the cell.

So-called mini-cells and midi-cells allow rapid analysis and are adequate for relatively uncomplicated samples. The design of these cells allows runs to be completed in as little as 35–45 min. Mini-gels are about 7 cm long by 8 cm wide while midi-gels are of the order of 15 × 10 cm. They are both very easy to handle. Some midi-gels can hold up to 26 samples.

All three size categories of gels can be used for nearly all purposes. The resolution that can be obtained between protein bands is the same for mini-, midi-, and maxi-gels. However, the separation between bands, as defined in Section 2.5, is greater with longer gels. Closely spaced bands are easier to distinguish from one another, their bandwidths are easier to measure, and they are more easily cut out from a large gel than from a small one.

2.4.1 Electrical considerations

Regulated direct current (DC) power supplies, designed for electrophoresis, allow control of every electrophoretic mode. Constant voltage, constant current, or constant power conditions can be selected. Many power supplies have timers and some have integrators allowing runs to be automatically terminated after a set time or number of volt-hours. All modes of operation can produce satisfactory results, but for best results and good reproducibility, some form of electrical control is important. The choice of which electrical parameter to control is almost a matter of preference. The major limitation is the ability of the chamber to dissipate the heat generated by the electrical current.

During an electrophoresis run, electrical energy is converted into heat, called Joule heat. This heat can have many deleterious consequences, such as band distortion, increased diffusion, enzyme inactivation, and protein denaturation. All good electrophoresis chambers are designed to transfer the heat generated in the gel to the outside environment. In general, electrophoresis should be carried out at voltage and current settings at which the run proceeds as rapidly as the ability of the chamber to draw off heat allows. That is, the run should be as fast as possible without exceeding desired resolution and distortion limits—and these can only be determined empirically for any given system. Each experiment will impose its own criteria on cooling efficiency. Nearly all electrophoresis runs can be carried out on the laboratory bench, but some delicate proteins may require that the runs be conducted in the cold room or with circulated coolant.

Electrical quantities are interrelated by fundamental laws. Each gel has an intrinsic resistance, R, determined by the ionic strength of its buffer (R changes with time in discontinuous systems). When a voltage V is impressed across the gel a current I flows through the gel and the external circuitry. Ohm's law relates these three quantities: $V = IR$, where V is expressed in volts, I in amperes, and R in ohms. In addition, power P, in watts, is given by $P = IV$. The generation of Joule heat, H, is related to power by the mechanical equivalent of heat, 4.18 J/cal, or $H = (P/4.18)$ cal/sec.

With continuous buffer systems, the resistance of the gel is essentially constant, although it will decrease a bit during a run as the buffer warms. With the discontinuous Ornstein–Davis or Laemmli buffers, R increases during the course of a run as the chloride ions are exchanged by glycinate. For runs at constant current in Laemmli gels, the voltage, power (I^2R), and consequently the heat generated in the gel chamber increase during the run. Under constant voltage conditions, current, power (V^2/R), and heat generation decrease during electrophoresis as R increases. Voltage and current should be set to keep H below the dissipation limit of the electrophoresis chamber. Follow the recommendations of the manufacturer for the proper electrical settings to use with any particular cell. Vertical cells are usually run at electric field strengths of 10–20 V/cm or currents in the range of 15–25 mA/mm of gel thickness.

The voltage applied to an electrophoresis cell is divided across three distinct resistance regions. The buffer paths from the open ends of the gel to the electrode wires form two of these regions. These two resistance regions are usually ignored but they should be kept in mind for electrical analysis when experimenting with electrode buffers having very high or very low conductivity. The gel buffer is the third resistance region. With the Laemmli SDS system, the buffers create two different resistive sections. The low resistance leading Cl^- ion forms a resistance segment that runs ahead of the higher resistance trailing glycinate ion segment. Taken together, the two gel segment resistors act as a voltage divider. The voltage across either one of the gel segments is proportional to the resistance of that segment. The voltage across the chloride section provides the force that pulls the ion front through the gel, whereas the voltage across the glycinate section pulls the proteins through the gel. This proportioning of the applied voltage can cause two gels of the same %T to run differently if their gel buffers are different. For example, the final band pattern in a 12%T Laemmli gel with a gel buffer at pH 8.6 looks like the band pattern of a 10%T Laemmli gel with a gel buffer at pH 8.8. The gel at pH 8.6 also takes about 20% longer to run than the gel at pH 8.8. The differences in the properties of the two gel types are due to the increased conductivity of the pH 8.6 gel relative to the pH 8.8 gel. The extra chloride needed to drop the pH of 0.375 M Tris from 8.8 to 8.6 brings about the increased conductivity (0.12 M Cl^- vs. 0.19 M Cl^-). A subtle electrochemical process related to the ionization of glycine accentuates this effect, which can sometimes be used to advantage (38).

2.5 Resolution and separation

Evaluations of electrophoretic data usually include mention of *separation* or *resolution*. Although the two terms are not synonymous they are often treated as such. Because of the way in which people visualize gels, separation may be the more important of the two terms for electrophoresis. Both large and small gels can have the same resolution, but there will be more space between the bands of larger gels. This added 'landscape' of larger gels is reassuring to researchers and makes the excision of bands from large gels easier to accomplish than from small gels.

Separation refers to the distance between two adjacent bands. The eye tends to see bands as being sharply defined with clearly evident blank spaces between adjacent bands. Thus, for practical purposes, separation should be taken to be the distance between the top of the faster running of two adjacent bands and the bottom of the slower one. It is the distance between the top of the bottom band and the bottom of the top band. This definition seems preferable in electrophoresis to defining separation as the distance between band centres.

Resolution, on the other hand, is a more technical term. It refers to the distance between adjacent bands relative to their bandwidths and acknowledges the fact that proteins are distributed in Gaussian profiles with overlapping distributions between bands. The numerical expression for resolution is obtained by dividing the distance between the centres of adjacent bands by some measure of their average bandwidths. It expresses the distance between band centres in units of bandwidth and gives a measure of the overlap between two adjacent bands. For preparative applications, when maximal purity is desired, the length of at least one bandwidth should separate two bands. In other applications, it may be sufficient to be able to simply discern that two bands are distinct. In this latter case, the bands can be less than a bandwidth apart. Several software packages are available that use Gaussian modelling to differentiate and quantify bands for the quantitative analysis of gel images.

2.6 Detergents

Detergents are employed in electrophoresis when it is necessary to disrupt protein–lipid and protein–protein interactions. A variety of detergents has been used for this purpose (39, 40).

SDS is the most common detergent used in PAGE analysis. Most proteins are readily soluble in SDS, making SDS–PAGE a generally applicable method. In SDS-PAGE, the quality of the SDS is of prime importance. The effects of impurities in SDS are unpredictable. Of the contaminants, the worst offenders are probably the alkyl sulfates other than dodecyl sulfate (C_{12}); especially decyl sulfate (C_{10}), tetradecyl sulfate (C_{14}), and hexadecyl sulfate (C_{16}) (41, 42). These bind to proteins with different affinities, thereby affecting mobilities. Lipophilic contaminants in SDS preparations, including dodecanol, can be trapped in SDS–protein complexes and SDS micelles, leading to loss of resolution. Only purified SDS should be used for electrophoresis, but even with pure SDS, various glycoproteins, lipoproteins, and nucleoproteins tend to bind the detergent irregularly. The resultant SDS–polypeptides then migrate 'anomalously' with respect to their molecular masses.

Several types of proteins do not behave as expected during SDS–PAGE (1, 34). Incomplete reduction, which leaves some intra- or intermolecular disulfide bonds intact, makes some SDS-binding domains unavailable to the detergent so that the proteins are not saturated with SDS. Glycoproteins and lipoproteins also migrate abnormally in SDS-PAGE, because their non-proteinaceous components do not bind the detergent uniformly. Proteins with unusual amino acid sequences, especially those with high lysine or proline content, very basic proteins, and very acidic proteins behave anomalously in SDS-PAGE, presumably because the charge-to-mass ratios of the SDS–polypeptide complexes are different than those that would be expected from size alone. Similarly, very large

SDS–proteins, with molecular masses in the several hundred kilodalton range, may have unusual conformations. Polypeptides smaller than about 12 000 Da are not resolved well in most SDS–PAGE systems. In most cases, they do not separate from the band of SDS micelles that forms behind the leading ion front. The Tricine buffer system was devised to separate these small polypeptides.

2.7 Choice of system

2.7.1 Native proteins

Continuous buffer systems are preferred for native work because of their simplicity. Furthermore, some native proteins may aggregate and precipitate at the very high protein concentrations reached during stacking in discontinuous electrophoresis. Consequently, either they might not enter the resolving gel or they might cause streaking as accumulated protein slowly dissolves during a run. If the proteins of interest behave in this manner, it is best to use some form of continuous buffer system.

The pH of the electrophoresis buffer must be in the range over which the proteins of interest are stable or where they retain their biological activity. The pH should also be properly chosen with respect to the isoelectric point (p*I*). The pH of the gel buffer should be far enough away from the p*I*s of the proteins of interest that they carry enough net charge to migrate through the gel in a reasonable time. On the other hand, separation of two proteins at a given gel concentration is best near one of their isoelectric points, because the isoelectric protein will barely move in that pH range. (Figure 4 shows how the buffer choice determines migration rates.) The choice of pH is often a compromise between considerations of resolution and stability. For best results with continuous systems, the concentrations of the proteins of interest should be at least 1 mg/ml to keep sample volume at a minimum. The sample should be loaded in gel buffer diluted to 0.5 to 0.2 strength (some form of buffer exchange may be required). The decreased ionic strength of diluted buffer causes a voltage to develop across the sample that assists in driving the proteins into the gel.

The choice of proper gel concentration (%T) is, of course, critical to the success of the separation, since it heavily influences separation. Too high %T can lead to exclusion of proteins from the gel and too low %T can decrease sieving (see Figure 3). One approach, useful with the McLellan continuous buffers (Table 1), is to use relatively large-pore gels (5%T to 7%T) and to alter mobilities with pH (Protocol 1 uses 6%T). An approach for discontinuous systems is to start with a 7.5%T gel, then, if that is not satisfactory, to try a number of gel concentrations between 5%T and 15%T. Pore-gradient gels can also be tried.

2.7.2 Denatured proteins

It is easier to choose suitable gel concentrations (%T) for SDS–PAGE than for native protein gels because the separation of SDS–polypeptides is dependent mainly on chain length. Laemmli gels with 7.5%T resolve proteins in the 40–200 kDa range, those with 10%T resolve 20–200 kDa proteins, 12%T gels separate proteins in the 15–100 kDa range, and 15% gels separate 6–90 kDa proteins (Figure 6).

2.8 Sample preparation

Samples for SDS–PAGE by the Laemmli procedure are prepared in diluted gel buffer containing SDS, 2-mercaptoethanol, glycerol, and bromophenol blue tracking dye (Table 3). It is best to prepare a stock solution of sample buffer containing everything but 2-mercaptoethanol and to add this reagent right before use. The glycerol provides density for applying the sample on the stacking gel under electrode buffer. The tracking dye allows both sample application and the electrophoretic run to be monitored (it migrates with the ion front). There is sufficient SDS present in the sample buffer to ensure saturation of most protein mixtures (31). Except in the rare instances when the sample is in a very high ionic strength solution (>0.2 M salts), it can be dissolved 1:1 (v/v) in stock sample buffer. It is much better, though, to dilute the sample at least 1:4 (v/v) with the sample buffer stock. The amount of sample protein to load on a gel depends on the detection method to be used (see Section 5). Enough of the protein of interest must be loaded on the gel for it to be subsequently located. Detection in gels requires on the order of 1 μg of protein for easy visibility of bands stained with anionic dyes such as Coomassie Brilliant Blue R-250 or 0.1 μg of protein with silver staining. Complete dissociation of most proteins is achieved by heating diluted samples to 95–100 °C for 2–5 min.

For native, discontinuous gels, upper gel buffer diluted two-fold to five-fold for sample application is commonly used. Tracking dye and glycerol are added to these samples also, and protein concentrations should fall within the same

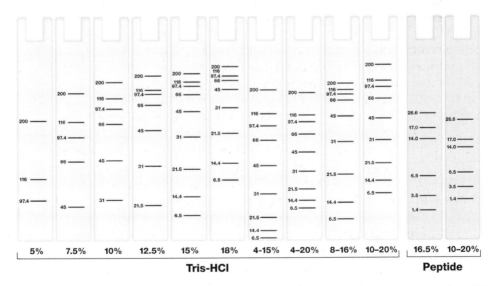

Figure 6 Protein migration charts. Relative positions of standard proteins are shown for several SDS–PAGE gels of the Laemmli type ('Tris–HCl') and two Tricine SDS–PAGE gels ('Peptide').

limits as for SDS–PAGE. With discontinuous systems, the volume of sample is not very important as long as the height of the stacking gel is at least twice the height of the sample volume loaded on the gel. Continuous systems require minimal sample volumes for best resolution.

Careful sample handling is important when sensitive detection methods are employed. Silver-stained SDS–PAGE gels sometimes show artefact bands in the 50–70 kDa molecular mass region and irregular but distinctive vertical streaking parallel to the direction of migration. The appearance of these artefacts has been attributed to the reduction of contaminant skin keratin inadvertently introduced into the samples (43). The best remedy for the keratin artefact is to avoid introducing it into the sample in the first place. Monomer solution, stock sample buffer, gel buffers, and upper electrode buffer should be filtered through nitrocellulose and stored in well-cleaned containers. It also helps to clean the gel apparatus thoroughly with detergent and to wear gloves while assembling the equipment.

Prepared samples are placed in sample wells of gels with microlitre syringes or micropipettes. Both types of liquid-handling device provide good control of sample volume. Syringes must be thoroughly rinsed between applications to avoid cross-contamination of different samples. Standard pipette tips are too wide to fit into narrow sample wells, but several thin tips, specifically designed for sample application, are available. The choice of sample-loading device is one of personal preference.

Protocol 4

Sample preparation and electrophoresis

Equipment and reagents

- Gel electrophoresis apparatus (e.g. the Bio-Rad Mini-PROTEAN 3)
- Electrophoresis power supply capable of delivering at least 300 V and 200 mA (e.g. the Bio-Rad PowerPac 300 or PowerPac 1000)
- Electrode buffer (see Table 2): consult the apparatus instructions for the volume required
- Sample buffer (see Table 3)

Method

1. Assemble the electrophoresis apparatus and insert the gels according to the manufacturer's instructions. With some electrophoresis cells, it is sometimes easier to load the wells before the gel assembly is placed in the buffer tank.

2. Remove the combs from the gels and rinse the wells with electrode buffer.

3. Fill the upper reservoir of the electrophoresis cell so that buffer fills the wells.

4. Mix 1 vol. of protein sample with at least 2 vol. of the appropriate sample buffer (see Table 3). Prepare protein standards or markers in the same manner as the samples.

Sample volumes depend on the types of proteins, the type of detection, and the size of the sample wells. Volumes loaded into the wells are often of the order of tens of microlitres.

5 Heat samples for SDS–PAGE (Laemmli or Tricine) at 95 °C for 2 min in a temperature block or water-bath. (Make sure that 2-mercaptoethanol is included in the sample buffer for SDS–PAGE; see Table 3.) Do NOT heat samples for non-denaturing conditions (McLellan or Ornstein–Davis systems).

6 Load the samples carefully in the wells of the gels with a microsyringe or micro-pipette; e.g. Bio-Rad Prot/Elec pipette tips. The glycerol in the sample buffers (Table 3) provides density for layering the samples under the buffer in the wells.

7 Place the gel assembly in the lower electrode buffer tank, if sample loading was done outside the tank. Add electrode buffer to the electrode chambers as necessary. It is often unnecessary to immerse the gels completely in buffer.

8 Put the lid on the tank and connect the leads to the power supply. For most gel types, the anode (+) (red lead) is at the bottom of the tank and the cathode (–) (black lead) is at the top. However, it is necessary to reverse the leads with CTAB gels and with the high pH McLellan gels. For safety, power supplies for electrophoresis should have isolated grounds.

9 Turn on the power supply and set it for the electrical conditions recommended by the manufacturer of the electrophoresis cell; e.g. 200 V constant voltage for Laemmli gels run in the Bio-Rad Mini-PROTEAN 3.

10 Allow the run to proceed until the blue tracking dye from the sample buffer reaches the end of the gel (about 40 min for Laemmli gels in the Mini-PROTEAN 3). It is not necessary to remove the gel immediately at the end of a run. Gels can be left for extended periods—at least overnight—without deterioration of the band patterns.

2.9 Molecular mass estimation

SDS–PAGE has become the most popular method of gel electrophoresis because it can be used to estimate molecular masses (1, 3, 8, 34, 44). To a first approximation, migration rates of SDS polypeptides are inversely proportional to the logarithms of their molecular masses. The larger the polypeptide, the slower it migrates in a gel.

Molecular masses are determined in SDS–PAGE by comparing the mobilities of test proteins to the mobilities of known protein markers. At one time, when samples for SDS–PAGE were run in individual tubes, it was necessary to normalize to a common parameter so that the different tube gels could be compared. This was because tube gels differ in length. The normalizing parameter that is still used is the relative mobility, R_f, defined as the mobility of a protein divided by the mobility of the ion front. In practice, when all gels are run for the same length of time, R_f is calculated as the quotient of the distance travelled by a protein from the top of the resolving gel divided by the distance migrated by the ion front. The distance to the ion front is usually taken as the distance to the tracking dye (measured or marked in some way before staining). With slab gels, this normalization is less important provided that a lane of standards is run in the same gel as the samples whose masses are to be determined. It is sufficient to compare migration distances of samples and standards. Plots of the logarithm of protein molecular mass ($\log M_r$) versus the migration distances fit reasonably straight lines.

In each gel, a lane of standard proteins of known molecular masses is run in parallel with the test proteins. After staining the gel to make the protein bands visible (Section 5), the migration distances are measured from the top of the resolving gel. The gel is calibrated with a plot of $\log M_r$ vs. migration distances for the standards. The migration distances of the test proteins are compared with those of the standards. Interpolation of the migration distances of test proteins into the standard curve gives the approximate molecular masses of the test proteins.

Pore-gradient SDS–PAGE gels can also be used to estimate molecular masses. In this case, $\log M_r$ is proportional to $\log (\%T)$. With linear gradients, %T is proportional to distance migrated, so that the data can be plotted as $\log M_r$ vs. log (migration distance).

Standard curves are actually sigmoid in shape (Figure 7). The apparent linearity of a standard curve may not cover the full-range of molecular masses for a given protein mixture in a particular gel. However, $\log M_r$ is sufficiently slow, in a mathematical sense, to allow fairly accurate molecular mass estimates to by made by interpolation, and even extrapolation, over relatively wide ranges. The approximate useful ranges of single-percentage SDS–PAGE gels for molecular mass estimations is as follows: 40 000–200 000 Da, 7.5%T; 30 000–100 000, 10%T; 15 000–90 000 Da, 12%T; 10 000–70 000 Da, 15%T. Mixtures of standard proteins with known molecular masses are available commercially for calibrating electrophoresis gels.

The semi-logarithmic plots used for molecular mass determinations are holdovers from the days when people had only graph paper and straight edges for curve fitting. [From a mathematical point of view, $\log M_r$ should be the independent variable (x-axis) and migration distance should be the dependent variable (y-axis). Not as usually drawn.] Several computer programs allow for standard curves to be fit with different mathematical functions than the semi-logarithmic model. Some other types of curves actually fit the data better than the semi-log function. Nevertheless, the semi-logarithmic model for standard curves is the accepted norm.

It is important to bear in mind that the molecular masses obtained using Laemmli SDS–PAGE are those of the polypeptide subunits and not those of native, oligomeric proteins. Moreover, proteins that are incompletely saturated with SDS, very small polypeptides, very large proteins, and proteins conjugated with sugars or lipids behave anomalously in SDS–PAGE, as mentioned above. Nevertheless, SDS–PAGE provides reasonable molecular mass estimates for most proteins.

224

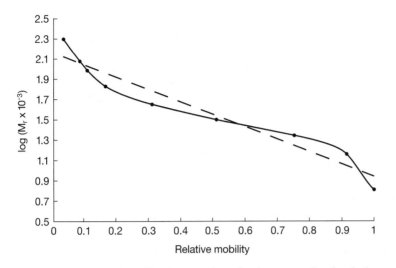

Figure 7 A representative calibration curve for molecular mass estimation. In the run that is plotted here (solid line), Bio-Rad SDS–PAGE standards with molecular masses of 200, 116.2, 97.4, 66.2, 45, 31, 21.5, 14.4, and 6.5 kDa (top to bottom, closed circles) were separated in a 15%T SDS–PAGE gel. The plot of $\log_{10} (M_r \times 10^{-3})$ vs. relative mobility (R_f) shows the inherent non-linearity of such curves. The straight line segment in the middle of the plot is the most accurate range for molecular mass estimations. Larger polypeptides experience greater sieving than do those in the middle range; the upper part of the curve has a different slope than in the middle. Small polypeptides experience less sieving than the others and also deviate from a strictly straight line dependence. It is customary to estimate molecular masses from a 'best fit' straight line (dashed). This is sufficient for some purposes and is acceptable because the logarithm function changes slowly with its argument.

2.10 Precast gels

The biggest change in gel electrophoresis since the advent of polyacrylamide gels in the 1960s is the commercial availability of precast gels. Since the early 1990s, several companies have made a wide variety of precast polyacrylamide gels available to the research community (Figure 8). Most of the precast gels offered are Laemmli SDS–PAGE gels of differing %T and numbers of wells. Because the Laemmli SDS–PAGE gel is so overwhelmingly popular, alternative types of electrophoresis gels have tended to be ignored by researchers and the companies have focused on this bias. In fact, it is probably safe to state that many people do not know that there is any other type of protein gel electrophoresis than the Laemmli method.

It took more than 20 years for manufacturers to devise production and distribution networks for delivering consistently high quality gels to customers. The problems that the companies faced stem from the limited shelf-life of polyacrylamide gels. Because polyacrylamide gels hydrolyse over time as shown in Figure 1, they are inherently unstable. The shelf-life of a gel cast in the Laemmli gel buffer (pH 8.8) is about three to four months. It is not possible to cast large volumes of gels and to hold them in a warehouse for long periods of time. A great deal of planning goes into the decisions of how many gels of each

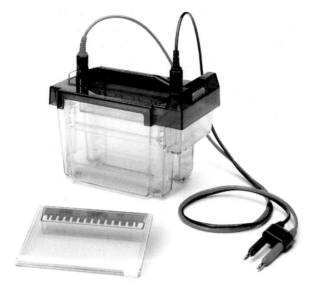

Figure 8 Precast gel system. A Bio-Rad Criterion™ Precast Gel and its electrophoresis cell are shown.

different type are to be cast at any particular time. Manufacturing and distribution issues have now been largely addressed and customers can now be guaranteed that they will receive gels that can be stored for several weeks before they are used. A limited number of precast gel products are available that are cast with neutral pH buffers (45). These gels have longer shelf-lives than gels made according to the Laemmli formulation. However, since the band patterns obtained with neutral pH gels are different than those obtained with Laemmli gels, they have not found universal acceptance.

People were initially drawn to precast gels by the gradients. It is more appealing to be able to buy gradient gels that are already made than to cast them. The ease in use of precast gradient gels led to the acceptance of single-percentage gel types as well. For all but the most demanding situations there is little reason to cast gels by hand. The gel types most in demand are 7.5%T, 10%T, 12%T, 4–15%T, and 4–20%T.

Precast gels differ from hand cast gels in three ways. They are cast with a single buffer throughout, without SDS, and without a sharp demarcation between the stacking and resolving gels. As pointed out previously, because SDS dominates the system, using different buffers in the stacking and resolving gels as in the original Laemmli formulation has no practical value. The two different buffers would mingle together on storage without elaborate means to keep them separate. In addition, during electrophoresis SDS from the cathode buffer sweeps past the proteins in the resolving gel and keeps them saturated with SDS even when there is no SDS in the gel when it is cast. Precast gels are thus made without SDS. This is beneficial to both the manufacturer and the user. SDS tends to form bubbles in the pumping systems used to deliver monomer solutions to gel cassettes, causing problems with monomer delivery. It also forms micelles that can trap acrylamide monomer and lead to heterogeneity of pore size.

When gels are cast by hand, it is customary to allow the resolving gel to harden before the stacking gel is placed on top of it. This practice is acceptable since hand cast gels are usually used within a short period of time. On the other hand, when gels are cast this way and stored, the stacking gel eventually begins to pull away from the resolving gel. The gap that forms between the two gels leads to lateral spreading of the stacked proteins and destruction of the stack as it leaves the upper gel. For this reason, precast gels are cast in a continuous manner with the stacking gel monomer solution added on top of the resolving gel monomer solution before gelation. This means that the separation between the two gel types is a gradual one rather than a sharp one. Even though the distance between the two gels is short, the transition between gel types exists as a short gradient of %T. Proteins 'unstack' gradually rather than abruptly. Because of this, the bands obtained with precast gels are not quite as sharp as those obtained with hand cast gels.

3 Isoelectric focusing

Isoelectric focusing is an electrophoretic method in which proteins are separated on the basis of their pIs (1-6, 46-51). It makes use of the property of proteins that their net charges are determined by the pH of their local environments. Proteins carry positive, negative, or zero net electrical charge, depending on the pH of their surroundings.

The net charge of any particular protein is the (signed) sum of all of its positive and negative charges. These are determined by the ionizable acidic and basic side chains of the constituent amino acids and prosthetic groups of the protein. If the number of acidic groups in a protein exceeds the number of basic groups, the pI of that protein will be at a low pH value and the protein is classified as being *acidic*. When the basic groups outnumber the acidic groups in a protein, the pI will be high with the protein classified as *basic*. Proteins show considerable variation in isoelectric points, but pI values usually fall in the range of pH 3-12 with a great many having pIs between pH 4 and pH 7 (52, 53).

Proteins are positively charged in solutions at pH values below their pI and negatively charged above their isoelectric points. Thus, at pH values below the pI of a particular protein, it will migrate toward the cathode during electrophoresis. At pH values above its pI, a protein will move toward the anode. A protein at its isoelectric point will not move in an electric field.

When a protein is placed in a medium with a linear pH gradient and subjected to an electric field, it will initially move toward the electrode with the opposite charge (Figure 9). During migration through the pH gradient, the protein will either pick up or lose protons. As it does, its net charge and mobility will decrease and the protein will slow down. Eventually, the protein will arrive at the point in the pH gradient equalling its pI. There, being uncharged, it will

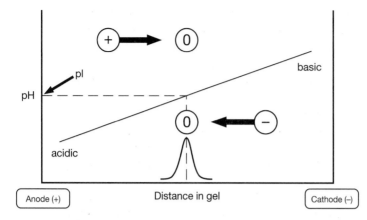

Figure 9 Isoelectric focusing. The motion of a protein undergoing isoelectric focusing is depicted (circles). The protein is shown near its p*I* in a pH gradient. Both the pH gradient and the motion of the protein are controlled by an electric field. At pH values lower than the p*I*, the protein is positively charged (+) and it is driven toward the cathode (arrow). Above the p*I*, the protein is negatively charged (−) and it moves toward the anode. There is no net electrical force on the protein at its p*I* (0). The protein focuses in a Gaussian distribution centred at the p*I*.

stop migrating. If a protein at its p*I* should happen to diffuse to a region of lower pH, it will become protonated and be forced toward the cathode by the electric field. If, on the other hand, it diffuses into a pH higher than its p*I*, the protein will become negatively charged and will be driven toward the anode. In this way, proteins condense, or focus, into sharp bands in the pH gradient at their individual, characteristic p*I* values.

Focusing is a steady-state mechanism with regard to pH. Proteins approach their respective p*I* values at differing rates but remain relatively fixed at those pH values for extended periods. This type of motion is in contrast to conventional electrophoresis in which proteins continue to move through the medium until the electric field is removed. Moreover, in IEF, proteins migrate to their steady-state positions from anywhere in the system. Thus, the sample application point is arbitrary. In fact, the sample can be initially distributed throughout the entire separation system.

3.1 Establishing pH gradients

Stable, linear pH gradients are the keys to successful IEF. Establishment of such gradients is accomplished in two ways with two different types of molecules, carrier ampholytes and acrylamido buffers.

Carrier ampholytes (*amphoteric electrolytes*) are mixtures of molecules containing multiple aliphatic amino and carboxylate groups. They are small (about 300–1000 Da in size) multi-charged organic buffer molecules with closely spaced p*I* values and high conductivity. Ampholytes are included directly in IEF gels. In electric fields, carrier ampholytes partition into smooth pH gradients that increase linearly from the anode to the cathode. The slope of a pH gradient is determined by the pH interval covered by the carrier ampholyte mixture and the distance between the electrodes. The use of carrier ampholytes is the most common and simplest means for forming pH gradients.

Acrylamido buffers are derivatives of acrylamide containing both reactive double bonds and buffering groups. Their general structure is $CH_2=CH-CO-NH-R$, where R contains either a carboxyl [$-COOH$] or a tertiary amino group [e.g. $-N(CH_3)_2$]. They are covalently incorporated into polyacrylamide gels at the time of casting. The key acrylamido buffers have pK values at pH 1, 3.6, 4.6, 6.2, 7.0, 8.5, 9.3, 10.3, and >12. They can be used to cast just about any conceivable pH gradients. In any given gradient, some of the acrylamido compounds act as buffers while others serve as titrants. Published formulations and methods are available for casting the most common gradients (48, 50). Because the buffering compounds are fixed in place in the separation medium, the gels are called 'immobilized pH gradients', or IPGs. IPGs offer the advantage of gradient stability over extended runs. They are, however, more cumbersome and expensive to cast than carrier ampholyte gels. IPGs are commercially available in sheet form in a few pH ranges. A greater variety of pH ranges are available in IPGs that have been cut into strips for the IEF first dimension of 2D PAGE.

IEF is a high resolution technique that can routinely resolve proteins differing in pI by less than 0.05 pH unit. Antibodies, antigens, and enzymes usually retain their activities during IEF. The proper choice of ampholyte or IPG range is very important to the success of a fractionation. Ideally, the pH range covered by an IEF gel should be centred on the pI of the proteins of interest. This ensures that the proteins of interest focus in the linear part of the gradient with many extraneous proteins excluded from the separation zone.

With carrier ampholytes, concentrations of about 2% (w/v) are best. Ampholyte concentrations below 1% (w/v) often result in unstable pH gradients. At concentrations above 3% (w/v) ampholytes are difficult to remove from gels and can interfere with protein staining. When casting IPGs, follow published recipes and use buffering powers of about 3 meq throughout the gradient (method not presented).

3.2 Gels for isoelectric focusing

As an analytical tool, IEF is carried out in large-pore polyacrylamide gels (5%T, 3%C) which serve mainly as anticonvective matrices. Polyacrylamide IEF gels are polymerized with an initiator system including riboflavin for photopolymerization. Photochemical initiation of polymerization with a combination of the three compounds riboflavin, ammonium persulfate, and TEMED, results in more complete polymerization of IEF gels than does chemical polymerization in gels containing low pH ampholytes (14). Suitable initiator concentrations are 0.015% ammonium persulfate, 0.05% TEMED, and 5 μg/ml riboflavin-5'-phosphate. Photochemical polymerization is allowed to continue for 2 h, with the second hour under direct lighting from a nearby fluorescent lamp.

DAVID E. GARFIN

The most common configuration for analytical IEF is the horizontal poly-acrylamide slab gel. Gels are cast with one exposed face on glass plates or specially treated plastic sheets. They are placed on cooling platforms and run with the exposed face upward. Electrolyte strips, saturated with 0.1–1 M phosphoric acid at the anode and 0.1–1 M sodium hydroxide at the cathode, are placed directly on the exposed surface of the IEF gel. Electrodes of platinum wire maintain contact between the electrical power supply and the electrolyte strips. In another possible configuration, the gel and its backing plate are inverted and suspended between two carbon rod electrodes without the use of electrolyte strips. IPG strips for 2D PAGE are often run with the gel facing down in dedicated IEF cells.

Protocol 5

Casting gels for isoelectric focusing

Equipment and reagents

- Gel casting apparatus for a horizontal electrophoresis cell (e.g. the Bio-Rad Model 111 or the Pharmacia MultiPhor)
- Fluorescent lamp
- 30%T, 3%C acrylamide stock solution[a]
- 50% glycerol[b]
- 0.1% riboflavin-5′-phosphate (FMN)[c]
- Carrier ampholytes with a pH range spanning the pIs of the proteins of interest[d]
- 10% APS and TEMED (see Protocol 1)

Method
The formulation given here is for 12 ml of gel solution containing 5%T (3%C) acrylamide, 2% carrier ampholytes, 5% glycerol. The volume needed depends on the casting apparatus that is used; adjust volumes accordingly. This recipe is sufficient for casting one gel of 100 × 125 × 0.8 mm (10 ml) or four 100 × 125 × 0.2 mm (10 ml total). 8 M urea can be substituted for the glycerol if desired (see ref. 50).

1 Assemble the casting apparatus according to the manufacturer's instructions. The use of gel support film for polyacrylamide is highly recommended.

2 Combine:

- 30%T, 3%C acrylamide stock   2.0 ml
- Carrier ampholytes (40%)   0.6 ml
- 50% glycerol   1.2 ml
- Water   8.2 ml
- 0.1% FMN   60 µl
- 10% APS   18 µl
- TEMED   4 µl

Swirl the solution gently to mix the components.

> **Protocol 5** continued
>
> **3** Transfer the gel solution to the casting apparatus with a pipette and bulb.
>
> **4** Position a fluorescent lamp about 3–4 cm from the gel solution and illuminate the solution for about 1 h.
>
> **5** Open the gel cassette or lift the gel from the casting tray to expose the face of the gel. Place the gel with the open face upward and illuminate it with the fluorescent lamp for an additional 30 min.
>
> **6** The gel may be used immediately or it can be covered with plastic wrap and stored at 4 °C for several days. Best results are sometimes obtained when IEF gels are left overnight at 4 °C before use.
>
> [a] To prepare 30%T, 3%C acrylamide solution, dissolve 29.1 g of acrylamide and 0.9 g of bisacrylamide in 72.5 ml of water. The final volume will be 100 ml.
>
> **Caution**: see note regarding acrylamide in Protocol 2.
>
> [b] To prepare 50% glycerol, mix 50 g of glycerol with 63 g of water (100 ml final volume).
>
> [c] To prepare 0.1% FMN, dissolve 50 mg of FMN in 50 ml of water.
>
> [d] Carrier ampholytes are usually supplied at 40% (w/v) concentration, but some are at 20% (w/v).

Ultrathin gels (<0.5 mm) allow the highest field strengths and, therefore, the highest resolution of the analytical methods. Electrofocusing can also be done in tubes, and this configuration once constituted the first dimension of 2D PAGE (54). Because of difficulties in handling and reproducibility with tube gels, IPG strips have largely replaced them.

Good visualization of individual bands generally requires a minimum of 0.5 μg each with dye staining or 50 ng each with silver staining (Section 5). One of the simplest methods for applying samples to thin polyacrylamide gels is to place filter paper strips impregnated with sample directly on the gel surface. Up to 25 μl of sample solution can be conveniently applied after absorption into 1 cm squares of filter paper. A convenient size for applicator papers is 0.2 × 1 cm, holding 5 μl of sample solution. Alternatively, 1–2 μl samples can be placed directly on the surface of the gel. In most cases, IPG strips (which are provided in dehydrated form) are rehydrated in sample-containing solution prior to electrophoresis (55). Rehydration loading allows higher protein loads to be applied to gels than do other methods. It is particularly popular because of its simplicity.

There are no fixed rules regarding the positioning of the sample on the gel. In general, samples should not be applied to areas where they are expected to focus. To protect the proteins from exposure to extreme pH, the samples should not be applied closer than 1 cm from either electrode. Forming the pH gradient before sample application also limits the exposure of proteins to pH extremes.

Precast IEF mini-gels (6 cm long by 8 cm wide and 1 mm thick) are available for carrying out carrier-ampholyte electrofocusing. A selection of IPG sheets is also available for horizontal IEF. Vertical IEF gels have the advantage that the electrophoresis equipment for running them is available in most laboratories and they can hold relatively large sample volumes. Because vertical electro-phoresis cells cannot tolerate very high voltages, this orientation is not capable of the ultrahigh resolution of horizontal cells. To protect the materials of the electrophoresis cells (mainly the gaskets) from caustic electrolytes alternative catholyte and anolyte solutions are substituted in vertical IEF runs. As catholyte, 20 mM arginine, 20 mM lysine is recommended in vertical slab systems (0.34 g arginine free base and 0.36 lysine free base in 100 ml of water). The recom-mended anolyte is 70 mM H_3PO_4, but it can be substituted with 20 mM aspartic acid, 20 mM glutamic acid (0.26 g aspartic acid and 0.29 g glutamic acid in 100 ml of water).

3.3 Power conditions and resolution in isoelectric focusing

The pH gradient and the applied electric field determine the resolution of an IEF run. According to both theory and experiment (1, 47, 51), the difference in p*I* between two resolved adjacent protein IEF bands (Δp*I*) is directly proportional to the square root of the pH gradient and inversely proportional to the square root of the voltage gradient (field strength) at the position of the bands:

$$\Delta pI \propto [(\text{pH gradient})/(\text{voltage gradient})]^{1/2}.$$

Thus, narrow pH ranges and high applied-voltages give high resolution (small Δp*I*) in IEF.

In addition to the effect on resolution, high electric fields also result in shortened run times. However, high voltages in electrophoresis are accompanied by large amounts of generated heat (Joule heating, Section 2.4.1). Thus, there are limitations on the magnitudes of the electric fields that can be applied and the ionic strengths of the solutions used in IEF. Because of their higher surface-to-volume ratio, thin gels are better able to dissipate heat than thick ones and are therefore capable of higher resolution (high voltage). Electric fields used in IEF are generally of the order of 100 V/cm. At focusing, currents drop to nearly zero since the current carriers have stopped moving by then.

3.4 Protein solubilization for isoelectric focusing

A fundamental problem with IEF is that some proteins tend to precipitate at their p*I* values. Carrier ampholytes sometimes help overcome p*I* precipitation and they are usually included in the sample solutions for IPG strips. In addition, non-ionic detergents or urea are often included in IEF runs to minimize protein precipitation.

Urea is a common solubilizing agent in gel electrophoresis. It is particularly useful in IEF, especially for those proteins that tend to aggregate at their p*I*s. Urea disrupts hydrogen bonds and is used in situations in which hydrogen bonding can cause unwanted aggregation or formation of secondary structures that affect mobilities. Dissociation of hydrogen bonds requires high urea concentrations (7–8 M). If complete denaturation of proteins is sought, samples must be treated with a thiol-reducing agent to break disulfide bridges (protein solutions in urea should not be heated above 30 °C to avoid carbamylation).

High concentrations of urea make gels behave as if they had reduced pore sizes. This is because of either viscosity effects or reductions in the effective size of water channels (pores). Urea must be present in the gels during electrophoresis, but, unlike SDS, urea does not affect the intrinsic charge of the sample polypeptides. Urea solutions should be used soon after they are made or treated with a mixed-bed ion exchange resin to avoid protein carbamylation by cyanate in old urea.

Some proteins, especially membrane proteins, require detergent solubilization during isolation. Ionic detergents, such as SDS, are not compatible with IEF, although non-ionic detergents, such as octylglucoside, and zwitterionic detergents, such as 2-[(3-cholamidopropyl)dimethylammonio]-1-propane-sulfonate (CHAPS) and its hydroxyl analogue CHAPSO, can be used. NP-40 and Triton X-100 sometimes perform satisfactorily, but some preparations may contain charged contaminants.

Concentrations of CHAPS and CHAPSO or of octylglucoside of 1–2% in the gel are recommended. Some proteins may require as high as 4% detergent for solubility. Even in the presence of detergents, some samples may have stringent salt requirements. Salt should be present in a sample only if it is an absolute requirement. Carrier ampholytes contribute to the ionic strength of the solution and can help to counteract a lack of salts in a sample. Small samples (1–10 μl) in typical biochemical buffers are usually tolerated, but better results can be obtained with solutions in deionized water, 2% ampholytes, or 1% glycine. Suitable samples can be prepared by dialysis or gel filtration.

Protocol 6

Isoelectric focusing

Equipment and reagents

- Flat-bed electrophoresis cell (e.g. the Bio-Rad Model 111 or the Pharmacia Multiphor)
- Power supply capable of delivering 2–3000 V and 6 W at high voltage (e.g. the Bio-Rad PowerPac 3000 or PowerPac 1000)
- Refrigerated water circulator if required

- 1 N NaOH[a] catholyte if required
- 1 N H$_3$PO$_4$[b] anolyte if required
- Electrode strips[c] if required
- Sample application strips[d]

Method

1   Set up the IEF cell as recommended by the manufacturer. This includes connecting a water circulator, if used, and preparing the cooling platform and electrode strips, if necessary.

2 Place sample application strips on a glass plate and pipette 5 μl of a protein sample to each strip. Place the application strips 1 cm from the anode end of the gel.

3 Position the gel in the IEF cell and make electrode contact as specified for the particular cell.

4 Close the electrophoresis cell and connect the leads to the power supply; the red lead is the anode and the black lead is the cathode.

5 Set the running conditions as recommended by the manufacturer of the electrophoresis cell.

[a] To prepare 1 N NaOH, dissolve 4 g of NaOH in 100 ml of water.

[b] To prepare 1 N H_3PO_4, dissolve 2.3 ml of 85% H_3PO_4 (14.6 M, 44 N) in 97.7 ml of water.

[c] For electrode strips, cut thick filter paper about 7 mm wide and about 4 mm shorter than the gel.

[d] For sample application strips, cut thin filter paper to 0.2 × 1 cm, one strip per sample.

4 Two-dimensional gel electrophoresis

2D PAGE provides the highest resolution separation method for proteins (54, 56). Following a first dimension IEF, proteins are subjected to SDS–PAGE in a perpendicular direction. The technique is a true orthogonal procedure in that the two separation mechanisms are based on different physical principles (they are orthogonal in that sense) and the two separations are done at right angles to one another (they are geometrically orthogonal). Thousands of polypeptides can be resolved in a single 2D PAGE slab gel. The technique works best with soluble proteins such as those from serum or cytoplasm. It is relatively labour-intensive for an electrophoresis technique, requiring a relatively high skill level for best results.

The best approach for 2D PAGE is to run the IEF first dimension using IPG strips, and the best approach to obtaining IPG strips is to purchase them already made. Following IEF, an IPG strip is inserted into the gel cassette on top of the SDS–PAGE slab gel. The SDS–PAGE gel is run and stained as with one-dimensional electrophoresis. The difference between 1D PAGE and 2D PAGE gels is that the protein patterns in 2D PAGE are spots rather than bands (Figure 10).

A large part of the success of a 2D PAGE run is determined by careful sample preparation. This topic and several of the nuances of 2D PAGE are outside the scope of this chapter. Those interested should consult refs 54–61.

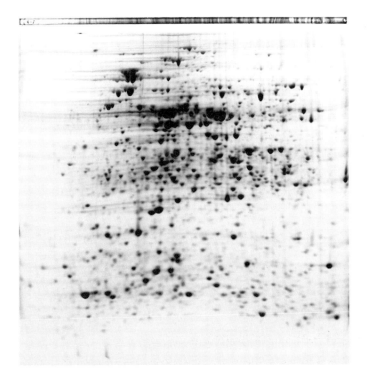

Figure 10 Two-dimensional polyacrylamide gel electrophoresis. Proteins from a lysate of *Escherichia coli* were subjected to IEF in a 17 cm IPG strip spanning the pH range of 4–7. The strip containing focused proteins was transferred to an 18 × 20 cm SDS–PAGE gel (8–16%T) and subjected to electrophoresis. Proteins in the gel were stained with SYPRO Ruby Gel Stain and the image shown was captured with a laser-based instrument. A second IPG strip was run in parallel and stained with colloidal CBB G-250. Its image is superimposed above that of the 2D PAGE gel for comparison.

Protocol 7
Two-dimensional polyacrylamide gel electrophoresis

Equipment and reagents

- Isoelectric focusing apparatus appropriate for use with IPGs (e.g. Bio-Rad PROTEAN IEF cell)
- Electrophoresis cell (e.g. Bio-Rad Mini-PROTEAN 3 or PROTEAN II cell)
- Power supply (e.g. Bio-Rad PowerPac 300)
- IPG trays or disposable 2 ml and 5 ml pipettes for treating IPG strips
- IPG strips in suitable pH range and length to fit the second dimension gel

- SDS–PAGE gel: commercial precast or made according to Protocol 2 (the well in the gel must match the length of the IPG strip)
- IPG sample solution[a]
- IPG equilibration solution[b]
- 0.5% agarose in Laemmli SDS–PAGE electrode buffer[c]

Protocol 7 continued

Method

1 Mix the sample proteins with IPG sample solution so that the proteins are at a final concentration of about 0.1–1 mg/ml, depending on the staining method to be used. The pH range of the carrier ampholytes in the sample solution should match the pH of the IPG strip. Concentrations of carrier ampholytes exceeding 0.2% (w/v) result in extended focusing times.

2 Carefully peel the protective plastic sheet from the (dehydrated) IPG strip. Rehydrate the IPG in protein solution (step 1) with the gel facing down in a tray or 2 ml disposable pipette. See the recommendations of the manufacturer of the IPG strip for the correct volume of solution to use in order to rehydrate the IPG strip properly. As a guide, 125 µl fully rehydrates a 7 cm strip, 200 µl rehydrates an 11 cm strip, and 300 µl rehydrates a 17 cm strip. After 1 h in a tray, cover the strip with light silicon oil. If a pipette is used rather than a tray, close off both of its ends with Parafilm. Allow at least 12 h for thorough rehydration. Use a fine-point forceps for all manipulations of IPG strips.

3 Transfer the IPG strip to the IEF cell and carry out the IEF according to the instructions provided by the manufacturer of the cell. The strip must be covered with light silicon oil during focusing. It might take some trial and error to arrive at proper focusing conditions for each different protein sample. Often 20–30 kVh (kilovolt hours) are sufficient for 7 cm IPG strips, 40–50 kVh for 11 cm strips, and 60–70 kVh for 17 cm strips.

4 Thaw one 10 ml aliquot of IPG protein equilibration solution for each IPG strip. Make sure that all components of the solution are thoroughly dissolved. Gentle warming may be required. Divide the aliquot into two 5 ml portions. To one of the 5 ml portions, add 50 mg of DTT (to 1%, or 65 mM). Add 75 mg of IAA to the other 5 ml portion (to 1.5% or 80 mM). Carry out this step while the IEF run is still in progress so that the IPG strip will not be subjected to drying out.

5 After IEF, place the strip with the plastic side down on a dry piece of filter paper and gently blot excess oil and other liquid from the gel with a piece of damp filter paper.

6 Transfer the IPG strip to a tray or 5 ml disposable pipette and incubate it with the 5 ml DTT solution prepared in step 4 for 15 min at room temperature with gentle rocking. This reduces disulfide bonds in the proteins in the IPG strip.

7 Remove the IPG strip from the reduction solution and blot off excess liquid as in step 5. Transfer the strip to the 5 ml IAA solution prepared in step 4 in a clean trough in a tray or a fresh pipette. Incubate the strip for 15 min at room temperature with gentle rocking to alkylate free sulfhydryl groups in the proteins in the IPG strip.

8 Remove the IPG strip from the alkylation solution and blot off excess liquid as in step 5. Transfer the strip to the top of the gel cassette holding the second dimension

gel. Use the forceps to place the plastic backing of the strip on the back plate of the gel cassette. Use a spatula to push the strip into the cassette above the gel so that it is within 1 mm of the bottom of the well.

9 Melt the agarose and let it cool until it is warm to the touch but still molten. Pipette the molten agarose into the gel cassette to seal the IPG strip to the gel. Hold the cassette at a slight angle as the agarose is being pipetted into it to provide a way for bubbles to escape. There must not be any bubbles trapped in the thin agarose layer between the IPG strip and the top of the gel. It may take some practice to become proficient in layering the agarose.

10 When the agarose has hardened, scrape excess agarose from the top of the cassette and transfer the cassette to the electrophoresis cell. Run the second dimension as for standard SDS–PAGE. Stain the completed gel as for SDS–PAGE.

[a] IPG sample solution is 8 M urea, 2% (w/v) CHAPS, 0.3% (w/v) dithiothreitol (DTT) (20 mM), 0.2% (w/v) carrier ampholytes. Dissolve 48 g of urea, 2 g of CHAPS, 0.3 g of DTT, and 0.5 ml of carrier ampholytes (assumed to be at 40%, w/v) in 60 ml of water. Adjust the final volume to 100 ml if necessary. Aliquot and freeze 5 ml portions. Mix this solution well upon thawing it for use so that all of the urea and CHAPS dissolve. Gentle warming may be required.

[b] IPG equilibration solution is 6 M urea, 75 mM Tris–HCl pH 8.8, 2% (w/v) SDS. Dissolve 36 g of urea, 5 ml of 1.5 M Tris–HCl pH 8.8 (Protocol 2), 2 g of SDS in 60 ml of water. Adjust the final volume to 100 ml. Aliquot and freeze 10 ml portions. Mix this solution well upon thawing it for use so that all of the urea and SDS dissolve.

[c] To prepare 0.5% agarose solution, add 0.5 g of low melting point agarose to 100 ml of Laemmli SDS–PAGE electrode buffer (see Table 2). Add a spatula-tip amount of bromophenol blue to impart colour. Melt the agarose on a hot plate or in a microwave oven. Mix the solution well and store the resultant gel at room temperature. The agarose must be remelted before each use.

5 Detection of proteins in gels

Gels are run for either analytical or preparative purposes. The intended use of the gel imposes restrictions on the amount of protein to be loaded and the means of detection. At one time, radioactive labelling of proteins was popular with detection of electrophoretic bands done by autoradiography. It is now more common to make protein bands in gels visible by staining them with dyes or metals (62–65). Each type of protein stain has its own characteristics and limitations with regard to the sensitivity of detection and the types of proteins that take up the stain the best (Table 4).

If the purpose of gel electrophoresis is to identify low-abundance proteins (e.g. low copy number proteins in a cell extract or contaminants in a purification scheme), then a high protein load (0.1–1 mg/ml) and a high sensitivity stain such as silver or fluorescence should be used. When the intention is to obtain enough protein for use as an antigen or for sequence analysis, then a high protein load

Table 4 Comparison of stains for proteins in gels

Stain[a]	Sensitivity (ng)[b]	Steps[c]	Time[d]	Gel types[e]
Coomassie stains				
CBB R-250	36–47	2	2.5 h	1D and 2D
Colloidal CBB G-250	8–28	3	2.5 h	1D, 2D, IPG, and blots
Silver stains				
Silver Stain Plus[f]	0.6–1.2	3	90 min	1D and 2D
Silver stain[g]	0.6–1.2	7	2 h	1D and 2D
Negative stains				
Copper stain[h]	6–12	3	10 min	1D
Zinc stain[i]	6–12	3	15 min	1D
Fluorescent stains				
SYPRO[j] Ruby (gel)	1–10	2	3 h	1D and 2D
SYPRO[j] Orange	4–8	1	45 min	1D
IEF stains				
SYPRO[j] Ruby (IEF)	2–8	2	2 h[k]	IEF and IPG
IEF stain	40–50	2	3 h	IEF and IPG
Blotting stains				
SYPRO[j] Ruby (blot)	2–8	3	50 min	NC and PVDF[l]
Colloidal Gold	1	3	2 h	NC and PVDF
Enhanced Colloidal Gold	0.01–0.1	4	3 h	NC and PVDF

[a] All stains listed are available from Bio-Rad in ready-to-use form.

[b] Sensitivities were determined with known masses of proteins run in 1D gels or blots.

[c] Minimum number of 'hands on' steps.

[d] Estimated staining time. Actual manipulations require considerably less time.

[e] Recommended types of gels. The stains may work for other types of gels as well.

[f] Silver staining kit based on the method of Gottlieb and Chavko (ref. 72).

[g] Silver staining kit based on the method of Merril et al. (ref. 71).

[h] Ref. 73.

[i] Ref. 74.

[j] SYPRO is a trademark of Molecular Probes, Inc., Eugene, Oregon.

[k] SYPRO Ruby IEF stain requires an overnight incubation.

[l] NC: nitrocellulose membrane. PVDF: polyvinylidene fluoride membrane.

should be applied to the gel and the proteins visualized with a staining procedure that does not fix the proteins in the gel. Quantitative comparisons require the use of stains with broad linear ranges of detection response.

The sensitivity that is achievable in staining is determined by:

(a) The amount of stain that binds to the proteins.

(b) The intensity of the coloration.

(c) The difference in coloration between stained proteins and the residual, background coloration in the body of the gel (signal-to-noise ratio). Unbound stain molecules can be washed out of the gels without removing much stain from the proteins.

Carry out all steps in gel staining at room temperature with gentle agitation (e.g. on an orbital shaker platform) in any convenient container, such as a glass casserole or a photography tray. Always wear gloves when staining gels, because fingerprints (and fingers) will stain. Permanent records of stained gels can be obtained by photographing them, drying them with the appropriate apparatus on filter paper or between sheets of cellophane, or by capturing electronic images of them.

All stains interact differently with different proteins. No stain is universal in that it will stain all proteins in a gel proportionally to their quantities. The only observation that seems to hold for most of the positive stains is that they interact best with basic amino acids. For critical analyses, replicate gels should be stained with two or more different kinds of positive stain. Of all the stains available, colloidal Coomassie Blue appears to stain the broadest spectrum of proteins. It is instructive, especially with 2D PAGE gels, to follow a colloidal Coomassie Blue stained gel with silver staining (66, 67) or to follow a fluorescence stain with colloidal Coomassie Blue or silver. Very often, this 'double staining' procedure will show a few differences in the two protein patterns. It is most common to stain gels first with Coomassie Blue or a fluorescent stain, then following that with a silver stain. However, the order in which the stains are used does not seem to be important (68, 69).

5.1 Dye staining

Coomassie Brilliant Blue R-250 (CBB R-250) is the standard stain for protein detection in polyacrylamide gels. It and the G-250 variety (CBB G-250) are wool dyes that have been adapted to the staining of proteins in gels. The 'R' and 'G' designations signify red and green hues, respectively. Easy visibility requires on the order of 0.1–1 μg of protein per band. The staining solution consists of 0.1% CBB R-250 (w/v) in 40% methanol (v/v), 10% acetic acid (v/v), which also fixes most proteins in gels (see Protocol 8). Absolute sensitivity and staining linearity depend on the proteins being stained.

Protocol 8

Gel staining with Coomassie Brilliant Blue R-250

Equipment and reagents

- Glass or plastic containers large enough to hold the gels to be stained
- Orbital shaker platform
- Staining solution[a]
- Destaining solution[b]

Method

1   Remove the gel from the cassette. It will most likely stick to one of the plates of the cassette. Invert the plate with the gel over a volume of staining solution and float the gel off of the plate into the staining solution. Several gels can be stained in the same container. They need only be covered with stain.

Protocol 8 continued

2   Place the staining container on the orbital shaker platform and agitate it gently for 30–60 min. Make sure that the staining solution reaches all parts of all the gels.

3   Pour off the staining solution and replace it with an excess of destaining solution. The staining solution can be reused several times.

4   Soak the gel in destaining solution under gentle agitation. Change the destaining solution several times, until the background has been satisfactorily removed.

5   The stained gel can be stored in water.

[a] Staining solution is 0.1% Coomassie Brilliant Blue R-250 (CBB R-250) in 40% methanol, 10% acetic acid. Mix 500 ml of water, 400 ml of methanol, and 100 ml of acetic acid, then add 1 g of CBB R-250. Filter the staining solution after the dye has dissolved. This solution can be purchased from Bio-Rad. Store it at room temperature.

[b] Destaining solution is 40% methanol, 10% acetic acid. Mix 500 ml of water, 400 ml of methanol, and 100 ml of acetic acid. Store the destaining solution at room temperature.

CBB G-250 is less soluble than the R-250 variety. In acidic solutions it forms colloidal particles that are too large to penetrate surface gel pores and can be formulated into a staining solution that requires little or no destaining. An environmentally safe formulation of CBB G-250 is available from Bio-Rad Laboratories under the name of 'Bio-Safe'. This stain is somewhat more sensitive than CBB R-250, in part because of increased signal-to-noise ratios since the bulk of the gel matrix does not pick up excess stain. Staining with colloidal CBB G-250 can be linear over two orders of magnitude of protein concentration for some proteins. Gels containing low molecular weight polypeptides ($\approx$1000 Da) can be stained with CBB G-250 with minimum fixation time (and minimal potential loss of material). The staining procedure is given in Protocol 9. IPGs can also be stained with colloidal CBB G-250 after IEF. To do this, blot excess oil from the strips and put them directly into the staining solution. Wash the strips with water after an hour of staining.

Protocol 9

Gel staining with colloidal Coomassie Brilliant Blue G-250

Equipment and reagents

• Staining container and shaker platform

• Bio-Safe™ Coomassie (Bio-Rad)

Method

1   Remove the gel from the cassette and wash it in an excess of water (see Protocol 8, step 1).

Protocol 9 continued

2   Wash the gel three times for 5 min each in water with gentle agitation.

3   Remove all the water from the staining container and add enough Bio-Safe stain solution to cover the gel. More than one gel can be stained in the same container.

4   Agitate gently for 1 h. If more than one gel is stained in the same container, make sure that the agitation is sufficient to ensure good coverage of all the gels. Protein bands will be visible within 20 min and reach maximum intensity in about 1 h. Longer incubations will not increase the background.

5   Wash the gel in an excess of water for at least 30 min. The water wash intensifies the band colour.

6   Stained gels can be stored in water.

7   For gels containing peptides such as Tricine SDS–PAGE gels (Protocol 3), fix them in 40% methanol, 10% acetic acid (Protocol 8) for 30 min. Then stain starting with step 3 above. Extend the water wash (step 5) to at least 2 h. Peptide bands may not be clearly visible until after the final water wash.

5.2 Silver staining

There are a number of different silver staining methods. Some are available in kit form from various manufactures. Others do not lend themselves to commercial kits. For a discussion of the mechanisms of silver staining see ref. 70. Merril and co-workers (71) developed a silver staining method that is available from Bio-Rad. It can be as much as 100 times more sensitive than CBB dye staining and allows visualization of heavily glycosylated proteins in gels. Bands containing 10–100 ng of protein can be easily seen. Proteins in gels are fixed with a solution containing methanol and acetic acid and the methanol is subsequently replaced with ethanol. Proteins are then oxidized in a solution of potassium dichromate in dilute nitric acid, washed with water, and treated with silver nitrate solution. Silver ions bind to the oxidized proteins and are subsequently reduced to metallic silver by treatment with alkaline formaldehyde. Colour development is stopped with acetic acid when the desired staining intensity has been achieved.

Protocol 10

Gel staining with Merril's silver stain[a]

Equipment and reagents

- Clean staining tray
- Orbital shaker platform
- Silver stain kit (Bio-Rad) or components identified below

Protocol 10 continued

Method

Reaction times vary with the thicknesses of the gels.

1 Fix the gel for from 60 min to overnight in enough 40% methanol, 10% acetic acid (v/v) to completely cover it. More than one gel can be fixed and stained in the same container. Agitation must be sufficient to allow good coverage of all gels.

2 Fix the gel twice in 10% ethanol, 5% acetic acid (v/v) for about 30 min each.

3 Soak the gel for 5–10 min in fresh oxidizer solution (0.0034 M potassium dichromate, 0.0032 N nitric acid).

4 Wash the gel three of four times with water for 5–10 min each, until the yellow colour has been washed out.

5 Soak the gel for 20–30 min in fresh silver reagent (0.012 M silver nitrate).

6 Wash the gel with water for 1–2 min.

7 Wash the gel for about 1 min in developer solution (0.28 M sodium carbonate, 0.0185% paraformaldehyde).

8 Replace the developer with fresh solution and incubate for 5 min.

9 Replace the developer a second time and allow development to continue until satisfactory staining has been achieved.

10 To stop development, pour off the developer and replace it with 5% acetic acid (v/v).

11 Stained gels can be stored in water.

[a] See ref. 71.

Note: Silvered staining containers can be cleaned with 50% nitric acid.

Another method that requires only one simultaneous staining and development step is that of Gottlieb and Chavko (72) (available from Bio-Rad as Silver Stain Plus). Proteins are fixed with a solution containing methanol, acetic acid, and glycerol and washed extensively with water. The gels are then soaked in a sol containing a silver–ammine complex bound to colloidal tungstosilicic acid. Silver ions transfer from the tungstosilicic acid to the proteins in the gel by means of an ion exchange or an electrophilic process. Formaldehyde in the alkaline solution reduces the silver ions to metallic silver to produce the images of the bands of macromolecules. The reaction is stopped with acetic acid when the desired intensity has been achieved. Because silver ions do not accumulate in the bodies of gels, background staining is light. Since this method lacks an oxidizing step, visualization of heavily glycosylated proteins and lipoproteins can be less sensitive than with the Merril stain.

Protocol 11

Gel staining with Silver Stain Plus[a]

Equipment and reagents

- Clean staining tray and shaker platform
- Silver Stain Plus Kit (Bio-Rad)

Method

1   Place the gel in Fixative Enhancer Solution (50% methanol, 10% acetic acid, 10% Fixative Enhancer Concentrate, 30% water). Use about 200 ml of Fixative Enhancer solution per gel. Fix for 20 min to overnight. Agitate gently. More than one gel can be stained per container.

2   Rinse the gel for 10 min with water. Repeat twice (three washes total).

3   During fixation or washing, prepare Development Accelerator Solution. Dissolve 5 g of Development Accelerator Reagent[b] per 100 ml water.

4   Place 35 ml of water in a beaker that can hold >100 ml and begin stirring. Add the following to the beaker in order: 5 ml Silver Complex Solution,[c] 5 ml Reduction Moderator Solution,[d] 5 ml Image Development Reagent.[e]

5   Immediately before use quickly add 50 ml of Development Accelerator Solution to the beaker (step 4). Stir well. Add the contents of the beaker to the staining tray.

6   Stain with gentle agitation for about 20 min or until the desired staining intensity is achieved. It may take 15 min before bands become visible.

7   To stop the staining reaction, pour off the staining solution and replace it with 5% acetic acid.

8   Stained gels can be stored in water.

[a] Based on the method of ref. 72.

[b] Development Accelerator Reagent contains Na_2CO_3.

[c] Silver Complex Solution contains NH_4NO_3 and $AgNO_3$.

[d] Reduction Moderator Solution contains tungstosilicic acid.

[e] Image Development Reagent contains formaldehyde.

Note: Silvered staining containers can be cleaned with 50% nitric acid.

5.3 Copper and zinc staining

Rapid negative staining of SDS–PAGE gels is achieved by incubating them for a short time in a copper chloride solution, then washing them with water (73). Blue–green precipitates of copper hydroxide form in the bodies of the gels except where there are high concentrations of SDS, such as that bound to the proteins. Clear protein bands can be easily seen against the blue–green backgrounds and photographed with the gels on black surfaces. Proteins are not permanently fixed by this method and can be quantitatively eluted after chelating the copper.

Protocol 12

Gel staining with copper[a]

Equipment and reagents

- Clean staining tray and shaker platform
- 10 × Copper Stain (Bio-Rad)[b]

Method

1   Remove the gel from the cassette and wash it briefly in water.

2   Dilute the 10 × Copper Stain ten-fold (to 1 ×).

3   Soak the gel in 1 × Copper Stain (0.3 M CuCl$_2$) for 5 min.

4   Wash the gel for 2–3 min in water.

5   The stained gel can be stored in water.

6   Visualize the bands by placing the gel on a black surface.

7   Excised bands can be destained with Tris–glycine buffer.

[a] See ref. 73.

[b] 10 × Copper Stain is 3 M CuCl$_2$. To make it, dissolve 51.2 g of CuCl$_2$·2H$_2$O in 100 ml of water.

A method using the combination of zinc and imidazole produces similar, negatively stained SDS–PAGE gels (74). Zinc imidazolate precipitates in gels except at the sites where precipitation is inhibited by SDS–protein bands (75). The resultant gels are opaque white with clear regions at the sites of the protein bands.

Protocol 13

Gel staining with zinc–imidazole[a]

Equipment and reagents

- Clean staining tray and orbital shaker
- Zinc Stain (Bio-Rad).

Method

1   Remove the gel from the cassette and wash it briefly with water.

2   Mix 1 part of Solution A[b] with 9 parts of water.

3   Mix 1 part of Solution B[c] with 9 parts of water.

4   Soak the gel for 10 min in diluted Solution A (step 2).

5   Immerse the gel in diluted Solution B (step 3) for 30–60 sec until the gel background becomes white.

6   Place the gel in water. The stained gel can be stored in water.

Protocol 13 continued

7   Visualize the bands with the gel on a black surface.

8   Excised bands can be destained with Tris–glycine buffer.

[a] See ref. 74.

[b] Solution A is 2 M imidazole. To make it, dissolve 13.6 g of imidazole in 100 ml of water.

[c] Solution B is 2 M zinc sulfate. To make it, dissolve 57.5 g of $ZnSO_4 \cdot 7H_2O$ in 100 ml of water.

With both the copper and zinc methods, the resultant negatively stained images of the electrophoresis patterns are intermediate in sensitivity between the Coomassie Brilliant Blue dyes and silver staining. The electrophoretic pattern is lost when copper or zinc stained gels are dried, so they must be photographed, re-stained with CBB, or stored in water. Neither of the negative staining methods works well for 2D PAGE gels because they do not give good discrimination of closely clustered spots. For preparative work (see Section 8.1), excess copper or zinc is removed by incubating gel slices in three changes (for 10 min each) of 0.25 M EDTA, 0.25 M Tris–HCl pH 9, or in Tris–glycine electrophoresis buffer.

5.4 Fluorescent stains

The rare earth chelate stains have desirable features that make them popular in high-throughput laboratories (76). They are end-point stains with little background staining (high signal-to-noise characteristics), and they are sensitive and easy to use. The rare earth chelate compounds possess three distinct domains. One domain binds a rare earth ion, such as ruthenium. A chromophoric domain is responsible for detection of the rare earth ion. And a third domain reversibly binds to proteins. Since these compounds are fluorescent, they require an imaging device capable of providing high intensity illumination at the excitation wavelength, bandpass filters for excitation and emission wavelengths, and a detector such as a photographic or CCD camera. Sensitivity varies from protein to protein but can exceed that of silver stain. Linearity can extend to three orders of magnitude. The most popular types of fluorescent protein stains are the SYPRO™ class of compounds. SYPRO Orange is recommended for 1D SDS–PAGE with SYPRO Ruby used for 1D and 2D SDS–PAGE and for native gels. (SYPRO is a trademark of Molecular Probes, Inc., Eugene, Oregon 97402, USA.)

Protocol 14

Staining with SYPRO™ Ruby Gel Stain

Equipment and reagents

- Polypropylene or polyethylene staining tray[a]
- Shaker platform
- Fluorescence imager (see Section 6)
- SYPRO Ruby Gel Stain (Bio-Rad)

Protocol 14 continued

Method

1 Remove the gel from the cassette and wash it for 30 min in 10% methanol, 7% acetic acid[b] under gentle agitation. Use a plastic (polypropylene or polyethylene) container.

2 Replace the wash solution with SYPRO Ruby Gel Stain. Use a volume of stain that is about 10 times the volume of the gel. Do not stain more than two gels in one tray.

3 Stain the gel with gentle agitation for a minimum of 3 h. For convenience, gels may be left in the stain solution overnight (16–18 h) without overstaining.

4 Rinse the gel for 30–60 min in 10% methanol, 7% acetic acid.[b] This step is optional. It can be valuable in removing 'speckles' that sometimes form on the surfaces of gels when dust particles provide nucleation sites for dye molecules.

5 Stained gels can be stored in water under a minimum of light.

6 Do not reuse the stain solution.

7 View the gel under in a fluorescence imager. SYPRO Ruby has two excitation peaks of 300 nm and 480 nm and an emission peak at 618 nm.

SYPRO Ruby is a trademark of Molecular Probes, Inc., Eugene, OR, the manufacturer of the dye.

[a] Glass staining containers are not recommended.

[b] Methanol at concentrations up to 40% and acetic acid at concentrations up to 10% can be used.

5.5 Detection of proteins in isoelectric focusing gels

The standard staining solution for proteins in IEF gels uses a combination of CBB R-250 and Crocein Scarlet in an ethanol–acetic acid solution containing cupric sulfate. The Crocein Scarlet binds rapidly to proteins and helps fix them in the large-pore IEF gels (77). The cupric sulfate enhances stain intensity (78). The procedures of staining and destaining are similar to those for CBB R-250 alone. IEF gels can also be silver stained for increased detection sensitivity.

Protocol 15 °

Staining IEF gels with CBB R-250

Equipment and reagents

- Staining tray and shaker platform
- IEF staining solution[a]
- IEF destaining solution[b]

Method

1 Remove the gel from the cassette and soak it for at least 1 h in IEF staining solution.

Protocol 15 continued

2 Destain the gel with a large excess of IEF destaining solution until a clear background is obtained. This will require several changes of destaining solution.

3 Soak the gel for 1 h in 12% ethanol, 7% acetic acid.[c]

4 The stained gel can be stored in water.

[a] IEF staining solution is 0.04% CBB R-250, 0.05% Crocein Scarlet, 0.5% $CuSO_4$ in 27% ethanol, 10% acetic acid. To make it dissolve 5 g of $CuSO_4$ in 630 ml of water. Then add 270 ml of ethanol and 100 ml of acetic acid. Dissolve 0.4 g of CBB R-250 and 0.5 g of Crocein Scarlet in the solution and filter it. This solution can be purchased from Bio-Rad.

[b] IEF destaining solution is 12% ethanol, 7% acetic acid, 0.5% $CuSO_4$. To make it dissolve 5 g of $CuSO_4$ in 810 ml of water and add 120 ml of ethanol and 70 ml of acetic acid.

[c] The wash in 12% ethanol, 7% acetic acid (810 ml water, 120 ml ethanol, 70 ml acetic acid) is to remove the cupric sulfate.

An easy way to stain IPGs is to immerse them for 1 h in colloidal CBB G-250 followed by two 10 min water washes. Some silver stains will turn the plastic backing sheets of IPGs into mirrors.

There is also a version of SYPRO Ruby Stain specifically formulated for use with both carrier ampholyte and IPG IEF gels.

Protocol 16

Staining IEF gels with SYPRO Ruby IEF Stain

Equipment and reagents

- Fluorescence imager (see Section 6)
- Polypropylene or polyethylene staining tray[a]
- Shaker platform
- SYPRO Ruby IEF Stain (Bio-Rad)

Method

1 Remove the gel from the cassette and soak it for 30 min in 10% methanol, 7% acetic acid[b] under gentle agitation.

2 Soak the gel overnight (12–18 h) in a volume of SYPRO Ruby IEF Stain that is about 10 times the volume of the gel. Do not stain more than two gels per container.

3 Wash the gel in water four times for 30 min each.

4 Stained gels can be stored in water with minimum light.

5 View the gel under in a fluorescence imager. SYPRO Ruby has two excitation peaks of 300 nm and 480 nm and an emission peak at 618 nm.

SYPRO is a trademark of Molecular Probes, Inc., Eugene, OR, the manufacturer of the dye.

[a] Glass staining containers are not recommended.

[b] Methanol concentrations of up to 40% and acetic acid concentration of up to 10% can be used.

6 Image acquisition and analysis

Once gels have been stained, it is often sufficient to examine them visually. The requirements of some types of experiments are satisfied by simple comparisons of the band patterns in relevant lanes in gels. In other situations, migration distances of selected proteins can be measured by hand with a ruler and standard curves drawn on graph paper. Simple analyses like these are usually done on the wet gels. Sometimes it is advantageous to photograph a stained gel and use the photographic image for measurements. In regards to archiving gel information, it is easier to tape photographs of gel images to notebook pages than to store wet gels or gels dried on cellophane or filter paper.

In laboratories where large numbers of gels are run on a routine basis, digital methods for image acquisition and data analysis have replaced wet gels and photographs (79, 80). Several types of imaging systems and associated software are commercially available for analysing gels stained with just about any kind of stain. These instruments greatly simplify data acquisition and analysis and the archiving of gel patterns.

The three categories of image acquisition devices used in with electrophoresis gels are:

- document scanners
- charge-coupled device (CCD) cameras
- laser-based detectors

Document scanners as configured for densitometry are for measurements on gels stained with one of the coloured materials, CBB or silver. They operate in visible light illumination, 400–750 nm, with dynamic ranges extending to 3 OD. The linear-array CCD detectors used with the better densitometers can distinguish adjacent features that are separated by 50 μm or greater (spatial resolution), which is more than adequate for most gel applications.

The better CCD cameras are cooled to increase their signal-to-noise ratios. They operate with illumination provided by either light boxes (UV or visible) for transmittance measurements or overhead lamps for epi-illumination. These devices are very versatile and they can acquire images from gels stained with coloured or fluorescent dyes or silver. The epi-illumination feature allows CCD cameras to capture images of blots on opaque membranes (see Section 7). The spatial resolution obtainable with the cameras is entirely dependent on the properties of the lenses used and the area being imaged, but is generally in the 100–200 μm range. Their dynamic ranges for quantification often exceed four orders of magnitude.

Laser devices are the most sophisticated image acquisition tools. They are particularly useful for gels labelled with fluorescent dyes, since the lasers can be matched to the excitation wavelengths of the fluorophores. Detection is generally with photomultiplier tubes. Some instruments incorporate storage phosphor screens for detection of radiolabelled and chemiluminescent compounds (not discussed). Resolution depends on the scanning speed of the illumination module and can be as low as 50 μm.

An imager is the most significant investment of all electrophoresis apparatus. As with all significant purchases, comparison shopping among the available products is highly recommended. In practice, researchers access the data in their gels through the analysis software and it is the software that should be the primary consideration in any imaging system. Good software will be able to use data from most imaging devices. However, dedicated software designed for use with particular instruments provides the desirable feature of controlling the imagers with the software.

Software for 1D gel analysis (Figure 11A) defines lanes and bands, quantifies bands, constructs standard curves, and determines molecular weights. Images can be adjusted for contrast, processed in various ways, annotated, and exported to other files for publication or document control. 2D analysis software (Figure 11B) defines and quantifies spots in 2D PAGE gels. Those programs that use Gaussian spot modelling are better able to quantify proteins in overlapping spots than are the programs that define spots by contours. Programs for 2D analysis include statistical software designed for quantitative comparisons of large numbers of gels. The programs are also set up for analysis of spot patterns derived from differentially expressed proteins and some can query databases to assist in protein identifications. They also can be used for image adjustments, annotation, and export in a variety of file forms.

7 Blotting

Certain synthetic membranes bind proteins tightly enough that they can be used as supports for solid-phase immunoassays. Bound proteins retain their anti-genicity and are accessible to probes. Several techniques have been developed for probing proteins bound to synthetic membranes. They are collectively known as 'blots'. Only the most common blotting technique is discussed here. In this technique, proteins are transferred from an electrophoresis gel to a support membrane and then probed with antibodies. This technique is called 'immuno-blotting' or, more popularly, 'Western blotting'. It combines the resolution of PAGE (1D or 2D) with the specificity of immunoassays allowing individual proteins in complex mixtures to be detected and analysed (81–84). A discussion of Western blotting is appropriate for this chapter because it complements electrophoretic separations and uses electrophoresis to transfer proteins from gels to membranes.

7.1 Immunoblotting

The immunoblotting procedure is as follows.

(a) Proteins are transferred from an electrophoresis gel to a membrane surface. The transferred proteins become immobilized on the surface of the mem-brane in a pattern that is an exact replica of the gel.

(A)

(B)

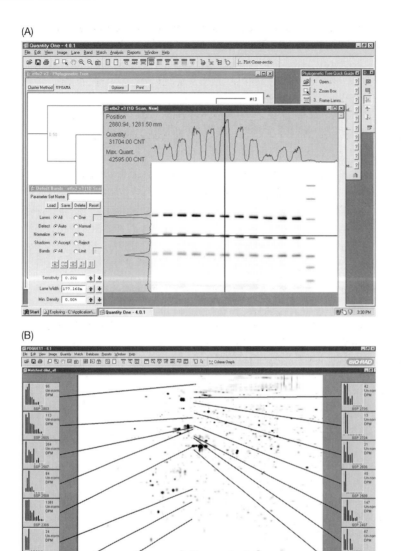

Figure 11 Image analysis software for 1D and 2D gels. (A) A 1D analysis software window is shown. Densitometric analyses of the column and row identified by the crosshairs appear above and at the left of the gel image. (B) In this 2D software window, bar graph annotations are shown representing the intensity patterns of selected protein spots in several related gels. This type of analysis is useful in comparing differing test samples of proteins.

(b) Unoccupied protein-binding sites on the membrane are saturated to prevent non-specific binding of antibodies. This step is called either 'blocking' or 'quenching'.

(c) The blot is probed for the protein of interest with a specific primary antibody.

(d) The blot is probed a second time. The second probe is an antibody that is specific for the primary antibody type and is conjugated to a detectable enzyme. The site of the protein of interest is thus tagged with an enzyme through the specificities of the primary and secondary antibodies.

(e) Enzyme substrates that are converted into insoluble, detectable products are incubated with the blot. The products leave a coloured trace at the site of the band or spot representing the protein of interest (Figure 12).

7.1.1 Apparatus for blotting

Electrotransfer from a gel to a membrane is done by directing an electric field across the thickness of the gel to drive proteins out of the gel and on to the membrane. There are two types of apparatus for electrotransfer: (A) buffer-filled tanks and (B) 'semidry' transfer devices (Figure 13).

Transfer tanks are made of plastic with two electrodes mounted near opposing tank walls. A non-conductive cassette holds the membrane in close contact with the gel. The cassette assembly is placed vertically into the tank parallel to the electrodes and submerged in electrophoresis buffer. A large volume of buffer in the tank dissipates the heat generated during the transfer.

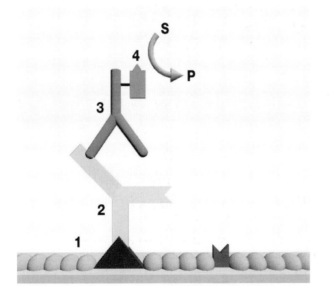

Figure 12 Specific enzymatic immunodetection of a blotted protein. Depicted are blocked binding sites on the membrane (1), a primary antibody (2) specifically bound to an antigenic protein, and a secondary antibody (3) bound to the primary antibody. The secondary antibody is conjugated to a reporter enzyme (4). Substrate (S) is converted into insoluble product (P).

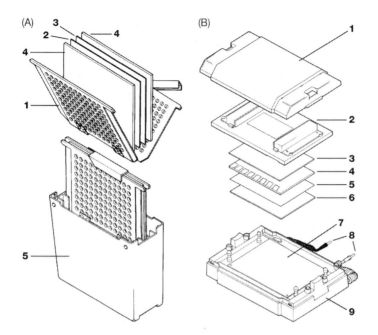

Figure 13 Two types of electrotransfer apparatus. (A) A tank transfer cell is shown in an exploded view. The cassette (1) holds the gel (2) and membrane (3) between buffer-saturated filter paper and fibre pads (4). The cassette is inserted vertically in the buffer-filled tank between the positive and negative electrodes (not shown). A lid with connectors and leads for applying electrical power is not shown. (B) An exploded view of a semidry transfer unit is shown. The gel (4) and membrane (5) are sandwiched between buffer-saturated stacks of filter paper (3 and 6) and placed between the cathode assembly (2) and anode plate (7). The safety lid (1) attaches to the base (9). Power is applied through cables (8).

In semidry blotting, the gel and membrane are sandwiched horizontally between two stacks of buffer-wetted filter papers in direct contact with two closely spaced solid plate electrodes. The close spacing of the semidry apparatus provides for high field strengths. The term 'semidry' refers to the limited amount of buffer that is confined to the stacks of filter paper.

Tanks rather than semidry apparatus should be used for most routine work. With tanks, transfers are somewhat more efficient than with semidry devices. Under semidry electrotransfer conditions, some low molecular weight proteins are driven through the membranes and because low buffer capacity limits run times, some high molecular weight proteins are poorly transferred.

7.1.2 Membranes and buffers for immunoblotting

The two membranes most used for protein work are nitrocellulose and polyvinylidene fluoride (PVDF). Both bind proteins at about 100 μg/cm^2. Nitrocellulose is the best membrane to use in the initial stages of an experiment. PVDF is used when proteins are to be sequenced. It can withstand the harsh chemicals of protein sequenators, whereas nitrocellulose cannot.

Tank transfers from SDS–PAGE gels are done in modified electrophoresis buffer, 25 mM Tris, 192 mM glycine, 20% (v/v) methanol pH 8.3. With semidry transfers from SDS–PAGE gels, the buffer is 48 mM Tris, 39 mM glycine, 20% methanol pH 9. The methanol in the buffers helps remove SDS from protein–detergent complexes and increases the affinity between proteins and the membranes. Methanol is not used in transfers from non-denaturing gels. Non-fat dry milk and Tween 20 detergent are used to block unoccupied sites in membranes and are included as carriers for the antibodies used to probe the membranes.

7.1.3 Immunodetection

Appropriate primary antibodies can be produced in any convenient animals, such as rabbits or mice. Antibodies to many important proteins can be purchased from a number of commercial vendors. Secondary antibodies (e.g. goat anti-rabbit immunoglobulin) conjugated to enzymes are also commercially available. The most common enzymes used in Western blotting are alkaline phosphatase and horseradish peroxidase. The preferred substrate for alkaline phosphatase is the mixture of 5-bromo-4-chloro-3-indolyl phosphate (BCIP) and nitroblue tetrazolium (NBT). The substrate BCIP is dephosphorylated by the enzyme and then oxidized in a reaction coupled to reduction of NBT. The resultant highly visible purple product is deposited on the protein bands or spots. With horseradish peroxidase, use 4-(chloro-1-naphthol) or diamino-benzidine as substrate (with added hydrogen peroxide). Chemiluminescent substrates for horseradish peroxidase are based on oxidation of luminol. The luminol substrate provides the most sensitive signal of the blotting substrates but requires photographic exposures or specially configured imaging devices. Protocol 17 gives a procedure for Western blotting. For alternative procedures and more detail than can be provided here, consult refs 81–83.

7.1.4 Total protein detection

For proper identification of the proteins of interest in a blot, immunodetected proteins must be compared to the total protein pattern of the gel. This requires the indiscriminant staining of all the proteins in the blot. Colloidal gold stain is a very sensitive reagent for total protein staining. It consists of a stabilized sol of colloidal gold particles. The gold particles bind to proteins on the surfaces of membranes. Detection limits are in the low hundreds of picogram range and can be enhanced by an order of magnitude by subsequent treatment with silver.

CBB G-250 is another popular total protein stain. Researchers blotting 2D PAGE gels particularly favour it since it is compatible with mass spectrometry. Stained blots provide good media for archiving 2D PAGE separations. A third version of SYPRO Ruby, formulated for blots, is a very sensitive total protein stain (51, 86, 87).

Protocol 17

Immunoblotting

Equipment and reagents

- Electrotransfer apparatus, tank or semidry, with filter papers, sponges, and power supply
- Blotting membrane, nitrocellulose or PVDF
- Transfer buffer[a]
- TBS[b] and TTBS[c]

- Primary antibody
- Secondary antibody–enzyme conjugate[d]
- Substrate for the enzyme conjugated to the secondary antibody[e]
- Total protein stain[f]

Method

1 Prepare transfer buffer appropriate to the electrotransfer apparatus. Refer to the recommendations of the manufacturer of the apparatus or use those given here. Make about 1 litre of buffer more than is required to fill the apparatus. Do not adjust the pH of transfer buffers; just confirm that they are close to the expected pH.

2 Remove the gel from the cassette and soak it in transfer buffer for about 10 min. It is helpful to cut off the stacking gel, if one was used, since the soft gel will stick to the transfer membrane.

3 Follow the manufacturer's instructions for setting up the transfer apparatus. Cut filter paper to size if necessary. Soak filter paper and sponge pads (if used) in transfer buffer.

4 Cut the transfer membrane to size with a clean, sharp scalpel or razor blade. Do not touch the membrane with bare hands. Use gloves and (or) blunt, flat-blade forceps to manipulate the membrane.

5 Completely wet the transfer membrane with transfer buffer. PVDF must be wetted in methanol prior to being placed in aqueous solutions. Avoid air bubbles in the membrane by slowly sliding it into buffer (or methanol) at a slight angle or by floating it on buffer. Immerse the membrane in buffer and let it soak for 15 min. Do not allow the membrane to dry out before beginning the transfer.

6 Place about 1 litre of transfer buffer in a large tray and assemble the transfer array in it. Use the buffer in the tray to keep all elements of the transfer array well wetted during the assembly process.

7 To avoid trapping air bubbles between the gel and the membrane, lay the membrane on the gel from the centre to the ends then gently roll a test-tube or pipette on top of the membrane to push out pockets of air.

8 Put the transfer array into the transfer apparatus. Follow the manufacturer's instructions for electrotransfer.

9 Wash the membrane for 5–10 min in TBS.

Protocol 17 continued

10 Incubate the membrane for 30 min to 1 h at room temperature in TBS containing 5% (w/v) non-fat dry milk to block excess protein binding sites on the membrane (5 g of non-fat dry milk per 100 ml of TBS).

11 Wash the membrane twice, for 5 min each time, with TTBS.

12 Incubate the membrane for 1–2 h at room temperature with primary antibody or antiserum diluted in TTBS containing 5% non-fat dry milk. Dilutions of primary antibody vary with the source, but are generally of the order of 1:100 to 1:3000.

13 Wash the membrane twice with TTBS, for 5 min each time.

14 Incubate the membrane for 1–2 h at room temperature with secondary antibody-enzyme conjugate appropriately diluted (e.g. 1:3000) in TTBS containing 5% non-fat dry milk.

15 Wash the membrane twice, for 5 min each time, with TTBS.

16 Wash the membrane with TBS to remove the Tween 20.

17 Incubate the membrane with substrate solution for about 1 h or until the desired intensity is obtained.

18 Wash the completed blot with water. Washed immunoblots can be stored dry.

19 For total protein staining with colloidal gold, do not block excess binding sites with non-fat milk. Rather, soak the membrane for 20 min in TBS containing 0.3% Tween 20, then wash it with an excess of water for 5 min. Immerse the membrane in colloidal gold stain (Bio-Rad) for 4 h or until desired intensity is obtained. Greater sensitivity is achievable by enhancing the gold stain with silver (Bio-Rad).

20 For total protein staining with SYPRO Ruby Blot Stain (Bio-Rad), do not block the membrane at all. Immerse (nitrocellulose) or float face down (dried PVDF) the membrane in 10% methanol, 7% acetic acid for 15 min, then wash it four times for 5 min each. Immerse (nitrocellulose) or float face down (PVDF) the membrane in stain for 15 min, then wash it with water. View the stain with epi-illumination.

[a] Transfer buffers. For tanks (25 mM Tris, 192 mM glycine, 20% methanol pH 8.3), dissolve 3.0 g of Tris and 14.4 g of glycine in 800 ml of water, then add 200 ml of methanol. For semidry apparatus (48 mM Tris, 39 mM glycine, 20% methanol pH 9), dissolve 5.8 g of Tris and 2.9 g of glycine in 800 ml of water, then add 200 ml of methanol.

[b] TBS is Tris-buffered saline (0.02 M Tris–HCl, 0.5 M NaCl pH 7.5). To make TBS, dissolve 2.4 g of Tris and 29.2 g of NaCl in approx. 800 ml of water. Adjust the pH to 7.5 with HCl and bring the volume to 1 litre with water.

[c] TTBS is TBS containing 0.05% Tween 20. Add 0.5 ml of Tween 20 to 1 litre of TBS.

[d] The usual enzymes conjugated to antibodies are alkaline phosphatase and horseradish peroxidase.

[e] Substrates. For alkaline phosphatase, the substrate is 0.15 mg of BCIP and 0.3 mg of NBT per ml of 0.1 ml Tris–HCl, 0.5 mM $MgCl_2$. The buffer consists of 1.2 g of Tris and 10 ml of 4.9 M $MgCl_2$ per 100 ml adjusted to pH 9.5 with HCl. Stock BCIP is 30 mg of BCIP (toluidine salt) per

Protocol 17 continued

ml of dimethylformamide, and stock NBT is 60 mg of NBT per ml of 70% dimethylformamide. To make the substrate solution, add 50 ml of stock BCIP and 50 ml of stock NBT to each 10 ml of buffer. For horseradish peroxidase, the substrate solution contains 0.015% hydrogen peroxide and 0.05% 4-(chloro-1-naphthol) in TBS containing 16.7% methanol. To make this substrate, dissolve 60 mg of 4-(chloro-1-naphthol) in 20 ml of methanol; protect this solution from light. Add 600 ml of 3% hydrogen peroxide to 100 ml of TBS. Mix the two solutions together and use the resultant solution immediately. An alternative substrate is prepared with 50 mg of diaminobenzidine and 100 ml of 3% hydrogen peroxide in 100 ml of TBS.

[f] For total protein stain use either colloidal gold stain or SYPRO Ruby Blot Stain (Bio-Rad).

8 Preparative gel electrophoresis

Gel electrophoresis and IEF have both proven useful as preparative methods. In many cases, electrophoretic methods are the only means for assuring that proteins are recovered at high purity. Devices for preparative electrophoresis are commercially available and have been reviewed (85). Since the best preparative IEF instruments do not employ gels in the way emphasized in this chapter, they are not discussed here. The topic of preparative IEF has been reviewed (51, 86, 87).

There are two general ways in which gel electrophoresis is used to purify proteins. In one scheme, proteins are first separated in a slab gel, then extracted from the gel. Depending on the intended use of the proteins, bands can be excised from a gel with a scalpel or razor blade and the proteins eluted either passively or actively. Alternatively, proteins can be electroeluted from intact gels into troughs in a special device. An entirely different preparative category is called 'continuous-elution electrophoresis'. With continuous-elution electro-phoresis, bands of separated proteins are run off the bottom of a gel and swept away to a fraction collector, as is done in column chromatography. Both the band excision and continuous elution methods retain the high resolution of gel electrophoresis. Elution from a whole gel into troughs loses some of the resolution of the gel but is useful for dividing complex samples into manageable fractions for subsequent assays or analyses.

The purity of proteins recovered from gels depends on how well the bands of interest can be identified and how cleanly the bands can be cut out of the gel slab. With a continuous elution device, the final purity of the recovered proteins depends mainly on the correct choice of gel and buffer.

8.1 Extraction from gel slices

Gels intended for protein extraction are often cast thicker than analytical gels (1.5–3 mm). A sample well spanning the width of a gel allows high sample loads to be applied. The maximum amount of sample that can be loaded on a gel depends on how well the molecules of interest can be separated from their neighbours in the sample mixture. Because bands become wider as the amount of material increases, the loss of resolution as sample loads are raised will eventually become unacceptable. Loads are easily tolerated that are 10- to 50-fold greater per unit of gel cross-section than are usually run in analytical gels. Thus with some large slab gels, proteins can be recovered in tens-of-milligram amounts.

High-throughput 2D PAGE laboratories employ automated cutting robots for removing pieces of gels (Figure 14). The cutting heads in these devices remove cylindrical plugs (1–1.5 mm diameter) from within stained bands or spots in gels or blots. Since most work with 2D gels is aimed at protein identification, the proteins in the gel plugs are subsequently treated for analysis in mass spectrometers (not discussed). In the case of 1D gels, the cutting devices can remove multiple plugs from a protein band until the entire band is taken. The cutting instruments offer the advantage over hand excision with blades in that the plugs are removed from the interiors of the spots or bands. Proteins fill the entire volumes of the cylindrical gel plugs, unlike manual cutting that can leave regions of empty gel at the periphery of the excised material. The reagents surrounding a gel plug have good access to the proteins themselves, especially along the lateral surfaces of the cylinders.

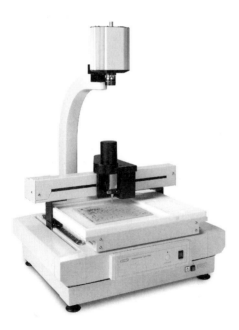

Figure 14 Gel cutting robot. The device shown was developed for excising spots from 2D PAGE gels. The gel is placed on a platform below an *x-y* mechanism that moves the cutting tool in the horizontal plane. A CCD camera records an image of the stained gel that is used by computer software to identify spots for excision. The *x-y* mechanism positions the cutting tool over individual spots to be excised. The cutting tool moves downward to remove cylindrical plugs from the gel (1–1.5 mm diameter). It then places the plugs in the wells of a microplate. The computer tabulates spot position and the location of the individual gel plugs in the microplate.

Several stains are compatible with mass spectrometry and protein sequencing. Manufacturers' specifications should be consulted if stained proteins are to be used for these kinds of analyses. Desired bands are located and cut from the stained gel. Gel slices are then incubated in the appropriate elution buffer.

8.1.1 Passive diffusion

Proteins are often extracted from gel slices by simple diffusion. Pieces of gel containing the molecules of interest are crushed with a mortar and pestle and are left covered with buffer for long enough for the proteins to diffuse into the supernatant fluid.

8.1.2 Electroelution

Electrophoretic elution is an efficient method for recovering proteins from gel slices (85, 88). In the simplest versions of this method, proteins are electro-

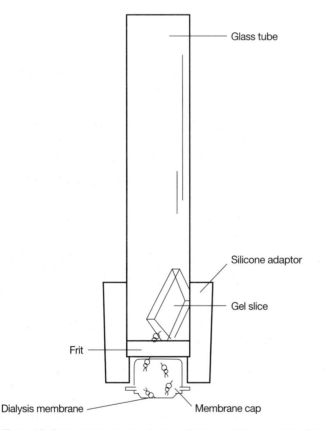

Figure 15 Electroelution device. A piece of gel containing a protein of interest is excised and placed on top of the porous frit in the bottom of the vertical glass tube. The tube is filled with elution buffer and a voltage is impressed across it. The voltage drives the protein out of the piece of gel, through the frit, and into the removable cap. The bottom of the cap is sealed with dialysis membrane. Proteins are retained in the cap by the dialysis membrane and can be recovered for subsequent use.

phoretically driven out of gel pieces into dialysis sacks in the type of apparatus used for running cylindrical gel rods. More sophisticated commercial devices are available for the rapid recovery of proteins in small volumes, with expected yields of greater than 70%. Elution takes about 3 h and is done in any suitable buffer. The Bio-Rad Model 422 Electro-Eluter was made specifically for this purpose (Figure 15).

Protocol 18

Protein recovery by electroelution

Equipment and reagents

- Model 422 Electro-Eluter and power supply (Bio-Rad)
- Elution buffer[a]

Method

1 Assemble the apparatus according to the manufacturer's instructions.

2 Soak the membrane caps for 1 h in elution buffer at 60 °C.

3 Fill the tubes with elution buffer and place a gel slice in each tube. Mincing the gel slices increases elution efficiency. Do not put more gel into a tube than will fill it to a depth of about 1 cm.

4 Place the tube module into the buffer chamber. Fill the chamber with elution buffer to a depth that will comfortably cover the silicon adapters (~600 ml). Add ~100 ml of elution buffer to the upper chamber. Stir the buffer vigorously with a stir bar.

5 Attach the lid and connect the leads to the power supply and carry out elution at 10 mA per tube for 4 h.

6 Drain the liquid from the upper chamber and remove the glass tubes. Remove the buffer from the tubes with a pipette down to the level of the frit.

7 Remove the silicon adapters with the membrane caps from the bottoms of the tubes and withdraw the liquid (containing the eluted protein) in the membrane cap. Rinse the membrane cap with ~200 ml of elution buffer and add the rinse to the protein solution.

8 Analyse the protein in the eluate by electrophoresis or other means. If necessary, remove SDS as recommended in the instruction manual.

[a] Elution buffer. Standard elution buffer is 1 × Laemmli SDS–PAGE electrode buffer (see Table 2). An alternative, volatile buffer is 50 mM ammonium bicarbonate, 0.1% SDS. To make this latter buffer, dissolve 4 g of NH_4HCO_3 and 1 g of SDS in 1 litre of water. Make this solution fresh.

8.2 The whole gel eluter

Screening complicated mixtures of proteins for antigenicity or activity can be a daunting prospect in terms of the number of proteins that must be tested. One approach to minimizing the workload is to separate the protein mixture into

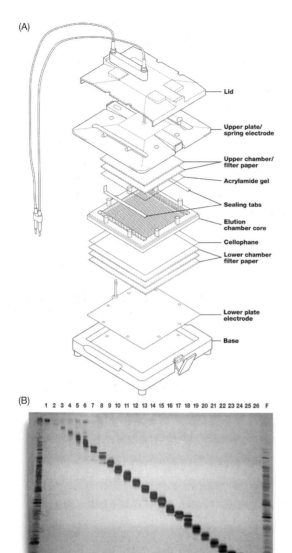

Figure 16 Whole gel elution. (A) The large gel version of the Whole Gel Eluter (Bio-Rad) is shown in an exploded view. A preparative gel is run with the protein bands extending its entire width. The gel is placed on the elution chamber core that contains 30 channels. A sheet of cellophane under the elution core converts the channels into 30 separate troughs. The troughs are filled with elution buffer. Electrical contact to the gel is made though the buffer in the troughs, buffer-saturated filter paper, and the cellophane. Proteins are driven out of the gel into the troughs. Each fraction contains a limited number of proteins. The protein fractions are collected by aspiration into test-tubes with a device that is not shown. (B) SDS–PAGE analysis of fractions collected with the Whole Gel Eluter. The protein sample was derived from *Mycobacterium tuberculosis*. The analytical gel was a silver stained 10–20%T gradient gel. Each eluter fraction was run in a separate lane. Only fractions 1–26 are shown in the analytical gel. Aliquots of the complete starting mixture were run in the left- and right-most lanes.

manageable fractions prior to testing. A device that can be used to divide crude mixtures of proteins for analysis is available as the Whole Gel Eluter (Bio-Rad) (Figure 16). The Whole Gel Eluter eliminates the need for cutting protein bands from gels (89). Instead, following native or SDS–PAGE the entire gel is placed in the elution device and subjected to an electric field directed across the thickness of the gel (as in blotting). The field drives proteins from the gel into buffer-filled troughs from which they can be collected for assay. There is much flexibility in the choice of trough buffer, so that proteins can be recovered ready for testing. Recovered proteins are not at single-band purity, but rather they are collected in small groupings of a few proteins each. The exact distribution of proteins in the individual fractions is dependent on the spacing of the bands in the gels and the placement of the gels on the collection plate of the device. An example protein distribution is shown in Figure 16B.

Protocol 19

Use of the Whole Gel Eluter

Equipment and reagents

- Whole Gel Eluter or Mini Whole Gel Eluter (Bio-Rad)
- Reagents for gel electrophoresis (Tables 1–3 and Protocols 1–4)

Method

1 Run an SDS–PAGE or native gel with a preparative comb.

2 Make 1 litre of elution buffer (500 ml for the Mini unit). Elution buffer is the same as one of the electrode buffers (Table 2). The buffer used for elution need not be the same as the one that was used for electrophoresis.

3 Following electrophoresis, remove the gel from the cassette and equilibrate it in elution buffer for about 15 min.

4 Assemble the base of the Whole Gel Eluter as described in the instruction manual, with buffer-moistened filter paper and cellophane on top of the electrode plate. Remove air bubbles from under the cellophane with the roller. Insert the core as shown and fill the chamber with elution buffer to cover all of the troughs.

5 Lay the gel on a clean flat surface and place the cutting template on top of it. Carefully excise the gel within the template using downward chopping motions with a single-edge razor blade.

6 Place the excised gel in the elution chamber. Avoid air bubbles. Place buffer-moistened filter paper on top of the gel and remove air bubbles with the roller. Seal the aspiration ports with the sealing tabs.

DAVID E. GARFIN

> **Protocol 19** continued

7 Place the upper electrode assembly on top of the chamber. Thread the sealing tabs through the cutouts in the upper assembly. Push down gently on the top of the unit and close the side clamps.

8 Place the lid on the assembled eluter and attach the leads to a power supply.

9 Carry out the elution as recommended in the manual. Proteins should elute in about 30 min.

10 After elution is complete, turn off the power supply, disconnect the leads, and remove the lid, but do not disassemble the eluter.

11 Gently remove the sealing tabs from the aspiration ports.

12 Fill the harvesting box with test-tubes, attach the lid, and connect the harvester to a vacuum source. Leave the vacuum source turned off.

13 Insert the needle array of the harvester into the aspiration ports on the eluter as directed in the manual.

14 Fully open the vacuum control valve on the harvesting box and turn on the vacuum source. Slowly close the control valve to achieve a gentle vacuum. Aspiration should be done slowly to avoid splattering and uneven collection. Harvesting can be done manually if a vacuum source is not available.

15 Run the collected fractions in individual wells of an analytical gel. Analyse the fractions for the proteins of interest by any suitable means.

8.3 Continuous elution electrophoresis

In continuous elution electrophoresis, separated bands migrate off the bottom of a gel and into an elution chamber (85, 90). A flow of buffer washes material out of the elution chamber and sweeps it away to a fraction collector. Purified molecules are recovered in test-tubes ready for analysis. It is important to realize that continuous elution electrophoresis is only meant to be used to purify a given protein from its nearest contaminants. It is designed as a last step in a purification protocol. Continuous elution electrophoresis is incapable of purifying all of the proteins in a sample in a single run. It may prove necessary to carry out separate purifications for different proteins in a sample.

The gels in the preparative gel electrophoresis device shown in Figure 17 are formed between a cylindrical outer tube and a cylindrical core. They can be thought of as either hollow cylinders or slab gels folded around so that their two lateral edges join. The device can hold gels that are 9 or 13 mm thick and up to 12 cm long. A companion device for smaller samples uses cylindrical gels that are 7 mm in diameter. The elution chambers of both devices consist of thin polyethylene frits. Dialysis membranes, directly beneath the elution frits, prevent macromolecules from being drawn out of the chambers by the electric

262

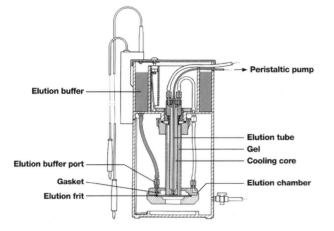

Figure 17 Continuous elution preparative electrophoresis. A cross-section of the Bio-Rad Model 491 Prep Cell is shown. The device is cylindrical in shape. The gel is cast in a glass tube surrounding the cylindrical cooling core. The pump draws elution buffer from the reservoir to the perimeter of the elution frit. From there, elution buffer is drawn radially inward to the centre of the elution frit and out through the elution tube in the centre of the cooling core. As proteins migrate off the bottom of the gel, the elution buffer sweeps them out of the chamber. The pump carries the elution buffer to a fraction collector.

field. Elution buffer enters the chambers around the perimeters of specially designed gaskets and is pumped out of the apparatus to a fraction collector.

8.3.1 Preparative SDS–PAGE continuous elution

The most important parameter in preparative SDS–PAGE is the pore size of the gel. The monomer concentration that best resolves two molecules varies with their molecular weights. Changing the gel composition from the optimal concentration by increasing or decreasing the monomer concentration ultimately decreases resolution. It is important to note that the appropriate monomer concentration for preparative electrophoresis is not necessarily the same as that used for analytical work. For reasons that are not entirely clear, optimum separation of two molecules occurs when the relative mobility of the protein of interest is around 0.55. Thus, the monomer concentration that provides sharp, well-resolved bands in the middle on an analytical gel will generally be the correct monomer concentration to use in a preparative gel. Molecules running with the ion front of an analytical gel will be best separated preparatively by a gel monomer concentration greater than that used in the analytical gel. Similarly, for preparative fractionation, molecules remaining near the top of an analytical gel will require a lower %T than that of the analytical gel. A series of analytical gels should always be run to ascertain the correct preparative gel concentration for SDS–PAGE. Optimal purity in preparative work requires that the gel to choose from among the analytical test gels is the one with the minimal bandwidths and the greatest separation between the protein of interest and its nearest contaminating band.

8.3.2 Preparative native PAGE continuous elution

Preparative native PAGE is a technique for high-yield purification of biologically active proteins. In contrast to SDS–PAGE where detergent–polypeptide complexes migrate according to size only, the mobilities of proteins in native PAGE systems depend on both their charges and their sizes. There is no single electrophoresis buffer system that will optimally purify all native proteins. When selecting conditions for the purification of a native protein, the pI of the protein under investigation and the pH of the electrophoresis system must both be considered.

The most important consideration in preparative native PAGE of proteins is the pH of the electrophoresis buffer. The pH of the electrophoresis buffer system must be within the pH range over which the protein under study is stable and retains its biological activity. In addition, the pH of the chosen buffer system must leave the protein with sufficient charge for it to move through the gel at a reasonable rate. Proteins move very slowly, if at all, in buffers with pH values near their pIs.

In native PAGE, protein mobilities are best varied by changing the pH of the buffer (Table 1) while holding the gel concentration constant at about 6%T. Theoretically, electrophoresis buffers with pH values close to the pI of the protein of interest will provide the best resolution. This is because other proteins will migrate away from the slow-moving protein of interest. However, the resultant migration rate will most likely be too slow for elution from the preparative gel column. Conversely, buffers with pH values far away from the pI of the protein of interest result in fast migration rates but with a loss of resolution. The choice of pH becomes a compromise between separation and speed (in the pH range of protein stability). Trial mini-gels should be run.

With dilute samples and (or) to obtain as much as possible of the protein of interest, the stacking effect of discontinuous buffer systems may be called for. In these cases, the Ornstein–Davis system should be the first non-denaturing gel system tried. The resolving gel of the Ornstein–Davis system becomes pH 9.5 once the glycinate ions displace the chloride in the gel. This pH may be outside the range of stability for some proteins. Alternative discontinuous buffer systems devised for preparative work, spanning the pH range from 3 to 10, can be found in refs 13 and 91. Protocols for using the alternative discontinuous buffers are analogous to those for the Ornstein–Davis buffer system.

Conditions for purification of native proteins should first be optimized on a small scale using mini-slab gels. The sample should be partially purified before gel electrophoresis. A method for identifying the protein of interest, such as by immunoblotting or enzyme activity, is essential. Gels used to analyse fractions from preparative electrophoresis should always be silver stained or stained with SYPRO Ruby to allow detection of trace contaminants that might not be visible after dye staining.

Protocol 20

Continuous elution electrophoresis

Equipment and reagents

- Model 491 Prep Cell or Mini Prep Cell (Bio-Rad)
- Reagents for gel electrophoresis (Tables 1–3 and Protocols 1–4)

Method

1 Determine the gel and buffer compositions appropriate to the sample according to the manufacturer's instructions. For SDS–PAGE, select a %T that gives an R_f between 0.5 and 0.6 for the protein of interest. With McLellan's buffers use 6%T.

2 Assemble the Prep Cell casting assembly as described in the instruction manual.

3 Cast the gel(s) as described in the instruction manual. Note that the catalyst concentrations are different than for slab gels and that it is advisable to cool the gel during polymerization as shown in the manual.

4 Transfer the gel assembly to the buffer chambers and connect the tubing to the pumps. Fill the chambers as instructed. Remember to purge the elution chamber of air as described in the manual.

5 Establish pump settings to give ~100 ml/min for the cooling line and 1 ml/min for elution (100 ml/min for the Mini Prep Cell).

6 Prepare the sample as appropriate to the requirements of the run (Protocol 4) and load it on top of the gel using the loading guide and syringe.

7 Attach the lid to the cell and plug the leads into the power supply. Run the cell at the power settings recommended in the instruction manual.

8 Begin collecting fractions when the tracking dye comes off the bottom of the gel (2–5 h depending on the cell type). The protein of interest elutes about 10 h after the start of the run for the full size gel.

9 Analyse the fractions as appropriate.

Acknowledgements

I thank my colleagues at Bio-Rad Laboratories for assistance with the illustrations. Dr Patti Taranto provided Figure 2, Adriana Harbers provided Figures 3, 4, and 7, and Mingde Zhu provided Figure 10.

References

1. Andrews, A. T. (1986). *Electrophoresis: theory, techniques, and biochemical and clinical applications* (2nd edn). Oxford University Press, Oxford.
2. Låås, T. (1989). In *Protein purification: principles, high resolution methods and applications* (ed. J.-C. Janson and L. Ryden), p. 349. VCH Press, Weinheim.

3. Dunn, M. J. (1993). *Gel electrophoresis: proteins*. BIOS Scientific Publishers, Oxford.

4. Allen, R. C. and Budowle, B. (1994). *Gel electrophoresis of proteins and nucleic acids*. De Gruyter, Berlin.

5. Garfin, D. E. (1995). In *Introduction to biophysical methods for protein and nucleic acid research* (ed. J. A. Glasel and M. P. Deutscher), p. 53. Academic Press, San Diego.

6. Westermeier, R. (1997). *Electrophoresis in practice: a guide to methods and applications of DNA and protein separations* (2nd edn). VCH Press, Weinheim.

7. Hames, B. D. (ed.) (1998). *Gel electrophoresis of proteins: a practical approach* (3rd edn). Oxford University Press, Oxford.

8. Garfin, D. E. (1990). In *Methods in enzymology* (ed. M. P. Deutscher), Vol. 182, p. 425. Academic Press, San Diego.

9. Makowski, G. S. and Ramsby, M. L. (1997). In *Protein structure: a practical approach* (2nd edn) (ed. T. E. Creighton), p. 1. Oxford University Press, Oxford.

10. Goldenberg, D. P. (1997). In *Protein structure: a practical approach* (2nd edn) (ed. T. E. Creighton), p. 187. Oxford University Press, Oxford.

11. Cantor, C. R. and Schimmel, P. R. (1980). *Biophysical chemistry, part 2: techniques for the study of biological structure and function*, p. 676. Freeman, San Francisco.

12. Righetti, P. G. (1989). *J. Biochem. Biophys. Methods*, **19**, 1.

13. Chrambach, A. (1985). *The practice of quantitative gel electrophoresis*. VCH Press, Weinheim.

14. Caglio, S. and Righetti, P. G. (1993). *Electrophoresis*, **14**, 554.

15. Bio-Rad Laboratories. (1993). *Bull. No. 1156*. Bio-Rad Laboratories, Hercules, CA.

16. Boschetti, E. (1989). *J. Biochem. Biophys. Methods*, **19**, 21.

17. Chen, B., Griffith, A., Catsimpoolas, N., Chrambach, A., and Rodbard, D. (1978). *Anal. Biochem.*, **89**, 609.

18. Chen, B., Rodbard, D., and Chrambach, A. (1978). *Anal. Biochem.*, **89**, 596.

19. Chrambach, A. and Jovin, T. M. (1983). *Electrophoresis*, **4**, 190.

20. Ornstein, L. (1964). *Ann. N. Y. Acad. Sci.*, **121**, 321.

21. Jovin, T. M. (1973). *Biochemistry*, **12**, 871, 879, and 890.

22. Kleparnik, K. and Bocek, P. (1991). *J. Chromatogr.*, **569**, 3.

23. Davis, B. (1964). *Ann. N. Y. Acad. Sci.*, **121**, 404.

24. McLellan, T. (1982). *Anal. Biochem.*, **126**, 94.

25. Margolis, J. and Kenrick, K. G. (1968). *Anal. Biochem.*, **25**, 347.

26. Fairbanks, G., Steck, T. L., and Wallach, D. F. H. (1971). *Biochemistry*, **10**, 2606.

27. Spiker, S. (1980). *Anal. Biochem.*, **108**, 263.

28. Reisfeld, R. A., Lewis, U. J., and Williams, D. E. (1962). *Nature*, **195**, 281.

29. Laemmli, U. K. (1970). *Nature*, **227**, 680.

30. Nielsen, T. B. and Reynolds, J. (1978). In *Methods in enzymology* (ed. C. H. W. Hirs and S. N. Timasheff), Vol. 48, p. 3. Academic Press, New York.

31. Wyckoff, M., Rodbard, D., and Chrambach, A. (1977). *Anal. Biochem.*, **78**, 459.

32. Reynolds, J. A. and Tanford, C. (1970). *Proc. Natl. Acad. Sci. USA*, **66**, 1002.

33. Weber, K. and Osborn, M. (1969). *J. Biol. Chem.*, **244**, 4406.

34. Shi, Q. and Jackowski, G. (1998). In *Gel electrophoresis of proteins: a practical approach* (3rd edn) (ed. B. D. Hames), p. 7. Oxford University Press, Oxford.

35. Neville, D. M., Jr. (1971). *J. Biol. Chem.*, **246**, 6328.

36. Akins, R. E., Levin, P. M., and Tuan, R. S. (1992). *Anal. Biochem.*, **202**, 172.

37. Schägger, H. and von Jagow, G. (1987). *Anal. Biochem.*, **166**, 368.

38. Makowski, G. S. and Ramsby, M. L. (1993). *Anal. Biochem.*, **212**, 283.

39. Neugebauer, J. M. (1990). In *Methods in enzymology* (ed. M. P. Deutscher), Vol. 182, p. 239. Academic Press, San Diego.

40. Hjelmeland, L. M. and Chrambach, A. (1981). *Electrophoresis*, **2**, 1.

41. Brown, E. G. (1988). *Anal. Biochem.*, **174**, 337.

42. Lopez, M. F., Patton, W. F., Utterback, B. L., Chung-Welch, N., Barry, P., Skea, W. M., *et al.* (1991). *Anal. Biochem.*, **199**, 35.

43. Ochs, D. (1983). *Anal. Biochem.*, **135**, 470.

44. Hames, B. D. (1990). In *Gel electrophoresis of proteins: a practical approach* (2nd edn) (ed. B. D. Hames and D. Rickwood), p. 1. IRL Press, Oxford.

45. Engelhorn, S. and Updyke, T. V. (1996). US Patent 5 578 180.

46. Låås, T. (1989). In *Protein purification: principals, high resolution methods, and applications* (ed. J.-C. Janson and L. Ryden), p. 376. VCH Press, Weinheim.

47. Righetti, P. G. (1983). *Isoelectric focusing: theory, methodology, and applications*. Elsevier, Amsterdam.

48. Righetti, P. G. (1990). *Immobilized pH gradients: theory and methodology*. Elsevier, Amsterdam.

49. Garfin, D. E. (1990). In *Methods in enzymology* (ed. M. P. Deutscher), Vol. 182, p. 459. Academic Press, San Diego.

50. Righetti, P. G., Bossi, A., and Gelfi, C. (1998). In *Gel electrophoresis of proteins: a practical approach* (3rd edn) (ed. B. D. Hames), p. 127. Oxford University Press, Oxford.

51. Garfin, D. E. (2000). In *Handbook of bioseparations* (ed. S. Ahuja), p. 263. Academic Press, San Diego.

52. Wilkins, M. R. and Gooley, A. A. (1997). In *Proteome research: new frontiers in functional genomics* (ed. M. R. Wilkins, K. L. Williams, R. D. Appel, and D. F. Hochstrasser), p. 35. Springer, Berlin.

53. Langen, H., Röder, D., Juranville, J.-F., and Fountoulakis, M. (1997). *Electrophoresis*, **18**, 2085.

54. Harrington, M. G., Gudeman, D., Zewert, T., Yun, M., and Hood, L. (1991). In *Methods: a companion to Methods in enzymology* (ed. M. G. Harrington), Vol. 3, No. 2, p. 98. Academic Press, San Diego.

55. Sanchez, J.-C., Rouge, V., Pisteur, M., Ravier, F., Tonella, L., Moosmayer, M., *et al.* (1997). *Electrophoresis*, **18**, 324.

56. Herbert, B. R., Sanchez, J.-C., and Bini, L. (1997). In *Proteome research: new frontiers in functional genomics* (ed. M. R. Wilkins, K. L. Williams, R. D. Appel, and D. F. Hochstrasser), p. 13. Springer, Berlin.

57. Hanash, S. M. (1998). In *Gel electrophoresis of proteins: a practical approach* (3rd edn) (ed. B. D. Hames), p. 189. Oxford University Press, Oxford.

58. Link, A. J. (ed.) (1999). *2-D proteome analysis protocols*. Human Press, Totowa, NJ.

59. Rabilloud, T. (ed.) (2000). *Proteome research: two-dimensional gel electrophoresis and identification methods*. Springer, Berlin.

60. Rabilloud, T. (1996). *Electrophoresis*, **17**, 813.

61. Molloy, M. P. (2000). *Anal. Biochem.*, **280**, 1.

62. Wirth, P. J. and Romano, A. (1995). *J. Chromatogr. A*, **698**, 123.

63. Merril, C. R. and Washart, K. M. (1998). In *Gel electrophoresis of proteins: a practical approach* (3rd edn) (ed. B. D. Hames), p. 53. Oxford University Press, Oxford.

64. Allen, R. C. and Budowle, B. (1999). *Protein staining and identification techniques*. BioTechniques Books, Natick, MA.

65. Rabilloud, T. (2000). *Anal. Biochem.*, **72**, 48A.

66. Irie, S., Sezaki, M., and Kato, Y. (1982). *Anal. Biochem.*, **126**, 350.

67. DeMoreno, M. R., Smith, J. F., and Smith, R. V. (1985). *Anal. Biochem.*, **151**, 466.

68. Dzandu, J. K., Deh, M. E., Barratt, D. L., and Wise, G. E. (1984). *Proc. Natl. Acad. Sci. USA*, **81**, 1733.

69. Vediyappan, G., Bikandi, J., Braley, R., and Chaffin, W. L. (2000). *Electrophoresis*, **21**, 956.

DAVID E. GARFIN

70. Rabilloud, T. (1990). *Electrophoresis*, **11**, 785.
71. Merril, C. R., Goldman, D., Sedman, S. A., and Ebert, M. H. (1981). *Science*, **211**, 1437.
72. Gottlieb, M. and Chavko, M. (1987). *Anal. Biochem.*, **165**, 33.
73. Lee, C., Levin, A., and Branton, D. (1987). *Anal. Biochem.*, **166**, 308.
74. Fernandez-Patron, C., Castellanos-Serra, L., and Rodriguez, P. (1992). *BioTechniques*, **12**, 564.
75. Fernandez-Patron, C., Castellano-Serra, L., Hardy, E., Guerra, M., Estevez, E., Mehl, E., et al. (1998). *Electrophoresis*, **19**, 2398.
76. Patton, W. F. (2000). *Electrophoresis*, **21**, 1123.
77. Crowle, A. J. and Cline, L. J. (1977). *J. Immunol. Methods*, **17**, 379.
78. Righetti, P. G. and Drysdale, J. W. (1974). *J. Chromatogr.*, **98**, 271.
79. Patton, W. F. (1995). *J. Chromatogr. A*, **698**, 55.
80. Patton, W. F. (2000). *BioTechniques*, **28**, 944.
81. Bjerrum, O. J. and Heegard, N. H. (ed.) (1988). *Handbook of immunoblotting of proteins*, Vol. 1 and 2. CRC Press, Boca Raton, FL.
82. Baldo, B. A. and Tovey, E. R. (ed.) (1989). *Protein blotting: methodology, research and diagnostic applications*. Karger, Basel.
83. Dunbar, B. S. (ed.) (1994). *Protein blotting: a practical approach*. Oxford University Press, Oxford.
84. Ledue, T. B. and Garfin, D. E. (1997). In *Manual of clinical laboratory immunology* (5th edn) (ed. N. R. Rose, E. Conway de Macario, J. D. Folds, H. C. Lane, and R. M. Nakamura), p. 54. ASM Press, Washington, DC.
85. Lee, K. H. and Harrington, M. G. (1998). In *Gel electrophoresis of proteins: a practical approach* (3rd edn) (ed. B. D. Hames), p. 93. Oxford University Press, Oxford.
86. Righetti, P. G., Bossi, A., Wenisch, E., and Orsini, G. (1997). *J. Chromatogr. B*, **699**, 105.
87. Bier, M. (1998). *Electrophoresis*, **19**, 1057.
88. Harrington, M. G. (1990). In *Methods in enzymology* (ed. M. P. Deutscher), Vol. 182, p. 488. Academic Press, San Diego.
89. Andersen, P. and Heron, I. (1993). *J. Immunol. Methods*, **161**, 29.
90. Chen, J.-H. (1989). US Patent 4 877 510.
91. Chrambach, A. and Jovin, T. M. (1983). *Electrophoresis*, **4**, 190.

Chapter 8
Biophysical methods in structural cell biology

Mavis Agbandje-McKenna, Arthur S. Edison, and Robert McKenna

Department of Biochemistry and Molecular Biology,
Center for Structural Biology, The Brain Institute, College of Medicine,
PO Box 100245, University of Florida, Gainesville, FL 32610-0245, USA.

1 Introduction

Over the past several decades biophysical methods have become essentials tools for studying biology at the molecular level, and have become an inter-disciplinary subject with interplay into biology, chemistry, and physics. It has also become very apparent that the instrumentation and methods used are key components for elucidating structure and function at the molecular level. Biological processes are governed by numerous intricate arrays of macro-molecular interactions. The role of the structural biologist is to visualize these interactions in three dimensions. This is essential for a full understanding of these interactions, as 'seeing is believing', and provides a means whereby strategies can be devised to 'correct' interactions that go wrong or 'prevent' undesirable interactions caused by diseases.

The range of biophysical methods used in structural biology is vast, ranging from hydrodynamical, optical, to scattering techniques, with advancements in instrumentation development resulting in constant updates in methodologies. In this chapter we focus on three techniques that have pushed forward the frontier of structural understanding of biological processes in recent years; nuclear magnetic resonance (NMR) spectroscopy, X-ray crystallography (XRC), and electron microscopy (EM). The choice of methodology for a particular structural elucidation endeavour is generally dictated by the size and complexity of the macromolecule under investigation, the amount of material available, its solubility in aqueous environments, and the type of interactions being visual-ized. For some studies, a combination of methodologies can be utilized in synergy to optimize the amount of information obtainable. Generally, NMR spectroscopy is utilized for small protein molecules that are flexible, XRC for medium size proteins and complexes that are compact, while very large macro-molecular assemblages or membranous protein structures are determined by such as COSY (Correlation Spectroscopy), TOCSY (Total Correlation Spectroscopy), and NOESY (Nuclear Overhauser effect Spectroscopy) can provide structures of proteins up to about 80 amino acids. Two-dimensional experiments have two frequency axes. The homonuclear (e.g. proton on each frequency axis) experiments described above contain two types of peaks: diagonal and cross-peaks. Diagonal peaks have the same chemical shift along each frequency axis and represent no transfer of magnetization. Cross-peaks have two different frequencies and represent transfer of magnetization from one atom to another (Figure 1A). When two atoms transfer magnetization from one to the other, they are said to be 'correlated' or 'coupled'.

For small proteins and peptides, 2D COSY or TOCSY spectra provide correlations of protons through chemical bonds. This information leads to characteristic patterns of each amino acid. However, because there is no proton–proton coupling across the peptide bond, COSY and TOCSY cannot provide sequential assignments. NOESY spectra provide correlations between protons that are close together in space (Figure 1A). As a result, NOESY spectra allow crossing of the peptide bond through interactions between alpha and beta protons and the following amide proton. By combining TOCSY or COSY with NOESY, complete assignments of all the resonances in small proteins can be made.

The NOESY data provide distance measurements between any protons within about 5 Å in space. Thus, the same data that allow sequential assignments also provides measurements of pairs of all close protons. This information can then be used as input for distance geometry, simulated annealing, or restrained molecular dynamics calculations. All of these methods are different approaches to use the experimental NMR distances to generate 3D structures of the macromolecule.

Because of signal overlap, 2D NOESY/TOCSY assignment strategies used for small proteins fail with larger proteins and nucleic acids. Instead, triple resonance experiments provide direct correlations through all the backbone and side chain atoms. These experiments are very efficient but require considerable NMR instrumentation, isotopically-labelled samples, and more operator knowledge than proton-based 2D methods. For proteins up to about 30 kDa, isotopic labelling with ^{13}C and ^{15}N allow structure and dynamics measurements through

---

Figure 1 (A) NMR data. Selected regions of a two-dimensional NOESY spectrum of a 16 amino acid peptide. NOESY experiments provide distances between two protons, and the intensity of the peaks are proportional to r^{-6}, where r is the distance between the two protons. Regions in the spectrum with interactions between different types of protons are indicated on the side. (B) Structures. Two different superpositions of a family of 30 structures of a 16 amino acid peptide. The structures were determined by using distances from NOESY data and other hydrogen bond restraints from pH titration data as restraints in molecular dynamics simulations, and all of the structures satisfy the available NMR data. The left and right sets of structures are superpositions of the same calculated structures of the C and N terminal halves of the peptide, respectively. Each half of the peptide is well-structured and has numerous NOESY restraints, but there are few restraints connecting the two halves, so they are disordered relative to each other.

moments, when placed into a static external magnetic field, become polarized with positive and negative energy level populations predicted by the Boltzmann distribution. In NMR, the difference in energy between the upper (positive) and lower (negative) energy levels is very small compared to thermal energy, and thus only a very small fraction of the magnetic moments contribute to the net polarization. Most of the polarized nuclear spins in the sample simply cancel each other out. Unfortunately, this makes NMR an insensitive technique and is one of the factors driving the development of higher magnetic fields. The small fraction of nuclear spins that contribute to the net polarization create a net magnetic moment, which behaves very much like a gyroscope precessing in Earth's gravitational field.

NMR transition energies are in the radio frequency range (e.g. 500 MHz), and if the proper frequency is applied to the net magnetization, the resonance condition is met: individual nuclear spins will change energy levels and the net magnetization will be tipped from the z-axis (parallel to the magnetic field) into the x-y plane. Just as a spinning gyroscope whose top is tipped toward the ground will precess with a frequency proportional to the strength of the gravitational field and spinning speed, so will the net magnetization in a magnetic field precess with a frequency proportional to the static magnetic field strength and the gyromagnetic ratio, an intrinsic property of each nucleus. The precession of the nuclear magnetization in the x-y plane of the magnetic field produces an electrical signal called a free induction decay (FID) that can be detected and amplified. This signal, which is recorded as a function of time, is converted by the Fourier transform into a mathematically equivalent signal that is a function of frequency. This Fourier transformed FID is the NMR spectrum.

A one-dimensional proton NMR spectrum of even a simple organic molecule contains peaks at several different frequencies. The different frequencies result from differences in the electronic and chemical environments of different types of atoms in the molecule. Therefore, the frequency of a particular NMR peak is called its 'chemical shift'. Chemical shifts are generally reported in parts per million (ppm) by dividing the frequency in Hertz (Hz) by the spectrometer frequency in megaHertz (MHz). This convention allows for direct comparison of NMR data at different magnetic field strengths. For example, if two peaks are

Table 1 Elements with non-zero spins routinely used for NMR

Nuclei	Nuclear spin	γ (MHz/T)
^{1}H	½	42.58
^{2}H	1	6.54
^{15}N	½	4.31
^{13}C	½	10.71
^{31}P	½	17.25
^{19}F	½	40.08
^{23}Na	1½	11.27

271

separated by 300 Hz in a 300 MHz (7 Tesla) magnet, they will be separated by 600 Hz in a 600 MHz (14 Tesla) magnet. In both cases, the peaks are separated by 1 ppm.

2.1 Sample preparation

NMR methods have been developed to record samples in the gas phase, solid state, and solution state. For biological applications, solution state NMR is most common, but solid state methods are being successfully applied to membrane-associated proteins. New methods are also being developed to create partially oriented solution state samples; these samples produce NMR spectra with important information on the orientation of bonds in space through dipolar couplings. For standard solution state NMR, samples are dissolved in about 600 µl of solvent that contains deuterium for a lock frequency. The most common solvent for macromolecules is 90% H_2O/10% D_2O with a salt concentration below about 100 mM. Sample concentrations need to be around 1 mM, and a 90% H_2O sample is about 100 M in water protons (e.g. about 100 000 times the concentration of the sample). Therefore, the solvent resonance needs to be experimentally attenuated to observe most samples. Amide protons that are not involved in internal hydrogen bonds will exchange rapidly with the solvent in a pH-dependent manner. If the large solvent signal is reduced by equalizing the solvent energy levels (called pre-saturation), exchangeable protons are also eliminated. Proton exchange can be catalysed by either basic or acidic conditions, and pH values between about 4 and 5 are optimal to minimize exchange. Gradient solvent suppression techniques allow for a wide range of pH values, and most standard buffers can be used, but they generally need to be deuterated to avoid contaminating the sample NMR signal. Sample concentrations should be close to 1 mM, but modern NMR spectrometers, higher field magnets, and better probes have made more dilute sample measurements possible. A useful 'rule of thumb' for an upper estimate for the number of milligrams needed for NMR studies is the molecular weight in kDa (e.g. 15 mg of a 15 kDa protein will produce 1 ml of a 1 mM sample). The dissolved samples are typically put into a 5 mm glass tube with an internal proton chemical shift standard for referencing. The preferred proton chemical shift standard for proteins in aqueous solution is DSS (sodium 2,2-dimethyl-2-silapentane-5-sulfonate). Referencing of ^{13}C and ^{15}N should be done relative to ^{1}H by multiplying the frequency in MHz of the ^{1}H DSS peak (referenced to 0.00 ppm) by known chemical shift ratios of ^{13}C (0.251449530) and ^{15}N (0.101329118). This will give the 0.00 ppm points for the ^{13}C and ^{15}N chemical shift scales. Current best values and temperature-dependent corrections for chemical shift ratios as well as a large amount of useful biological NMR information is available at the BioMagResBank (BMRB) database (www.bmrb.wisc.edu).

For proteins larger than about 80 amino acids, isotopic labelling is required. Several methods have been developed to label proteins and nucleic acids, but the most common is to express the protein of interest in *E. coli* using a minimal media with defined carbon and nitrogen sources. Thus, to produce a 'double labelled' $^{13}C/^{15}N$ protein sample, [^{13}C]glucose and [^{15}N]ammonium chloride (or similar compounds) are added to the growth media. Doubled labelled samples are expensive to produce, so it is important to first optimize growth and purification conditions with unlabelled samples. For larger proteins, in addition to ^{13}C and ^{15}N, partial labelling with ^2H is sometimes necessary to reduce line broadening caused by ^1H. These 'triple labelled' samples are very expensive and are often more difficult to produce because of large isotope effects in the growth of the bacteria.

2.2 Instrumentation hardware

In their most basic form, NMR spectrometers consist of a static magnet, a probe which sits in the magnet and transmits radio frequency signals to and from the sample, frequency generators, amplifiers, and computer control. Today's magnets are almost always superconducting and generate magnetic fields ranging from about 7 Tesla (300 MHz) to 21 Tesla (900 MHz). The proper unit for field strength is Tesla, but it is most common to refer to magnets by the frequency that protons resonate in that field (e.g. a 600 MHz magnet is the resonance frequency of protons in a 14 T magnet). Probes are the 'heart' of an NMR spectrometer, because they allow communication with the sample. There are a multitude of probes available, but the most common probe for biological NMR is a 5 mm 'triple resonance' probe. Despite its name, a triple resonance probe has four channels: ^1H, ^{13}C, ^{15}N, and ^2H for a frequency lock. Substantial developments in probe design have significantly improved the sensitivity of NMR experiments over the last decade. New probe developments include cold metal cryoprobes and high temperature superconducting (HTS) probes. Under optimal conditions, these probes can produce about a four-fold increase in signal-to-noise (S/N) over a conventional probe. This increased S/N will lead to shorter acquisition times and the ability to detect less concentrated samples. Another important development has been the pulsed field gradient (PFG). PFGs change the strength of the applied magnetic field along either just the z-axis or the x-, y-, and z-axes. Among other things, PFG capability allows for significantly better water suppression and reduces the need for extensive and time-consuming phase cycling.

2.3 Data acquisition and structure determination

The most basic form of NMR spectroscopy utilizes the chemical shifts of the protons in the protein as probes of molecular structure and dynamics. In NMR, the term 'dimension' refers to the number of different frequency axes recorded in the experiment. There is no theoretical upper limit on the number of dimensions, but a current practical limit is four and the most common for macromolecules is two or three. One-dimensional (1D) spectra, while often satisfactory for small organic molecules, are generally too overlapped in proteins to yield useful information. More complicated two-dimensional (2D) proton experiments such as COSY (Correlation Spectroscopy), TOCSY (Total Correlation Spectroscopy), and NOESY (Nuclear Overhauser effect Spectroscopy) can provide structures of proteins up to about 80 amino acids. Two-dimensional experiments have two frequency axes. The homonuclear (e.g. proton on each frequency axis) experiments described above contain two types of peaks: diagonal and cross-peaks. Diagonal peaks have the same chemical shift along each frequency axis and represent no transfer of magnetization. Cross-peaks have two different frequencies and represent transfer of magnetization from one atom to another (Figure 1A). When two atoms transfer magnetization from one to the other, they are said to be 'correlated' or 'coupled'.

For small proteins and peptides, 2D COSY or TOCSY spectra provide correlations of protons through chemical bonds. This information leads to characteristic patterns of each amino acid. However, because there is no proton–proton coupling across the peptide bond, COSY and TOCSY cannot provide sequential assignments. NOESY spectra provide correlations between protons that are close together in space (Figure 1A). As a result, NOESY spectra allow crossing of the peptide bond through interactions between alpha and beta protons and the following amide proton. By combining TOCSY or COSY with NOESY, complete assignments of all the resonances in small proteins can be made.

The NOESY data provide distance measurements between any protons within about 5 Å in space. Thus, the same data that allow sequential assignments also provides measurements of pairs of all close protons. This information can then be used as input for distance geometry, simulated annealing, or restrained molecular dynamics calculations. All of these methods are different approaches to use the experimental NMR distances to generate 3D structures of the macromolecule.

Because of signal overlap, 2D NOESY/TOCSY assignment strategies used for small proteins fail with larger proteins and nucleic acids. Instead, triple resonance experiments provide direct correlations through all the backbone and side chain atoms. These experiments are very efficient but require considerable NMR instrumentation, isotopically-labelled samples, and more operator knowledge than proton-based 2D methods. For proteins up to about 30 kDa, isotopic labelling with ^{13}C and ^{15}N allow structure and dynamics measurements through

---

Figure 1 (A) NMR data. Selected regions of a two-dimensional NOESY spectrum of a 16 amino acid peptide. NOESY experiments provide distances between two protons, and the intensity of the peaks are proportional to r^{-6}, where r is the distance between the two protons. Regions in the spectrum with interactions between different types of protons are indicated on the side. (B) Structures. Two different superpositions of a family of 30 structures of a 16 amino acid peptide. The structures were determined by using distances from NOESY data and other hydrogen bond restraints from pH titration data as restraints in molecular dynamics simulations, and all of the structures satisfy the available NMR data. The left and right sets of structures are superpositions of the same calculated structures of the C and N terminal halves of the peptide, respectively. Each half of the peptide is well-structured and has numerous NOESY restraints, but there are few restraints connecting the two halves, so they are disordered relative to each other.

a variety of experiments that provide correlations through atoms of the backbone and side chains. These multidimensional 'triple resonance' experiments often are named according to the atoms they correlate. For example, HNCO is a 3D experiment that produces a peak at the intersection of three different frequencies: the amide proton, the amide nitrogen, and the carbonyl carbon of a peptide bond. Combinations of many different types of triple resonance experi-

(A)

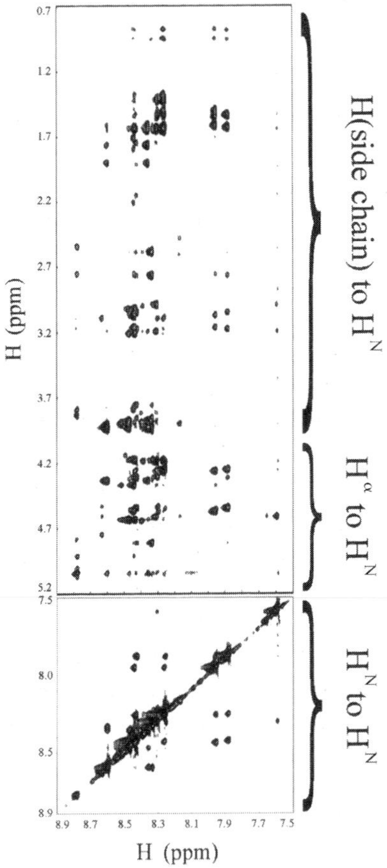

(B)

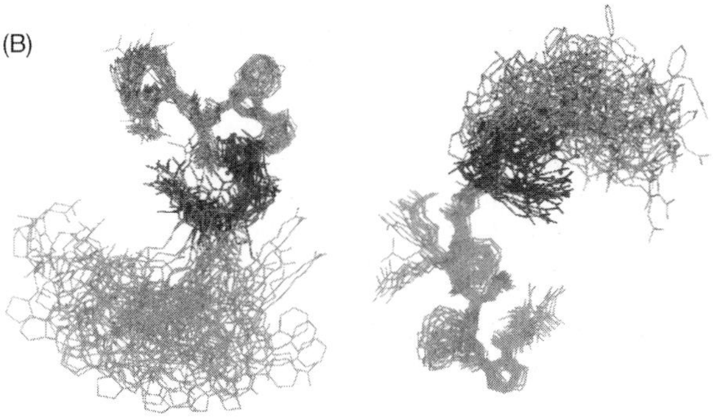

275

ments can provide data for complete assignments and structural parameters for structure determination and dynamics measurements of medium to large proteins, nucleic acids, and macromolecular complexes. Higher magnetic fields and new techniques such as TROSY (Transverse Relaxation Optimized Spectroscopy) will expand the practical molecular weight limit in the near future.

Protein structures larger than 80 amino acids are also solved with NOESY data in which third or fourth dimensions correlating frequencies of ^{13}C or ^{15}N nuclei. Distances from NOESY data continue to be the most important experimental constraint in structure determination, but many other parameters are also widely used, including J-coupling to identify rotations about dihedral angles, dipolar coupling to identify orientations of bonds in space, and chemical shifts to identify regions of secondary structure in proteins.

NMR structures using experimental distances, dihedral angles, and bond orientations are generated with one of the methods mentioned above and presented as families of structures, each satisfying the experimental constraints equally well. NMR families of structures typically have regions that are very well-defined and others that are 'fuzzy' or ill-defined. An example of a family of conformations from a flexible peptide with a hinge region is shown in Figure 1B. The 'fuzzy' regions can arise through lack of NMR data, because of molecular flexibility, or both. Proof of flexibility requires extensive NMR relaxation measurements, which can be used to measure the time scales of motion at different sites in a macromolecule.

3 X-ray crystallography

For the last 50 years, X-ray crystallography (XRC) has dominated structural biology, from the study of single globular proteins to large multiple complexes involving proteins, RNA, and DNA. XRC enables the precise mapping of macromolecular surfaces, enzymatic active sites, and interactions at interfaces. The technique requires the formation of crystals, a solid regular array of identical molecules of the sample under investigation. This is often the most difficult or limiting step in XRC. XRC is the method of choice for high resolution structure determinations if crystals can be obtained. The methodology exploits the interaction of electrons surrounding each atom in the crystal with a directed X-ray beam to obtain data that can be manipulated to produce the three-dimensional (3D) structure of the sample under consideration.

Electrons interact with X-rays in the crystal, and oscillate, giving rise to a new source of radiation that is emitted in all directions. The experimental data recorded from crystals are the resultant scatter arising from X-rays emitted in a certain direction which add together to give a diffraction pattern or diffraction spots (Bragg's Law). Each spot in the diffraction pattern is made up from scattered X-rays from each atom in the protein molecule that formed the crystal. The patterns are generally recorded on a photographic film or some other form of detector (see Section 3.4). This diffraction pattern can then be used to calculate an electron density map of the protein or sample using the Fourier transform function (the mathematical equivalent of a glass lens in an optical microscope). Obtaining the phase information that is required for the electron density map calculation is the second obstacle in XRC. This is the so-called 'phase problem' that all crystallographers must solve before they can calculate an electron density map.

Each recorded spot in a diffraction pattern has four properties, an amplitude (square root of its intensity), wavelength, frequency, and phase. The amplitude is derived from the intensity of the spot and is the experimentally measured size of the wave, the wavelength is set by the X-ray source monochromatic radiation, and the frequency is derived from the Bragg plane from which it originated in the crystal lattice. The phase is the only property that cannot be measured directly experimentally. Calculation of the position of all the atoms in the molecule requires knowledge of all four properties. Crystallographers utilize a number of methodologies to solve the 'phase problem' that are discussed in Section 3.5. Once phases are obtained, the calculated electron density map allows for the structural mapping of all the non-hydrogen atoms that formed the crystal lattice. In rare cases, where atomic resolution diffraction data is available, the hydrogen atoms can also be assigned.

3.1 Sample preparation for crystallization experiments

Crystal growth is achieved by the slow dehydration of the water of solvation from the sample in a controlled manner that prevents precipitation and takes the sample out of solution and into a crystalline state. However, the success of this process is highly dependent on sample purity. The sample must be at least 95% pure, and preferably 99%, because the greater the percentage of impurities, the less likely the sample is to crystallize. The purity of the sample should always be checked before starting a crystallization experiment or trial. For example, SDS–PAGE developed by silver staining should be used to check purity before crystallization. In addition, the concentration of the sample should be at least 5 mg/ml, preferably 10 mg/ml or higher. The greater the concentration of the sample, the greater the chance of success.

The sample should also be stable in the buffer solution in which it is dissolved for at least two weeks (the optimal time required for protein crystal growth) and should not be denatured by the solution at the temperature (generally ranging from 4 °C to room temperature) of the experiment. The buffer solution should contain as minimum an amount of additives (counter ions, detergents, etc.) as possible. Ideally, the sample should be allowed to buffer itself in deionized water, although in practice low molarity buffers are used. Table 2 gives a list of standard buffers and their pH range used in crystallization trials. The precipitant used is also of great importance and Table 3 gives a list of commonly used organic and inorganic precipitants. Many successful crystallization trials also require the use of additional additives that stabilize the sample during crystallization. Table 4 gives a list of the types and concentrations of common additives used.

Table 2 Commonly used buffers in crystallization trials

Buffer	pH range
Acetate	3.6–5.6
Bis–Tris	5.8–7.2
Cacodylate	5.0–6.0
Citrate	2.6–7.0
HEPES	6.8–8.2
Imidazole–HCl	6.2–7.8
MES	5.5–7.0
MOPS	6.5–7.9
PIPES	6.1–7.5
Phosphate	5.9–8.0
Sodium–acetate	3.6–5.6
Sodium–cacodylate	5.0–7.4
Sodium–citrate	3.0–6.2
Succinate	5.0–6.0
Tris–HCl	7.2–9.0

Table 3 Commonly used precipitants in crystallization trials

Polymeric precipitants

Polyethylene glycol (PEG) 1000–20 000 molecular weight

Organic solvents

Acetone

Ethanol

Isopropanol

Methanol

2-methyl-2,4-pentanediol (MPD)

Polyethylene glycol (PEG) 400

1,3-propanediol

Tert-butanol

Salts

Acetates (ammonium, sodium)

Chlorides (ammonium, calcium, lithium, potassium, sodium)

Citrate (ammonium, sodium)

Halides (NaCl, NaBr, NaI, NaF)

Nitrates (ammonium, potassium, sodium)

Phosphates (ammonium, potassium, sodium)

Sulfates (ammonium, calcium, cadmium, lithium, magnesium, sodium)

Sulfonates (*p*-toluene, propane, benzoate, betaine)

Tartrates (potassium, sodium–potassium)

Thiocyanates (ammonium, potassium, sodium)

Table 4 Commonly used additives in crystallization trials

Additive	Concentration
Divalent cation	
Chloride dihydrates (barium, cadmium, calcium, cupric)	1–10 mM
Chloride hexahydrates (cobaltous, magnesium, strontium)	
Organic, non-volatile	
Ethylene glycol, glycerol anhydrous, 1-6 hexanediol,	2–5% (v/v)
2-methyl-2,4-pentanediol, polyethylene glycol 400	
Organic, volatile	
Acetone, dioxane, ethanol, isopropanol, methanol, propanol	2–5% (v/v)
Salt	
Ammonium sulfate, caesium chloride, potassium chloride,	0.05–0.2 M
lithium chloride, sodium chloride, sodium fluoride	
Reducing agent	
L-cysteine, 1,4-dithio-DL-threitol	10 mM
Carbohydrate	
D(+)-glucose, D(+)-sucrose, xylitol	2–5% (w/v)
Detergent	
β-Octylglucoside, Triton X-100	0.1–1 mM
Co-factor	
ATP, NAD	5–20 mM
Polyamine	
Spermidine, spermine tetra-HCl	5–20 mM

Finally, before setting up any crystallization experiment, microcentrifuge the sample for a few minutes to remove insoluble sample and other debris.

3.2 Crystal growth

The practical development of methodologies that allow growth of sample crystals of X-ray diffraction quality is the most fundamental in XRC. The forces governing crystal nucleation are difficult to understand and pinpoint because of the many factors that affect the solubility of the sample in the solvent in which it is dissolved. Factors include the buffer, the pH of the solution, the concentration of the sample and counter ions, the type and concentration of the precipitant used to bring the sample out of solution, the temperature, the surface area of droplet, and the gravity of the system used. In general, when no homologous proteins have been crystallized, all these factors have to be explored before suitable conditions are identified. If a homologous protein has been crystallized, the conditions for the new protein should explore those published and be expanded from this starting point. As a rule of thumb, the optimum size for a crystal is when at least two of the three crystal lengths measure 0.1–0.4 mm. However, crystals as small as 0.05 mm in all dimensions have been successfully used to determine structures when data is collected at microfocused, highly intense, synchrotron beamlines.

There are many methods for growing crystals and some of the more common methodologies are described below. There are also fundamental tips that are essential for successful crystal growth:

(a) Choose a solvent/buffer in which your sample is moderately soluble. Supersaturated solutions tend to give crystals that are too small in size.

(b) It is important to minimize dust or other extraneous particulate matter in the crystal-growing vessel.

(c) Allow crystals to grow with a minimum of disturbance. Vessels should be placed in a quiet place and left undisturbed for at least a week.

(d) Moving the vessel everyday to check on how your crystals are growing will almost certainly precipitate the sample or produce micro crystals.

3.2.1 Sitting and hanging drop

These are the most popular techniques used for crystallization. They are easy to perform and require only a small amount of sample. In addition, both methods allow a wide range of crystallization conditions to be screened and are therefore the methods of choice when trying to identify crystallization conditions for a new protein. The crystallization experiments are set up in linbro plates (referred to as trays), which can contain 24 or 96 individual crystallization compartments or wells. A different condition can be tested in each well.

In the sitting drop technique, a 2–40 μl droplet of sample is mixed with an equal amount of precipitant solution (also referred to as mother solution, mother liquor, or reservoir solution) and is placed on a bridge/post sitting inside a well, in vapour equilibration with the precipitant at the bottom of the well. With the hanging drop technique, the droplet, also mixed 1:1 with the precipitant, is suspended from a coverslip over the precipitant at the bottom of the well. In both methods the sealing of each crystallization well is essential to prevent air evaporation of the drop. The wells are sealed by creating an interface between the coverslip and the rim of each well on the linbro plate using vacuum grease, oil, or sealing tape in the case of the sitting drops.

The initial precipitant concentration in the droplet is less than that in the reservoir, thus, over time, the reservoir solution will draw water from the droplet in a vapour phase such that an equilibrium will exist between the droplet and the reservoir. During this equilibration process the sample is also concentrated, increasing its relative supersaturation, thus slowly bringing the sample out of solution and into a crystalline state.

The major advantages of the sitting and hanging drop techniques are speed and simplicity. The disadvantage of the sitting drop technique is that crystals can sometimes adhere to the sitting drop surface, making removal difficult. The hanging drop method avoids the problems of surface crystal adherence, but droplet volume is limited compared to sitting drops. Both methods are excellent for crystallization condition screening and optimization.

3.2.2 Sandwich drops

This method follows the same protocol as the sitting and hanging drop method, but the drops of sample mixed with precipitant solution are sandwiched between two coverslips. This method is rarely used, but does have the advantage of reducing the exposed surface area of the drop. This reduces the rate at which the precipitant reservoir solution draws water from the droplet and slows down the equilibration process compared to the sitting and hanging drop methods. This could be advantageous if the crystals are forming too fast and are thus too small for X-ray diffraction data collection.

3.2.3 Batch

Here the sample is mixed with the precipitant and appropriate additives to create a homogeneous crystallization medium requiring no equilibration with a reservoir. Therefore supersaturation is achieved directly rather than by diffusion. The advantages are speed and simplicity, but a major disadvantage is that only a narrow range of precipitant/sample concentration can be sampled in a single experiment. A batch experiment can be readily performed in a capillary, with, for example, 10 μl of solution, in a test-tube with 300 μl of solution, or plate with a small reservoir with <500 μl of solution. This technique is ideal when the crystallization conditions are known.

3.2.4 Microbatch

This is the same as the batch method, but uses extremely small drop volumes (<2 μl) and the droplet is covered by oil to prevent evaporation.

3.2.5 Free interface diffusion

This method of crystallization is less frequently used than the sitting or hanging drop vapour diffusion methods, but it is one of the methods used by NASA in microgravity crystallization experiments. The sample, in solution, is placed in direct contact with the precipitant. This creates a boundary interface between the sample and precipitant. Over time the sample and precipitant diffuse into one another and crystallization occurs either at the interface or on the side of the droplet containing high sample/low precipitant or low sample/high precipitant ratios. The technique allows the screening of crystallization conditions with respect to a gradient of sample/precipitant concentration combinations. The technique can readily be performed in small capillaries and uses only small amounts of sample.

3.2.6 Dialysis

Dialysis crystallization is one of the first methods used by crystallographers. It requires placing the sample in a dialysis cell, this can be as simple as a small length of narrow dialysis tubing tied at each end or a button (of defined volume typically between 10–1000 μl), which is sealed with a dialysis membrane at one or both sides. The dialysis cell is placed in a suitable container (typically a ~250 ml glass beaker) holding the precipitant or crystallization medium. The beaker is then sealed with Parafilm to prevent evaporation. The dialysis cell permits water and some precipitants to exchange while retaining the sample in the cell. The technique allows for salting-in and salting-out of the sample, as well as using a pH change to induce crystallization. For example, a sample requiring a high ionic strength for solubility can be dialysed against a solution of low ionic strength to salt it out.

3.3 Cryo-cooling of crystals for data collection

The cooling of crystals to near liquid nitrogen temperatures before data collection has become an essential technique in all areas of crystallography. Originally pioneered in the late 1960s for small molecule applications, low temperature techniques are used routinely in the field of macromolecular crystallography. A modern X-ray laboratory is not considered complete without a reliable cryo-cooling system because frozen crystals are much less sensitive to radiation damage in an X-ray beam than crystals cooled to 4°C, which decay because of radiation induced damage after initial exposure.

Freezing the sample crystal before X-ray diffraction data collection is done by removing the crystal from the drop it was grown in using a loop made of nylon or hair of a size similar to that of the crystal, and placing it in a cryogenic protective solution. This protective solution is usually the same as the precipitant solution used to grow the crystal but with an additional additive, which is a freezing protectant (more commonly referred to as cryoprotectant). Commonly used additives are polyethylene glycol 400 and glycerol. The solution can be placed in a watch glass or on a glass coverslip (usually 50 μl of solution is enough to immerse the crystal completely). This protectant will prevent the crystal from cracking as it is flash frozen in the pathway of a gaseous stream of liquid nitrogen with a temperature of approximately −173°C (100 K). Table 5 gives a list of cryoprotectants and useful concentration ranges.

Most crystals are frozen by the 'quick dip' method, which means that the crystal is dipped into the cryoprotectant solution for a few seconds and then flash frozen by being placed in the liquid nitrogen stream. In general, this is sufficient to protect the crystal from radiation damage, but a slower approach is sometimes necessary where the crystal is allowed to soak up the cryoprotectant for several minutes before being frozen. In the most extreme cases where no cryoprotectant can be found in which the crystal is stable (i.e. the crystal cracks or dissolves in the solution), the crystal may have to be grown in drops that contain cryoprotectant. This normally requires new crystallization conditions.

Table 5 Commonly used cryoprotectants

Cryoprotectant	Concentration
Glycerol	10–30% (v/v)
Vacuum pump oils	100% v
Ethylene glycol	20–40% (v/v)
Polyethylene glycol 400	10–30% (v/v)

3.4 Instrumentation hardware

There are four components to the collection of data in XRC.

3.4.1 The X-ray source

This can be in the form of a sealed tube (these are generally only used in laboratories for routine small molecule structures); a rotating anode (this is the X-ray generator of choice found in most structural biology laboratories); or a synchrotron source (these are large ring systems, usually in the order of a ¼ of a mile in diameter, that can produce very intense X-rays). The most common X-ray radiation used in the laboratory is CuKα with its characteristic wavelength, $\lambda = 1.5418$ Å, created by accelerating electrons to collide with a copper target.

3.4.2 The optics system

This is used to select the wavelength of the X-rays as well as to direct and focus the X-ray beam onto the sample crystal. At synchrotron sources special crystal prisms are used, which is especially useful for multi-wavelength data collection.

3.4.3 The X-ray detector

The major advancements in instrumentation for XRC have been in the area of detectors. For many years after the discovery of X-rays, photographic film had been used to detect the scattered X-rays from crystals, but this has changed. The detectors currently used are multi-wire systems, image plates, and with the onset of silicon chip technology, the charge-couple device (CCD). Within the next ten years the CCD will become the detector of choice because of its greater sensitivity, large dynamic range, and increased speed of data collection. At present, the best quality CCD detectors are found only at synchrotron sources, but with increased demand and use, prices will become affordable for structural biology laboratories.

3.4.4 The computer/graphics system

Developments in computer technology (storage space and memory capacity) have made it possible for the crystallographer to develop software to solve structures very quickly. These enable parallel sampling of many methods for determining the diffracted X-ray data phases, create electron density maps, and interpret structures in real time.

3.5 Phasing of X-ray diffraction data

3.5.1 Heavy atom soaks

One technique for phasing X-ray diffraction data is the heavy atom method, which exploits the ability to identify the position of electron dense atoms attached to specific amino acids. Several X-ray diffraction data sets of different heavy atom-soaked crystals can be used to determine the phases of the X-ray diffraction pattern using Patterson function searches. A native data set, plus data sets of native crystals soaked with heavy atoms, are collected. These heavy atoms (or heavy atom salt solutions), that bind to various surface accessible residues of the sample, are soaked into the crystals by incubation for 6–48 h prior to data collection (see Table 6 for a list of commonly used heavy atom salts). The crystals are usually transferred using a nylon loop or syringe into a new drop containing a solution of the crystallization precipitant or stabilizing solution containing a higher concentration of the precipitant plus 1–10 mM concentration of the heavy atom salt. The new drops are sealed to prevent air evaporation. In certain situations, crystals do not survive the transfer to the new drops and thus crack. For these crystals, attempts may be made to introduce low dilution heavy atom solutions, in small volumes (1–2 µl), to the side of the original drops to allow slow diffusion. Most heavy atom solutions are light sensitive, so the stock solutions should be covered to prevent exposure to direct light.

3.5.2 Selenium–methionine substitution

The most practical way to phase X-ray diffraction data of proteins is to use the properties of anomalous scattering that atoms containing large electrons have. This can be achieved by substituting all the methionine residues in the protein with selenium–methionine, as the selenium exhibits anomalous scattering which the sulfur atom does not. Hence, this method only works if there are available methionine sites to substitute and the protein can be expressed in an *E. coli* expression system where the source of methionine is replaced by selenium–methinoine. If selenium–methinoine-substituted crystals can be produced the X-ray diffraction data can be phased using the method of multi-anomalous dispersion (MAD) phasing. This method requires a synchrotron X-ray source for data collection because of the need to collect several data sets from the same crystal with different X-ray wavelengths. Thus, for this method to be successful, the crystals must be frozen to prevent radiation damage.

3.5.3 Molecular replacement

This method of phase determination uses information from a previously determined homologous protein structure. The method is dependent on knowledge of the orientation and position of the protein molecule of interest in the

Table 6 Commonly used heavy atom salts in the determination of protein structures

Au	Hg
Gold chloride	Mercury (II) chloride
Gold (III) chloride	Mercury (II) acetate
Gold potassium bromide	Mercury (II) bromide
Gold (I) potassium cyanide	Mercury (II) cyanide
Pt	**Other examples**
Platinum potassium cyanide	Lathanum (III) nitrate hexahydrate
Potassium tetrachloroplatinate (II)	Thallium (I) chloride
Ammonium tetrachloroplatinate (II)	Lead (II) chloride
Potassium hexabromoplatinate (IV)	Osmium (III) chloride hydrate

crystal unit cell. In general, a model is built for the unknown structure based on a sequence alignment with the known structure, which is then placed at the appropriate position in the crystal cell. Phases are then calculated for the model based on calculating scattering factors for the atoms in the model that can be compared and combined with the observed diffraction data.

3.6 Structure determination

It would be impossible, in the context of this chapter, to detail how to determine a crystal structure and no attempt will be made to do so. Essentially, the crystallographer requires the observed X-ray diffraction data and then, from one of many methods, must determine the phase of each of those waves. Some of these methods of phasing such as the Patterson and MAD method require special preparation of the crystals to collect additional experimental data compared to that of the native sample crystal, as mentioned above. Other phasing methods, such as molecular replacement (MR), only need an approximate model of what the structure might be. The rationale of all these methods is to obtain a set of phases to be assigned to each of the observed X-ray diffraction data thus permitting the calculation of an electron density map that can be used to interpret the structure. Figure 2 shows an example of the steps involved in structure determination using XRC.

4 Electron microscopy

Ever since the transmission electron microscope (TEM) was invented, biologists have used it to image cellular structures. The TEM provides a direct way to

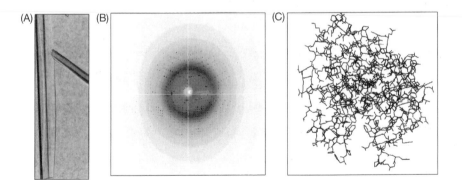

(A) (B) (C)

Figure 2 (A) Crystal. An optical photograph of a protein crystal. The crystal has dimensions of $2.0 \times 0.1 \times 0.05$ mm. These crystals were obtained by the hanging drop method. (B) X-ray diffraction pattern. An X-ray oscillation diffraction photograph taken of the crystal shown in Figure 2A. Many of these diffraction patterns would have to be collected, for different orientations of the crystals, to obtain all the data needed to solve the structure. A diffraction data set combined with phase information allows the calculation of an electron density map. (C) Structure. Stick diagram of the amino acid fold of the protein that formed the crystals shown in Figure 2A. The protein model was built into the electron density map generated from many experimentally collected diffraction patterns, just like the one shown in Figure 2B.

visualize the cell either by staining, shadowing, or sectioning the sample under study. Electron microscopy also provides a means for determining the structure of large macromolecular assemblages or complexes that are beyond the scope of NMR and XRC. It is also an invaluable tool for obtaining information on small complexes that cannot be crystallized. Recent advances in the methods of specimen preparation, such as frozen-hydration and 2D crystallization, have resulted in high resolution structural study applications using state-of-the-art TEMs. This section will deal with these two methods of structure determination by EM.

EM uses electrons to image the protein under investigation. An electron beam is passed through a specimen (prepared on a grid) and the resultant scattered beam is focused (using magnets) or collected (for data analysis) to produce an image, all carried out in a vacuum to prevent diffusion of the electron beam.

4.1 Sample preparation

4.1.1 Negative staining

Negative staining is used to check samples before more detailed analysis using specimens prepared by frozen-hydration. The sample preparation is to enhance the contrast of the protein/DNA sample relative to the background solvent, by embedding it in an electron dense medium. The stain covers the sample, and hence its size and shape can be determined from the stain mould, which scatters the electron beam more than the background.

The sample should usually be between 1–2 mg/ml concentration in a buffer containing low salt concentrations. Samples with high salt (e.g. CsCl) or sugar (e.g. sucrose) concentrations must be washed first to avoid high background signal.

Forceps (reversible with a fine tip) are used to hold the edge of the EM grid so that it is horizontal. Then, 5 μl of sample is applied to the grid for approx. 1 min. Excess sample is removed by touching filter paper to the edge of the grid. Then, 5 μl of stain (e.g. 2% uranyl acetate or phosphotungisitc acid in water, pH approximately 4–5) is placed on the grid for 10 sec and the excess drawn off. The grid should be dried for 5–10 min at room temperature before placing it in the EM. An over-stained grid will look black when viewed in the EM.

Negative staining is quick and easy to do. It allows for the rapid screening of samples and is inexpensive. Disadvantages include the fact that the stain shows only the surface features of the sample and lacks fine detail. Furthermore the technique allows for the possibilities of structural distortion and artefacts, because of quick sample dehydration, exposure to high concentrations of heavy metal salts, and abrupt pH changes.

4.1.2 Frozen-hydration (vitrification)

This technique of specimen preparation is the method of choice for observing an undistorted sample suspended in buffer. For this method to work, the sample needs to be concentrated to between 1–5 mg/ml in a buffer containing little or no salts or sugars, as these may substantially lower the freezing point of the buffer or cause dehydration of the sample.

Forceps (reversible with a fine tip) should be used to hold the edge of the EM grid so that it is vertically positioned in a sample plunging device (these can be commercially purchased but most are custom-built by the microscopist). The purpose of the device is to immerse the sample grid into liquid ethane or nitrogen so as to flash freeze the sample on the grid. The temperature must be less than $-150\,^{\circ}C$. This has to be done instantaneously after the excess sample is drawn off the grid, and is achieved in a guillotine-like fashion, as the grid is plunged into the liquid ethane.

Approximately 5 μl of sample should be applied to the grid for approx. 10 sec, and excess sample should be removed rapidly by pressing filter paper over the entire grid. This removal of excess sample has to be precise, removing enough to leave only a thin layer of sample, and prevent drying of the grid. As soon as the grid is prepared it has to be instantly flash frozen.

The prepared grid must be transported to the EM while maintaining a temperature less than $-150\,^{\circ}C$. Commercially available cryoholders cooled by liquid nitrogen can maintain the sample temperature at $-180\,^{\circ}C$. The EM itself has to have a special sample stage holder to accommodate the cryoholder. A few EM are now built that can operate at liquid helium temperatures $(-269\,^{\circ}C)$, which reduces thermal vibrations and confers some additional radiation protection to the sample grid when it is exposed to an energy intense electron beam.

Provided that the sample layer remaining is thin and the freezing is fast enough to prevent ice formation, the sample and buffer will be immobilized in an amorphous vitreous state, thus avoiding the problems of sample dehydration and distortions observed with negative staining and/or shadowing techniques.

The major advantage to this method is that the sample is not dehydrated, but is preserved in a thin film of ice (<100 nm) in random orientations. The sample is maintained at $-180\,^{\circ}C$ and prevents sublimation and crystallization.

4.1.3 Two-dimensional crystals

Another method for preparing samples for the EM is to form 2D crystals on the grid. This method is ideal for insoluble proteins, which are not amenable for study by the techniques of crystallography and NMR spectroscopy. The method not only allows for the imaging of the sample in an orderly crystalline array but also the collection of the electron scatter from the crystal itself. Hence, just as with crystallography, it allows the mathematical Fourier transform of this scatter to be calculated to form the 3D image. The advantage with the EM scattered data compared to that of crystallography is that the phases can also be directly calculated from the image of the crystalline arrays, therefore eliminating the phase problem that crystallography has to contend with.

Two approaches can be used; both based on the concept of keeping the crystalline sample to a 2D surface while maintaining hydration:

(a) The first is the formation of crystal from a gel liquid phase. In this method, after purification and concentration, the sample has to be reinserted into a lipid bilayer. Thus, when concentrating the sample, the choice of detergent used to maintain the solubility of the sample is as important as the exogenous lipid used for reinsertion, as is the concentration of the components. These parameters will vary widely from sample to sample. The 2D crystals are formed using the temperature dependence of lipid mobility, that is, as the sample temperature is decreased below the gel/liquid, threshold the sample can be 'frozen out' of the ordered lipid phase to form a 2D crystal.

(b) An alternative method of sample preparation is to take the saturated sample (2D crystal) from a flat lipid monolayer formed at the air/solvent interface. Such monolayers present suitable surfaces for ordered aggregation of the sample. The sample should have an affinity for the polar surface of the monolayer, this affinity may be a natural property of the sample or be induced by chemical modification of the sample or the lipid head group. Each sample will attach to the monolayer in an ordered manner and thus will pack adjacent to each other to form the 2D crystalline array.

4.2 Instrumentation hardware

The TEM comes in many forms, from the low end of the market; manually operated, tungsten source, photographic film image detector, which can be found at most biological related research institutes, to the high end of the market; remote computer operated, field emission gun (FEG) source, CCD image detector, with sample cooling holders that can go down to liquid helium temperatures. High end microscopes are now becoming available for use in the study of biological samples which can generate data with greater than 3 Å resolution.

4.3 Recording images, analysis, and reconstruction

Just as with NMR and XRC, the data analysis to achieve a high resolution image of a sample requires years of training, as each data set will bring its own unique problems.

The concepts themselves are simple; an electron micrograph from frozen-hydrated sample produce random orientations of 2D projections of three-dimensional objects. Therefore, by calculating the orientation of these projections and their centre of rotation (many biological sample have centres of symmetry which makes this process simpler), it is possible to reconstruct the three-dimensional image. This is achieved by adding the 2D Fourier transform of the individual images to create a 3D Fourier transform that can then be calculated back to the 3D sample that produced the images. The more images collected, the better the distribution of orientations of the sample, and the better the quality and resolution of the final image reconstruction. Figure 3 shows the random 2D projections of a sample and the subsequent 3D image produced when these images are combined.

The data collection from a 2D crystal sample is performed in a manner similar to crystallography; the electron diffraction data is collected as a series of tilt images of the sample diffraction. This data can be image analysed and then each

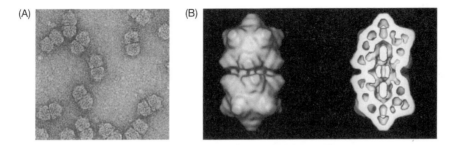

Figure 3 (A) Frozen-hydrated particles. Vitrified particles in a thin layer (<30 nm) of buffer at 100 K. The image shows the particles in a variety of random orientations. (B) Cryo-electron microscopy reconstruction. Surface shaded representation of the three-dimensional structure of the particles shown in Figure 3A. View of the exterior (left) and interior (right) surface. This particle reconstruction was produced by the summation of many (~2000) different two-dimensional projections, some of these are shown in Figure 3A.

of the individual diffraction data sets can be scaled and combined together with the phase information of the images from the crystal to compute an electron density map using Fourier transforms.

The major source of error from both of these methods comes from the optics of the microscope, the stability of the stage where the sample sits (the more movement the worse the resolution of the individual images), and the radiation damage of the sample as the image is collected. The resolution of the single particle reconstructed image is also limited by the number and variation of orientation of the images and the 2D electron diffraction by the degree of order of the sample crystal itself.

5 The protein database

With the continuing development and accelerating success of all three of the techniques discussed, a wealth of structural data has now been obtained and will continue to be amassed at an ever increasing rate. The completion of the human genome project will have a significant effect in the field of structural biology with the initiation of the determination of structures on a genomic scale in a high-throughput mode. In an effort to store and make accessible this information, the Research Collaboratory for Structural Bioinformatics (RCSB; http://www.rcsb.org) was created in 1998. This database is the most complete collection of structural biology and is updated on a daily basis.

Suggested reading

Nuclear magnetic resonance spectroscopy

General

1. Wüthrich, K. (1986). *NMR of proteins and nucleic acids*. Wiley, New York.
2. Ernst, R. R., Bodenhausen, G., and Wokaun, A. (1987). *Principles of nuclear magnetic resonance in one and two dimensions*. Oxford University Press, Oxford.

3. Evans, J. N. S. (1995). *Biomolecular NMR spectroscopy*. Oxford University Press, Oxford.
4. Cavanagh, J., Fairbrother, W. J., Palmer III, A. G., and Skelton, N. J. (1996). *Protein NMR spectroscopy: principles and practice*. Academic Press, London.

Protein multinuclear triple resonance

1. Edison, A. S., Abildgaard, F., Westler, W. M., Mooberry, E. S., and Markley, J. L. (1994). In *Methods in enzymology* (ed. N. J. Oppenheimer and T. L. James), Vol. 239, p. 1. Academic Press, London.
2. Clore, G. M. and Gronenborn, A. M. (1998). *Curr. Opin. Chem. Biol.*, **2**, 564.

Relaxation and NOE

1. Neuhaus, D. and Williamson, M. (2000). *The nuclear Overhauser effect in structural and conformational analysis*, second edition. Wiley-VCH, New York.
2. Peng, J. W. and Wagner, G. (1994). In *Methods in enzymology* (ed. N. J. Oppenheimer and T. L. James), Vol. 239, p. 563. Academic Press, London.

TROSY

1. Pervushin, K., Riek, R., Wider, G., and Wüthrich, K. (1997). *Proc. Natl. Acad. Sci. USA*, **94**, 12366.

Partially oriented samples

1. Tjandra, N. and Bax, A. (1997). *Science*, **278**, 1111.

Unstructured proteins

1. Wright, P. E. and Dyson, H. J. (1999). *J. Mol. Biol.*, **293**, 321.

X-ray crystallography
General

1. McRee, D. E. (1993). *Practical protein crystallography*. Academic Press, London.
2. Rhodes, G. (2000). *Crystallography made crystal clear: A guide for users of macromolecular models*. Second edition. Academic Press, London.

Growing crystals

1. Bergfors, T. M. (ed.) (1999). *Protein crystallization*. International university line, California.

Solving crystal structures

1. Wyckoff, H. W., Hirs, C. H. W., and Timasheff, S. N. (ed.) (1985). *Diffraction methods for biological macromolecules (Part A)*. Vol. 114. Academic Press, London.
2. Wyckoff, H. W., Hirs, C. H. W., and Timasheff, S. N. (ed.) (1985). *Diffraction methods for biological macromolecules (Part B)*. Vol. 115. Academic Press, London.

Electron microscopy (single particle reconstruction)
General

1. Toyoshima, C. and Unwin, N. (1988). *Ultramicroscopy*, **25**, 279.
2. Baker, T. S. and Cheng, R. H. (1996). *J. Struct. Biol.*, **116**, 120.
3. Fuller, S. D., Butcher, S. J., Cheng, R. H., and Baker, T. S. (1996). *J. Struct. Biol.*, **116**, 48.
4. Conway, J. F., Trus, B. L., Booy, F. P., Newcomb, W. W., Brown, J. C., and Steven, A. C. (1996). *J. Struct. Biol.*, **116**, 200.

5. Mancini, E. J., de Haas, F., and Fuller, S. D. (1997). *Structure*, **5**, 741.
6. Baker, T. S., Olson, N. H., and Fuller, S. D. (1999). *Microbiol. Mol. Biol. Rev.*, **63**, 862.

Electron microscopy (diffraction)

Two-dimensional crystals

1. Kühlbrandt, W. and Downing, K. H. (1989). *J. Mol. Biol.*, **207**, 823.
2. Wang, D. N. and Kühlbrandt, W. (1991). *J. Mol. Biol.*, **217**, 691.
3. Kühlbrandt, W. (1992). *Q. Rev. Biophys.*, **25**, 1.

Data collection to obtain structural amplitudes

1. Wang, D. N. and Kühlbrandt, W. (1992). *Biophys. J.*, **61**, 287.

Phase determination

1. Unwin, P. N. T. and Henderson, R. (1975). *J. Mol. Biol.*, **94**, 425.
2. Henderson, R., Baldwin, J. M., Downing, K. H., Lepault, J., and Zemlin, F. (1986). *Ultramicroscopy*, **19**, 147.
3. Henderson, R., Baldwin, J. M., Cseka, T. A., Zemlin, F., Beckmann, E., and Downing, K. H. (1990). *J. Mol. Biol.*, **213**, 899.

Databases

1. Glusker, J. P. (ed.) (1998). *Acta crystallographica*, **D54**, 1065.
2. BioMagResBank (BMRB) database (www.bmrb.wisc.edu)
3. Research Collaboratory for Structural Bioinformatics (RCSB) database (www.rcsb.org)

Chapter 9
Nucleic acids

T. A. Brown

Department of Biomolecular Sciences, University of Manchester
Institute of Science and Technology, Manchester M60 1QD, UK.

1 Introduction

A full description of the nucleic acid techniques relevant to cell biology would require hundreds of pages of text. This chapter is therefore restricted to the key procedures involved in the extraction of nucleic acids from living tissues and the study of those nucleic acids by the polymerase chain reaction (PCR), agarose gel electrophoresis, hybridization analysis, and DNA sequencing. The reader is referred to the relevant parts of larger and more comprehensive manuals (e.g. refs 1 and 2) for further information on these techniques and for descriptions of procedures that are beyond the scope of this chapter.

The problems faced by a newcomer to nucleic acid analysis are eased greatly by the availability of a range of commercial kits that provide all the reagents and other materials needed for a particular experiment. These kits usually give reasonably good results and in most cases these results are easier to obtain than by the traditional method of purchasing the materials separately, making up one's own buffers, and suchlike. Cost is also a factor because kits are usually cheaper than the sum of the individual materials they contain. Unfortunately, kits rarely provide the researcher with any flexibility with regards to the design of a procedure, which means that the traditional route, though slower, will usually give better results in the long run, as experience gained along the way can be used to modify steps and buffer compositions to suit the particular DNA molecules being studied. Kits are an immense help for those whose involvement with nucleic acids will be little more than a brief flirtation, but if you plan a longstanding relationship then the best advice is to start by learning and understanding each technique. Once you have mastered the techniques then determine if you can obtain equally good results with kits, and if so then use the kits for repetitive experiments.

2 Purification of nucleic acids

Purification of DNA and RNA from various sources is central to all work with nucleic acids. As well as purification of genomic DNA, total RNA, and poly(A^+)

RNA from animal and plant material, techniques are also needed for preparation of plasmid and bacteriophage clones from *Escherichia coli* cells, purification of individual DNA bands from agarose gels, and clean-up of PCR products prior to cloning or sequencing. Until recently, this aspect of molecular biology was complex with many different procedures available for each application, most of these being messy, time-consuming, and difficult. Over the last five years, the entire technology has been revolutionized by the introduction of purification procedures based on the selective binding of nucleic acids to silica gel membranes and to anion exchange resins. Both matrices selectively bind nucleic acids and can purify DNA and RNA from complex mixtures without the need for the lengthy organic extractions and selective precipitations that characterized the earlier techniques. The specificity of the procedure for DNA or RNA is determined by the binding conditions, and the parameters can be directed at DNA molecules of different sizes, enabling plasmid and phage clones to be separated from bacterial genomic DNA, and unused primers to be removed from PCR products. Resins with surface oligo(dT) groups are used for poly(A$^+$) RNA purification prior to cDNA cloning. The preparative steps prior to gel/resin binding depend on the nature of the source material. Many protocols make use of the chaotropic reagent guanidinium thiocyanate, which dissolves tissues and leave nucleic acids intact and in a form suitable for binding to silica gel (3). Animal tissues and microbial cultures can be treated directly with guanidinium thiocyanate; plant material may need to be disrupted by grinding in liquid nitrogen.

It is possible to carry out this type of nucleic acid purification from first principles, using silica gel or an anion exchange resin in suspension or in home-made columns (4). There may be advantages in doing this if the source material is unusual and the extraction parameters need to be monitored in order to achieve satisfactory results. For most applications it is easier to use the nucleic acid purification systems marketed by various companies (e.g. Qiagen, Promega). In these systems the binding matrix is enclosed within a filtration unit such as a spin-column, and the procedure simply involves application of the sample to the filtration unit followed by washing and elution of the nucleic acid. Different systems are available for purification of DNA and RNA of various types from various sources, including agarose gels and PCR mixtures.

3 The polymerase chain reaction

PCR is a deceptively simple but remarkably versatile procedure that has applications in virtually all areas of molecular biology research. All that a PCR does is make multiple copies of a target DNA sequence, the amplification process enabling microgram quantities of DNA to be obtained from a single starting molecule. Any segment, up to 40 kb in length, of any DNA or RNA molecule can be amplified, providing the sequences to either side of the targeted region are known. If genomic DNA is used as the starting material the PCR can be used in an analogous fashion to cloning, to provide a pure sample of a single gene or other DNA sequence, with the advantage that the gene is obtained after a few hours work, rather than the weeks needed to prepare and screen a clone library. If RNA is the starting material then PCR can achieve the same objective as cDNA cloning, again much more rapidly. PCR of RNA samples from different tissues can indicate the expression pattern of a gene, and specialized techniques such as rapid amplification of cDNA ends (RACE) (5) can map the 5' or 3' ends of an RNA on to a DNA sequence, enabling the initiation and termination positions for transcription of a gene to be identified. *In situ* PCR (6), like *in situ* hybridization, can be used to locate transcripts within cells, but being more sensitive it can detect rare transcripts as well as ones that are highly abundant.

3.1 PCR in outline

PCR is carried out by mixing together the appropriate reagents and incubating them in a thermal cycler, a piece of equipment that enables the incubation temperature to be varied over time in a pre-programmed manner. The basic steps in a PCR experiment are as follows (Figure 1):

(a) DNA is prepared from the organism being studied and denatured by heating to 94°C.

(b) A pair of oligonucleotides is added to the DNA, the sequences of these oligonucleotides enabling them to anneal either side of the gene or other DNA segment that is to be amplified, and the mixture is cooled to 50–60°C so these oligonucleotides attach to their target sites.

(c) A thermostable DNA polymerase is added together with a supply of deoxyribonucleotides and the mixture is heated to the optimal temperature for DNA synthesis, 74°C if *Taq* DNA polymerase (the DNA polymerase I enzyme from the thermotolerant bacterium *Thermus aquaticus*) is used. The annealed oligonucleotides now act as primers for synthesis of new polynucleotides complementary to the template strands.

(d) The cycle of denaturation–annealing–extension is repeated 25–30 times, with the number of newly synthesized DNA molecules doubling during each cycle. This exponential amplification results in synthesis of a large number of copies of the DNA sequence flanked by the pair of oligonucleotides.

3.2 Planning and carrying out a PCR

The primers are the critical component of a PCR. They must anneal to either side of the region targeted for amplification, which means that the sequences of these boundary regions must be known. If the starting material is genomic DNA then each primer should be at least 15 nucleotides in length, to reduce the possibility that binding sites other than the targets are present in the template DNA. The primers should be less than 20 nucleotides because long primers anneal slowly reducing the efficiency of amplification. The primers should not be able to anneal to one another, and the design should avoid the possibility of a primer forming intramolecular base pairs and becoming tied up in a hairpin

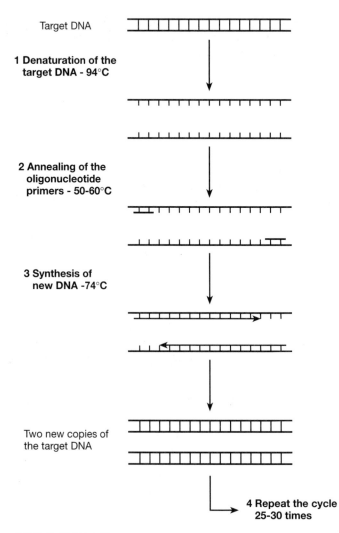

Target DNA

1 Denaturation of the
target DNA - 94°C

2 Annealing of the
oligonucleotide
primers - 50-60°C

3 Synthesis of
new DNA -74°C

Two new copies of
the target DNA

4 Repeat the cycle
25-30 times

Figure 1 PCR in outline.

structure. Either of these events will result in 'primer–dimers' that will pre-
dominate over the desired amplification product at the end of the PCR. The
primers should also have similar melting temperatures, to ensure that they have
similar annealing characteristics. The melting temperature of a short oligonucleotide can be estimated from the equation:

$$\text{melting temperature} = 4 \times (\text{number of GC base pairs}) + 2 \times (\text{number of AT base pairs})$$

The annealing temperature used in the PCR should be 2–4°C below the lower
of the two melting temperatures for the primer pair. Several computer pro-
grammes are available to help in the design of suitable pairs of primers.

Protocol 1 gives a basic PCR procedure. It uses a 'hot-start', which means that addition of one key component of the reaction ($MgCl_2$ in this version) is delayed until the mix has been heated to the denaturation temperature in the first thermal cycle. This strategy improves specificity by preventing DNA synthesis from occurring before the reaction has begun, when the tubes are on ice or at room temperature before being placed in the cycler. At these low temperatures, primers can anneal to non-specific sites in the template DNA and be extended by the polymerase into short products which re-anneal at those non-target sites during the PCR, leading to spurious products. In Protocol 1 the hot-start requires that the tubes be re-opened, which could result in cross-contamination. Two alternative procedures avoid this problem. The first involves constructing a wax barrier over the main reaction mix, with the $MgCl_2$ pipetted on top. When the reaction heats up to the first denaturation temperature the wax melts and the supplement enters the reaction. The second method is to use a complex consisting of *Taq* polymerase and a heat-sensitive antibody that prevents activity of the enzyme. When the first denaturation temperature is reached the antibody is degraded and the enzyme is activated (7). Commercial systems for carrying out both procedures are available from a variety of sources.

Protocol 1

Basic procedure for PCR[a]

Equipment and reagents

- Thermal cycler with a hot lid (available from many suppliers)
- 0.5 ml microcentrifuge tubes
- 10 × PCR buffer: 100 mM Tris–HCl pH 9.0, 500 mM KCl, 0.1% Triton X-100 (prepared in sterile water)
- 20 mM $MgCl_2$ (prepared in sterile water)
- 40 mM dNTP solution: 10 mM dATP, 10 mM dCTP, 10 mM dGTP, 10 mM dTTP (prepared in sterile water)

- Primers 1 and 2: 10 μM solutions of each primer in sterile water
- *Taq* DNA polymerase
- Template DNA[b]
- Gel loading buffer: 15% Ficoll 400, 0.25% bromophenol blue, 0.25% xylene cyanol

Method

1 Prepare a 'master mix' containing the following components. The amount of reagents to add depends on the number of reactions that are being carried out. The following recipe is sufficient for a single reaction and should be scaled up accordingly.[c]

   - 10 × PCR buffer   5 μl
   - 40 mM dNTP solution   1 μl
   - Primer 1   2.5 μl

Protocol 1 continued	

- Primer 2   2.5 µl
- Sufficient water to bring the volume to 49 µl x µl
- *Taq* DNA polymerase   2.5 U
- Total volume   44 µl

2   Transfer 44 µl of master mix to a 0.5 ml microcentrifuge tube and add 1 µl template DNA.

3   Place each tube in a thermal cycler that has been programmed to provide the following heating regime:

- 2 min at 94 °C
- 30 cycles of 1 min at the annealing temperature,[d] 1 min at 74 °C, 1 min at 94 °C
- 8 min at 74 °C

4   After 1 min at 94 °C, pause the thermal cycling and add 5 µl of 20 mM $MgCl_2$ to each tube.[e] Continue the thermal cycling.

5   When the thermal cycling is complete, add 10 µl gel loading buffer to each tube.

6   Analyse the results by agarose gel electrophoresis (Section 4). If the PCR has worked efficiently then a clear band should be visible after electrophoresis of a 10 µl aliquot of the reaction.

7   PCR products can be stored at 4 °C for several months.

[a] This protocol is suitable for amplification of fragments up to 3 kb in length. Longer fragments—up to 40 kb—are amplified by a modified procedure (8). RNA can also be used as a substrate for PCR providing that the amplification protocol is preceded by a step in which the RNA is converted to cDNA by an enzyme with reverse transcriptase activity. This is most easily achieved by substituting *Tth* DNA polymerase for *Taq* DNA polymerase, the *Tth* enzyme having both RNA- and DNA-dependent DNA polymerase activities (8).

[b] The size of the template DNA dictates how much should be added to a PCR. For human genomic DNA, use 1 µg per reaction. For a recombinant plasmid, 10 pg is sufficient. However, the amount of template DNA that is added often depends on availability and successful PCRs are possible with substantially less than the amounts stated.

[c] Good controls are essential in all PCRs because the sensitivity of the reaction makes it very susceptible to contamination with extraneous DNA or PCR products from past experiments, which may persist as aerosols in laboratory air (8). Always include a 'water blank', which is a PCR set up with sterile water instead of template DNA and which tests for the absence of contamination in the PCR reagents, and an 'extract blank', which checks for contamination of reagents used in preparation of the template DNA. An extract blank can be obtained by carrying out the procedure for template DNA preparation with no starting material or by starting with equivalent material that does not contain the primer annealing sites. A positive control, with template DNA that has proven ability to yield the correct PCR product, is optional. Its main purpose is to confirm that the reagents are active and the cycling parameters are appropriate, but this will be apparent if the test PCRs give the anticipated product. After electrophoresis, the sizes of the test products can be checked by comparison with the positive control, but this can equally well be done by comparison with DNA size markers.

[d] The annealing temperature is determined as described in the text. With a new pair of primers it might be necessary to optimize this parameter by setting up a series of PCRs with annealing temperatures ranging from 8 °C below to 4 °C above the estimated value, in 2 °C increments.

[e] This addition gives a final Mg^{2+} concentration of 2 mM. The optimum value depends on the sequence characteristics of the region to be amplified and should be determined experimentally whenever a new PCR is designed. The optimum can be at any point between 1.0–5.0 mM Mg^{2+}, so test PCRs should be set up covering this range in 0.5 mM increments by changing the amount of $MgCl_2$ added at this step, with the water content of the master mix altered to compensate. The optimum Mg^{2+} concentration is assessed by comparing product yields on an agarose gel. It is not necessary to optimize any of the other components of the PCR, but improvements in yield can be obtained by adding a compound that reduces decay of the polymerase during the thermal cycling (e.g. final concentrations of 10 mM 2-mercaptoethanol, 10 mM dithiothreitol, 10% glycerol, 5 μg/ml *E. coli* single-stranded binding protein, or 50 mM tetramethyl ammonium chloride, the last particularly useful if the template DNA is GC-rich). The accumulation of pyrophosphates (a product of dNTP polymerization) in the latter stages of a PCR, which reduce yield by promoting depolymerization of DNA, can be avoided by including a thermostable pyrophosphatase (Roche) in the reaction. Yield and/or specificity can also be improved by addition of PCR enhancers available from a number of companies (8).

4 Agarose gel electrophoresis

The standard method for monitoring the results of a nucleic acid technique is to examine a sample of the resulting material by agarose gel electrophoresis. The appearance of the nucleic acid in the gel indicates if a genomic DNA or total RNA preparation has been successful, whether a PCR has yielded the desired product, and whether manipulations such as restriction and ligation (Section 7) have worked.

4.1 DNA gels

DNA molecules are separated in non-denaturing slab gels, usually less that 50 ml in volume, prepared in one of two buffers. TBE (89 mM Tris base, 89 mM boric acid, 2 mM EDTA) is suitable for most purposes, but TAE (40 mM Tris base, 40 mM acetic acid, 1 mM EDTA) is preferred if the DNA fragments of interest are >15 kb or if one or more fragments are to be recovered from the gel. Electrophoresis is at 1–2 V/cm for large molecules or 5–10 V/cm for fragments <15 kb, the latter corresponding to an applied voltage of up to 120 V. The specification of the agarose, and the concentration at which it is used, determines the range of DNA molecules that can be resolved. A 1% gel in standard agarose allows molecules of 400 bp to 8 kb to be separated. Resolution in the 10–500 bp range requires a 4–6% gel made with one of the higher-performance agaroses that are available from many companies (9). Detection limits of 10 ng DNA per band are possible with ethidium bromide and other stains.

Protocol 2 describes a standard method for running and staining a DNA gel. It omits what is probably the most critical step with regards to the clarity of the resulting gel, this step being preparation of the dissolved agarose. Agarose under-goes a series of processes when it is dissolved: dispersion, hydration, and melting/dissolution (9). Problems are sometimes encountered with hydration when using a microwave oven to dissolve agarose. In part, this occurs because hydration is time-dependent and microwave ovens bring the temperature up rapidly. The problem is exacerbated by the fact that the agarose is not being agitated to help dilute the highly concentrated solution around each particle. The recommended procedure is to use a beaker that is two to four times the volume of the gel to be made, and to add the agarose powder to the buffer (not the other way round) by sprinkling while the mixture is being rapidly stirred. Then take out the stir bar and heat in a microwave oven at high power until bubbles appear. Remove the beaker and swirl gently to resuspend any settled powder, taking care to avoid scalding as the solution might be superheated and foam over when agitated, then replace in the oven and boil for 1 min. Add enough hot distilled water to return to the desired volume and allow to cool to 50–60 °C prior to casting the gel. If the gel concentration is >2%, then the hydration step needs to be extended by leaving the buffer–agarose mix to soak at room temperature for 15 min before dissolving on medium power, possibly with a second 15 min rest period halfway through dissolution if the concentration is 4% or more (9).

Protocol 2

Agarose gel electrophoresis of DNA[a]

Equipment and reagents

- Horizontal electrophoresis apparatus plus accessories (e.g. power supply, gel plate, comb)
- Staining vessels larger than the gel (e.g. glass dishes)
- 254 nm wavelength UV transilluminator
- Agarose solution prepared in 1 × TBE or 1 × TAE
- 1 × TBE or 1 × TAE electrophoresis buffer 5 × TBE: dissolve 54.0 g Tris base, 27.5 g boric acid, 3.72 g EDTA in 750 ml distilled water, then add distilled water to a final volume of 1 litre. 50 × TAE: mix 242 g Tris base, 57.1 ml glacial acetic acid, 18.61 g EDTA in 750 ml

distilled water, then add distilled water to bring to a volume of 1 litre
- Gel loading buffer (see Protocol 1)
- Ethidium bromide stock solution (10 mg/ml): add 1 g ethidium bromide to 100 ml distilled water and stir on a magnetic stirrer for several hours. Transfer the solution to a dark bottle and store at room temperature. **CAUTION: Ethidium bromide is a powerful mutagen. Gloves should be worn when handling solutions of this dye and stained gels. Appropriate eye and skin protection should be worn when observing gels on a UV transilluminator.**

A. Preparing the gel

1 Allow the agarose solution to cool to 50–60 °C.

2 While the solution is cooling, assemble the gel casting tray. Level the casting tray and check the teeth of the comb for residual dried agarose. Dried agarose can be

removed by scrubbing the comb with a lint-free tissue soaked in hot distilled water. Allow a small space (approx. 0.5–1.0 mm) between the bottom of the comb teeth and the casting tray.

3   Pour the agarose solution into the gel tray. Position the comb and allow the agarose to gel at room temperature for 30 min.[b]

4   Once the gel is set, flood with electrophoresis buffer (either 1 × TBE or 1 × TAE, whichever was used to prepare the gel) and slowly remove the comb. Place the gel casting tray into the electrophoresis chamber and fill the chamber with electrophoresis buffer until the buffer reaches 3–5 mm over the surface of the gel. Gently flush out the wells with electrophoresis buffer using a Pasteur pipette to remove loose gel fragments prior to loading the samples.

B. Loading the gel and electrophoresis

1   Add a one-fifth volume of gel loading buffer to each DNA sample and pipette each sample carefully into the wells.[c] Use the outermost lanes for DNA size markers (available from many suppliers).

2   Place the lid on the gel apparatus and attach to the power supply. For standard mini-gels in TBE buffer, an applied voltage of up to 120 V (5–10 V/cm) is suitable for 1–2 h run. For TAE gels use 1–2 V/cm.

3   Stop the electrophoresis when the dyes have reached the desired position within the gel.[d]

C. Staining and visualizing the gel[e]

1   Prepare enough working solution of ethidium bromide (0.5–1.0 mg/ml of ethidium bromide in distilled water or electrophoresis buffer) to cover the surface of the agarose gel.

2   Remove the gel from the electrophoresis chamber and submerge for 20 min[f] in the ethidium bromide solution.

3   Remove the gel and destain by submerging for 20 min in a new container filled with distilled water.

4   Repeat step 3, using fresh distilled water.[g]

5   View on a UV transilluminator.

[a] Modified from ref. 9.

[b] Low-melting- and intermediate-melting-temperature agaroses require an additional 30 min of gelling at 4 °C to obtain the best gel handling.

[c] The volume of sample depends on the sizes of the wells. Usually, volumes of 10–20 μl are loaded. If DNA is to be recovered from the gel then a larger loading might be necessary to give a satisfactory yield, but the volume should still be kept as small as possible.

[d] The migration rates of the bromophenol blue and xylene cyanol dyes can be used to determine the approximate positions within the gel of DNA molecules of different sizes. Their migration rates are dependent on agarose type, concentration, and running buffer (see ref. 9).

Protocol 2 continued

[e] To avoid the hazards of ethidium bromide staining, use an alternative commercial stain such as SYBR Green—follow the manufacturer's instructions. As an alternative to post-staining, ethidium bromide can also be included in the electrophoresis buffer, but this is not recommended because it increases the hazard. Ethidium bromide solutions must be treated before disposal by diluting to <100 µg/ml, adding 0.05 g/ml Amberlite XAD-16, stirring for 2 h per 10 ml of solution, and filtering to recover the resin. The resin must be treated as hazardous waste; the filtrate is non-hazardous and can be disposed of in the normal way.

[f] Double this time if the gel concentration is 4% or greater.

[g] The destaining steps can be repeated more than twice if the background is still too high.

4.2 RNA gels

RNA is single-stranded and can form secondary structures, and so must be electrophoresed in a denaturing system. Various denaturants can be used (9) but a combination of glyoxal and DMSO (Protocol 3) is recommended as it is adequate for most purposes and presents a much less severe health hazard than formaldehyde or methylmercuric hydroxide.

In all manipulations involving RNA, care must be taken to avoid contamination of samples with ribonucleases that are present in glassware and plasticware and which will rapidly degrade the RNA being studied. Electrophoresis equipment should be cleaned in detergent, rinsed with distilled water, and dried with ethanol before soaking in 3% hydrogen peroxide for 15 min, followed by a thorough rinse with DEPC-treated water. The latter is prepared by adding DEPC to double-distilled water to a final concentration of 0.1%, autoclaving for 1 h, and then leaving overnight in the autoclave after releasing some of the steam, the lengthy heat treatment being needed to decompose residual DEPC. For more detailed precautions for avoiding ribonuclease contamination see ref. 10.

Protocol 3

Electrophoresis of RNA by the glyoxal–DMSO method[a]

Equipment and reagents

- Horizontal electrophoresis apparatus plus accessories (e.g. power supply, gel plate, comb)
- Apparatus for recirculating the electrophoresis buffer
- DEPC-treated water (see text)
- 100 mM sodium phosphate pH 7.0: mix 5.77 ml of 1 M Na_2HPO_4 with 4.23 ml of 1 M NaH_2PO_4 and bring the volume up to 100 ml with RNase-free water

- DMSO
- 6 M glyoxal (40% solution), deionized immediately before use: pass the solution through a small column of mixed-bed ion exchange resin (e.g. Bio-Rad AG501-X8 or X8(D) resins) until the pH is >5.0
- Glyoxal loading buffer: 10 mM sodium phosphate pH 7.0, 0.25% bromophenol blue, 0.25% xylene cyanol, 50% glycerol

Protocol 3 continued

A. Denaturation of samples

1   Bring the RNA volume up to 11 μl with DEPC-treated water.

2   Add 4.5 μl of 100 mM sodium phosphate pH 7.0, 22.5 μl of DMSO, and 6.6 μl of deionized glyoxal.

3   Mix thoroughly then heat at 50 °C for 1 h.

4   Chill on ice for at least 1 min before loading.

B. Electrophoresis

1   For a 1% gel, dissolve 1.0 g of agarose in 100 ml of 10 mM sodium phosphate pH 7.0. Adjust the amounts for different per cent gels. Cool to 60 °C in a water-bath.

2   Cast the gel using the procedure described in Protocol 2A. Gels should be cast to a thickness where the wells can accommodate 60 μl.

3   Remove the comb and cover the gel with 10 mM sodium phosphate to a depth of 1 mm.

4   Add 12 μl of glyoxal loading buffer per 45 μl of sample and mix thoroughly.

5   Load the samples[b] and electrophorese at a maximum of 4 V/cm while the running buffer is being recirculated.[c] Continue the electrophoresis until the bromophenol blue (the darker blue and more rapidly migrating dye) reaches the end of the gel.

[a] Modified from ref. 9.

[b] Load 0.5–1.0 μg of RNA per lane.

[c] Recirculation of the buffer is needed to prevent formation of a pH gradient. If no recirculation apparatus is available, pause the electrophoresis every 30 min and remix the buffer in the apparatus. Alternatively, one of two electrophoresis buffers that do not require recirculation can be used (though with less good results): 10 mM PIPES, 30 mM bis-Tris; or 20 mM MOPS, 5 mM sodium acetate, 1 mM EDTA, 1 mM EGTA.

5 Southern hybridization

DNA molecules that have been separated by agarose gel electrophoresis can be examined by transfer to a membrane support followed by hybridization analysis with a labelled probe. The probe might be a cloned gene, the results showing which restriction fragments in a genomic DNA digest contain the gene of interest, or it could be an oligonucleotide used to confirm that a PCR has resulted in the expected amplification product. The procedure is called Southern hybridization (northern hybridization is the equivalent technique with electrophoresed RNA). DNA and RNA can also be directly dot blotted on to a membrane prior to examination by hybridization probing, and special methods enable cloned DNA within plasmid colonies or bacteriophage plaques to be blotted. For all of these techniques, the reader should consult ref. 11. In this section the description is limited to the Southern method, which illustrates the key issues involved in hybridization analysis.

5.1 The Southern blot

The objective of the Southern blot (Protocol 4) is to transfer the DNA from the gel to a membrane in such a way that the relative positioning of DNA bands in the gel is reproduced on the membrane. The membrane provides a suitable format for hybridization probing whereas the gel matrix does not. Several types of membrane can be used (11), including nitrocellulose filters, which give low backgrounds but are fragile and less suitable for re-probing, and positively-charged and neutral nylon membranes, which bind DNA molecules more tightly. Transfer from gel to membrane is usually achieved by capillary action, using the apparatus illustrated in Figure 2. Buffer soaks through the wicks, gel, and filter or membrane to the paper towels, the overall architecture of the blot ensuring that band positioning remains unaltered during the transfer. More rapid blotting can be achieved with a vacuum device. Prior to transfer the DNA is depurinated by soaking the gel in HCl. This treatment results in strand breakage, enhancing transfer efficiency because short molecules diffuse more quickly than long ones. The subsequent hybridization analysis requires that the DNA be single-stranded, so the gel is next soaked in alkali before being neutralized. The blot is then assembled and transfer achieved overnight.

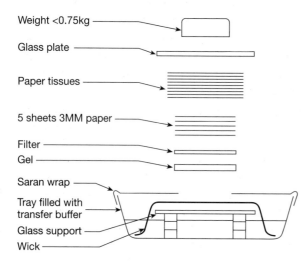

Figure 2 A typical set-up for nucleic acid blotting.

Protocol 4

Southern blotting on to a nitrocellulose filter[a]

Equipment and reagents

- Shaking platform
- Small weight (approx. 500 g)
- Vacuum oven
- Nitrocellulose filter (e.g. Amersham-Pharmacia Hybond-C)
- Whatman 3MM paper
- Paper towels

- 0.25 M HCl
- Denaturing solution: 1.5 M NaCl, 0.5 M NaOH
- Neutralizing solution: 1.5 M NaCl, 0.5 M Tris–HCl pH 7.0
- 20 × SSC: 3 M NaCl, 0.3 M trisodium citrate pH 7.0

Method

1   **Depurination**. Rinse the gel[b] with distilled water and place it in a glass dish with 0.25 M HCl. Use between five and ten times the gel volume for this and subsequent washing steps. Leave the gel for 30 min at room temperature on a shaking platform set at low speed.

2   **Denaturation**. Remove the HCl and rinse the gel with distilled water. Add the appropriate volume (see step 1) of denaturing solution and incubate at room temperature with shaking for 20 min. Pour off the solution and replace with fresh denaturing solution. Incubate for a further 20 min, shaking gently.

3   **Neutralization**. Remove the denaturing solution and rinse the gel with distilled water. Add the appropriate volume of neutralizing solution and incubate at room temperature with shaking for 20 min. Remove the solution and replace with fresh neutralizing solution. Incubate for a further 20 min, shaking gently.

4   **Preparing the blotting apparatus**. A typical blotting apparatus is shown in Figure 2. Fill the reservoir with at least 500 ml of 20 × SSC and set up a flat platform. Wet three sheets of 3MM paper in 20 × SSC and use to cover the platform. The ends of the 3MM paper should dip into the 20 × SSC reservoir and serve as a wick. Smooth the 3MM paper over the platform with a gloved hand by rolling a clean pipette over it. This removes any air bubbles that may be trapped. Measure the gel and cut a piece of nitrocellulose to the exact size, using a pair of sharp scissors or a scalpel blade. Wet the filter by laying it on the surface of distilled water in a glass container. The filter should wet quickly and evenly. Shake the container to submerge the filter. Pour off the water and soak the filter in 20 × SSC. Cut five sheets of 3MM paper to the same size as the filter.

5   **Assembling the blot**. Place the neutralized gel on the saturated 3MM paper on the level platform. Remove any air bubbles that may be trapped between the paper and the gel by rolling a clean pipette over the surface of the gel or by gently pressing the gel with a gloved hand. Take the wetted filter from the 20 × SSC container and place it on the top of the gel. Remove any bubbles that may be trapped between the filter

---

Protocol 4 continued

and the gel. Place Saran wrap or Parafilm around the edges of the gel so that no part of the nitrocellulose filter can directly contact the 3MM paper wick. Place the five sheets of dry 3MM paper on top of the filter and stack paper towels to a thickness of 4 cm on top of the 3MM paper. Put a glass plate on top of the paper towels and place a small weight (approx. 500 g) on top of the plate. Leave the blot overnight.

6   **Disassembling the blot**. Remove the weight, paper towels, and 3MM paper. The gel will have compressed leaving a thin gel/filter sandwich. With a pencil, mark the position of the wells on the filter and cut one corner to allow orientation of the gel. Carefully peel off the filter and gently submerge it in a solution of 6 × SSC in a glass container. Leave for 3 min, then lay the filter on a clean piece of 3MM paper to dry. The gel should be re-stained with ethidium bromide to check that most of the DNA has been transferred.

7   **Immobilizing the DNA**. After allowing the filter to air dry, place it between two sheets of 3MM paper and bake at 80 °C in a vacuum oven for 2 h. The filter is ready for hybridization, but may also be stored in a dark, dry place.

[a] Taken from ref. 11.

[b] The quantity of DNA that needs to be loaded on the gel will depend on the relative abundance of the target sequence within the sample. Using radioactive probes of high specific activity (10^8 to 10^9 d.p.m./µg), between 1 and 100 pg of target DNA is routinely detectable in an overnight exposure. Note that a 1 kb sequence present in single copy in human genomic DNA comprises 3 pg of a 10 µg sample.

Southern blotting to different membrane types follows the same basic procedure but with certain modifications. For example, with a positively-charged nylon membrane (e.g. Amersham-Pharmacia Hybond-N+, Bio-Rad ZetaProbe), an alkaline transfer buffer is used (0.4 M NaOH). Denaturation is carried out by soaking the gel in this buffer for 20 min; neutralization is unnecessary. The alkaline transfer results in covalent attachment of the DNA to the membrane, which means that after hybridization the probe can be stripped off by rinsing with hot 0.1% SDS (11), enabling the membrane to be probed a second time, up to ten re-probings often being possible with a single membrane. In contrast, DNA attaches to nitrocellulose by non-covalent interactions, so probe stripping has to be done under more gentle conditions which still lead to some loss of bound DNA. The DNA loss, combined with the fragility of a nitrocellulose membrane, means that these blots can be reused only one or two times.

5.2 Hybridization analysis of a Southern blot

The sequence of interest on a Southern or other type of blot is detected by hybridization with a labelled probe. Several methods exist for labelling DNA and RNA with radioactive nucleotides (12), the most generally-applicable procedures being nick translation, in which nicks in a double-stranded DNA molecule act as initiation sites for synthesis of new DNA strands by DNA polymerase I, and random primer labelling, which involves denaturation of double-stranded DNA followed by annealing of short random oligonucleotides which prime new strand synthesis. In both cases, inclusion of a ^{32}P-labelled nucleotide in the reaction mix results in the new strands becoming labelled to a specific activity of 10^8 to 10^9 d.p.m./μg. Kits for carrying out both procedures are available from various suppliers. Many companies also market non-radioactive labelling systems, in which detection of the hybridized probe is by a colorimetric or chemiluminescent method, rather than by exposure of X-ray film. Non-radioactive probes are suitable for most purposes but may not give the sensitivity needed for detection of a single-copy sequence in a Southern blot of genomic DNA.

The basic hybridization method given in Protocol 5 describes the use of a radioactive, double-stranded DNA probe produced by nick translation or random primer labelling. Slight modifications are needed if the probe is an oligonucleotide or RNA molecule (11), and more extensive changes might be needed for a non-radioactive probe, in particular with regards to the procedure used to detect the bound probe (follow the supplier's instructions).

Protocol 5

Basic method for filter hybridization[a]

Equipment and reagents

- Hybridization oven
- Boiling water-bath
- Nitrocellulose filter carrying a Southern blot
- 20 × SSC (see Protocol 4)
- 100 × Denhardt's: 2% bovine serum albumin, 2% Ficoll 400, 2% polyvinylpyrrolidone (mol. wt. 400 000)
- Non-homologous DNA (10 mg/ml calf thymus, salmon sperm, or herring sperm DNA, in water): prepare a 10 mg/ml solution of salmon sperm DNA in water, vigorously pass through a 17 G syringe needle approx. 20 times, place in a boiling water-bath for 10 min, chill, and use immediately
- Hybridization solution: 5 × SSC, 5 × Denhardt's, 1% SDS, 10% dextran sulfate

(mol. wt. 500 000), 0.3% tetrasodium pyrophosphate, 100 μg/ml non-homologous DNA. Mix together: 25 ml of 20 × SSC, 5 ml of 100 × Denhardt's, 5 ml of 20% SDS, 10 g dextran sulfate, 0.3 g tetrasodium pyrophosphate, 1 ml of 10 mg/ml non-homologous DNA, plus 60 ml of water. Allow the dextran sulfate to dissolve and then adjust the volume to 100 ml with water. Do not add SDS directly to SSC because it will precipitate
- Radioactively labelled, double-stranded probe DNA
- 2 × SSC, 0.1% SDS
- 0.2 × SSC, 0.1% SDS
- 0.1 × SSC, 0.1% SDS

Method

1 Warm the aqueous hybridization solution to 68 °C.

2 Wet the filter in 5 × SSC and place the damp filter in one of the cylinders of the hybridization oven.[b]

Protocol 5 continued

3 Add at least 1 ml of warmed aqueous hybridization solution per 10 cm^2 of filter.[c] Carry out the pre-hybridization by rotating the cylinder in the hybridization oven for 3 h at 68 °C. The filter should be able to move freely in the cylinder and the solution should sweep gently over the surface of the filter.

4 Denature the double-stranded probe by incubating in a boiling water-bath and snap-cooling on ice.[d] Add the denatured probe to the pre-warmed hybridization solution. There should be at least 1 ml of hybridization solution per 10 cm^2 of filter, and final probe concentration should be 10 ng/ml (for a probe of specific activity 10^8 d.p.m./μg) or 2 ng/ml (for a probe with specific activity 1×10^9 d.p.m./μg). Mix well.[e]

5 Remove the pre-hybridization solution from the cylinder and replace with the hybridization solution plus labelled probe.[f]

6 Rotate the cylinder overnight at 68 °C.

7 Remove the hybridization solution and dispose in the appropriate way. Add 5–10 ml of low stringency wash solution (2 × SSC, 0.1% SDS) per 10 cm^2 of filter. Rotate the cylinder for 5 min at room temperature.

8 Repeat step 7 twice with fresh aliquots of low stringency wash solution.

9 Increase the stringency of the wash by washing the filter twice in 0.2 × SSC, 0.1% SDS at room temperature for 5 min.

10 The stringency of the washes can be increased further by increasing the washing temperature and/or lowering the salt concentration. Washing twice for 15 min at 42 °C in 0.2 × SSC, 0.1% SDS is a moderately stringent wash. Washing twice for 15 min at 68 °C in 0.2 × SSC, 0.1% SDS or 0.1 × SSC, 0.1% SDS is a high stringency wash. It is not necessary to wash for greater than 30 min for each level of stringency. Washing solutions should be pre-warmed to the appropriate temperature before use.

11 Finally rinse the filter in 2 × SSC at room temperature and lay the filter on blotting paper to remove excess liquid. Do not allow the filter to dry if it is to be re-washed or stripped for re-probing.

12 Autoradiograph the filter.

[a] Modified from ref. 11.

[b] If a hybridization oven is not available then it is possible, though inadvisable, to carry out the hybridization in a heat-sealed plastic bag. This procedure is inadvisable because of the difficulty in repeatedly sealing and opening the bag without contaminating the work area with radioactive material. If a bag is used then it should be double-sealed along all sides approx. 1 cm from the edge of the filter. Use the same volumes of hybridization solution as described in the protocol. Ensure that all air bubbles are removed. A simple way to do this is to seal the fourth side of the bag 2 cm from the edge of the filter, then snip off a corner and displace the bubbles from the bag through the small hole by running a ruler along the bag or running the bag against the edge of a bench. Then reseal the bag. The filter should be able to move freely in the bag and the solution should sweep gently over the surface of the filter: the latter will require that the bag is placed on a rocking platform in an incubator. The filter should be removed from the bag and placed in a glass dish in order to carry out the washes.

Protocol 5 continued

c More than one filter can be hybridized at one time. Up to ten filters can be hybridized in a single cylinder or bag, but ensure that air bubbles are not trapped between the filters and that the filters do not stick together during the incubations.

d DNA probes can also be denatured by treatment with alkali. Add one-tenth the original volume of 3 M NaOH. After 5 min at room temperature add one-fifth the original volume of 1 M Tris–HCl pH 7.0 and one-tenth the original volume of 3 M HCl.

e Radioactive probes must be handled in accordance with local safety procedures. Hybridization and wash solution must be disposed of in the appropriate way. Note that all wash solutions will be radioactive.

f Do not allow the filter to dry out at any time during the procedure.

6 DNA sequencing

Virtually all DNA sequencing is performed using the chain termination method first devised by Sanger *et al.* (13) in the 1970s. This technique involves a DNA synthesis reaction in which a polymerase enzyme copies a DNA template into new strands of DNA, each of which is terminated with a modified nucleotide that blocks further strand extension. The modified nucleotides are 2′,3′-dideoxynucleoside triphosphates (ddNTPs) that cause chain termination because they lack the 2′-hydroxyl that normally participates in phosphodiester bond synthesis. Therefore, when the reaction is carried out in the presence of a ddNTP, chain termination occurs at positions opposite complementary nucleotides in the template DNA (e.g. opposite thymidines if ddA is used). The strand synthesis reaction is carried out four times in parallel, once with each of ddA, ddC, ddG, and ddT, and the products examined to determine the lengths of the chain terminated molecules that have been synthesized. This can be achieved by electrophoresis in thin, denaturing polyacrylamide gels, in which molecules differing in length by just a single nucleotide can be separated into a series of bands. The products of the four reactions are run in four adjacent lanes and the sequence read from the bottom of the gel upwards. First the band that has moved the furthest is located, this representing the strand that has been terminated by incorporation of the dideoxynucleotide at the first position in the template. The track in which this band occurs is noted because this indicates the identity of the first nucleotide in the sequence. The next band up corresponds to a DNA molecule that is one nucleotide longer and hence indicates the identity of the second nucleotide in the sequence. The process is continued until the individual bands become too close together to be distinguished.

6.1 Sequencing PCR products

PCR products are sequenced by the method called thermal cycle sequencing. The version given in Protocol 6 is based on ref. 14 and is designed for the thermostable polymerase called Vent (exo⁻) (New England Biolabs). Any equivalent thermostable polymerase can be used if the necessary changes are made to the buffer conditions.

Protocol 6

Thermal cycle sequencing

Equipment and reagents

- Thermal cycler
- Oligonucleotide primer (one of the primers used in the PCR that generated the product that is to be sequenced)
- 10 × kinase buffer: 0.5 M Tris–HCl pH 7.6, 0.1 M MgCl$_2$, 50 mM dithiothreitol, 1 mM spermidine
- [γ-^{32}P]ATP at 3000 Ci/mmol and 10 mCi/ml
- T4 polynucleotide kinase
- 0.5 M EDTA pH 8.0
- A sequencing mix: 30 μM dATP, 100 μM dCTP, 100 μM dGTP, 100 μM dTTP, 900 μM ddATP, prepared in 1 × Vent buffer. Prepare termination mixes from 20 mM stock solutions. Both the stock solutions and the termination mixes can be stored at −20°C for several months.
- C sequencing mix: 30 μM dATP, 40 μM dCTP, 100 μM dGTP, 100 μM dTTP, 500 μM ddCTP, prepared in 1 × Vent buffer
- G sequencing mix: 30 μM dATP, 100 μM dCTP, 40 μM dGTP, 100 μM dTTP, 400 μM ddGTP, prepared in 1 × Vent buffer
- T sequencing mix: 30 μM dATP, 100 μM dCTP, 100 μM dGTP, 30 μM dTTP, 700 μM ddTTP, prepared in 1 × Vent buffer
- 10 × Vent buffer: 200 mM Tris–HCl pH 8.8, 100 mM KCl, 50 mM MgSO$_4$, 100 mM (NH$_4$)$_2$SO$_4$
- PCR product to be sequenced [a]
- Vent (exo$^-$) DNA polymerase (New England Biolabs)
- Formamide dye mix: 0.1% bromophenol blue, 0.1% xylene cyanol, 10 mM EDTA pH 8.0, prepared in deionized formamide. To deionize formamide, add 5 g of mixed-bed ion exchange resin (e.g. Bio-Rad AG 501-X8 or X8(D) resins) to 100 ml formamide, stir at room temperature for 1 h, and filter twice through Whatman No. 1 paper. **CAUTION: Formamide is a teratogen and an irritant. Avoid exposure and take appropriate safety precautions.**

Method

1 End-label the oligonucleotide primer by the forward reaction with T4 poly-nucleotide kinase.[b] To do this, set up the following mix in a microcentrifuge tube on ice:

   - Oligonucleotide primer   10 pmol ends[c]
   - 10 × kinase buffer   5 μl
   - [γ-^{32}P]ATP (3000 Ci/mmol)   200 μCi (65 pmol)
   - T4 polynucleotide kinase   10 U
   - Sufficient water to bring the volume to 50 μl   x μl

   Incubate at 37°C for 30–45 min. Terminate the reaction by addition of 2 μl of 0.5 M EDTA pH 8.0.

Protocol 6 continued

2 Add 3 µl of the appropriate sequencing mix to each of four 0.5 ml microcentrifuge tubes labelled A, C, G, and T.

3 Set up the following mix in a microcentrifuge tube on ice:

- PCR product   5 µl[d]
- End-labelled primer (from step 1)   5 µl
- 10 × Vent buffer   1.5 µl
- Vent (exo⁻) DNA polymerase   2 U
- Sufficient water to bring the volume to 15 µl   x µl

4 Transfer 3 µl of the mix from step 3 into each of the tubes prepared in step 2.

5 Place each tube in a thermal cycler that has been programmed for 20 cycles of 20 sec at 95 °C, 20 sec at 55 °C, and 20 sec at 72 °C.

6 Terminate the reaction by adding 4 µl formamide dye mix to each tube.

7 Prepare a denaturing polyacrylamide gel[e] and load and run the samples. Set up an autoradiograph.

[a] Before sequencing a PCR product it must be purified. If the PCR product to be sequenced is the only visible band on an agarose gel, then it can be purified directly from the PCR mixture (e.g. by using the Qiagen QIAquick PCR Purification Kit). If additional bands are seen in the gel then cut out the desired band and purify the product from the gel slice (e.g. by using the Qiagen QIAquick Gel Extraction Kit).

[b] For more details, see ref. 12.

[c] 10 pmol ends = 0.065 µg of a 20-mer oligonucleotide.

[d] The volume of PCR product may need to be adjusted to provide a clear sequence.

[e] Use a commercial sequencing gel apparatus in accordance with the manufacturer's instructions. See ref. 15.

6.2 Thermal cycle sequencing of cloned DNA

Thermal cycle sequencing is not limited to sequence analysis of PCR products—the same approach can be taken for an any DNA template. For a single-stranded M13 clone you should use 0.05 pmol of template DNA, corresponding to 125 ng of an M13 clone with a 500 bp insert, and for a plasmid clone you should use 0.1 pmol, which is 100 ng of a pUC18 clone with a 500 bp insert. This template DNA is substituted for the PCR product in the mixture prepared in Protocol 6, step 3. A universal sequencing primer is used (e.g. New England Biolabs, Sigma, Stratagene, Promega). The thermal cycle method is also the basis for automated DNA sequencing. In the automated system, fluorescent labels are used, either a single label attached to the primer, in which case the sequencing reactions are carried out as described in Protocol 6, or four different labels, one for each ddNTP. The latter strategy necessitates a few modifications to Protocol 6, but the principle of the technique is the same. If an automated sequencer is used then the components of the sequencing reactions and the cycle times used for generation of the chain terminated molecules must be strictly compatible with the specifications of the sequencer. The manufacturer's instructions should therefore be followed.

7 Other DNA and RNA manipulations

The above sections have described the most important basic techniques for studying DNA and RNA. Many other procedures are also used in nucleic acid analysis. Restriction (Protocol 7) is the process by which DNA molecules are cut at specific nucleotide sequences. Along with ligation (Protocol 8), it forms the basis to the construction of recombinant DNA molecules prior to cloning. The latter requires that recombinant molecules be introduced into E. coli cells that have been made competent for DNA uptake (16), suitable preparations usually being purchased along with a specialized cloning vector (17) from any one of several commercial suppliers. Other important general procedures are ethanol precipitation, used to concentrate DNA samples (Protocol 9) and the measurement of DNA concentrations by spectrophotometry (Protocol 10).

More advanced techniques enable the start and end positions of transcripts to be mapped onto DNA sequences (18). DNA–protein interactions can be studied by gel retardation analysis, in which a restriction fragment that binds a protein is identified by virtue of its changed mobility in an agarose gel, and by footprint analysis, which delineates the specific region of a DNA molecule that participates in protein binding (19).

Protocol 7

Restriction enzyme digestion[a]

Reagents

- Restriction enzyme plus appropriate buffer

Method

1 Make up a 20 μl reaction mixture by adding the reagents in order into a sterile 1.5 ml microcentrifuge tube:

- 2 μl of 10 × restriction enzyme buffer[b]
- x μl DNA
- y μl water

The value of x will depend on the concentration of the DNA solution. The value of y should bring the volume up to 19 μl.

2 Add 1 μl of the restriction enzyme[c] and mix gently with the pipette tip.

Protocol 7 continued

3 Incubate in a water-bath at the appropriate temperature for the desired length of time. The temperature and time depends on the enzyme and will be stated in the product guide provided by the supplier.

4 Store the reaction on ice while analysing the digestion by gel electrophoresis.

[a] Modified from ref. 20.

[b] Use the buffer provided by the suppliers of the restriction enzyme, or make a buffer according to their instructions.

[c] Use 1 U of enzyme per μg of DNA. If necessary dilute the enzyme stock with the storage buffer described in the supplier's product guide. It is acceptable to add more than 1 U/μg of DNA, but large excess amounts should be avoided.

Protocol 8

DNA ligation[a]

Reagents

- 10 × ligation buffer: 0.66 M Tris–HCl pH 7.6, 50 mM $MgCl_2$, 50 mM dithiothreitol, 10 mM ATP
- T4 DNA ligase

Method

1 The volume of the ligation mixture and the DNA concentration depend on the type of ligation experiment. Use a 10 μl reaction with DNA at a concentration of >100 ng/μl for concatamer ligation products, or a 10 μl reaction with DNA at <10 ng/μl for circular ligation products.

2 Add T4 DNA ligase. For a cohesive-end ligation add 0.25 U of enzyme per μg of DNA, and for a blunt-end ligation add 2.5 U/μg.

3 Incubate the reaction mixture at 15 °C for 1–16 h. Simple cohesive-end ligations are usually complete in 1–2 h.

4 Analyse for correct and complete ligation by gel electrophoresis with unligated material as a marker, or by transformation of *E. coli* if the resulting DNA contains vector sequences.

[a] Taken from ref. 20.

Protocol 9

Precipitation of DNA with ethanol[a]

Reagents

- 3 M potassium acetate solution pH 5.6, prepared by adding glacial acetic acid to 3 M potassium acetate until this pH is obtained
- Absolute ethanol[b]
- 70% ethanol
- TE buffer pH 8.0: 10 mM Tris–HCl pH 8.0, 1 mM EDTA

Method

1 Place a microcentrifuge tube containing the DNA solution on ice and add 0.1 vol. of 3 M sodium acetate solution pH 5.6, and mix.

2 Add 2.5 vol. of cold absolute ethanol (−20 °C) and mix by inversion. Incubate at −20 °C for 30 min or −70 °C for 10 min.[c]

3 Centrifuge at 15 000 g for 30 min. Remove the supernatant, replace with 0.5 ml of 70% ethanol, mix by inversion, and centrifuge at 15 000 g for 5 min.

4 Pour away the supernatant and dry the mouth of the inverted tube with tissue. Place the tube in a horizontal position at room temperature and allow the traces of ethanol to evaporate. Redissolve the DNA in water or TE pH 8.0.

[a] Modified from ref. 4.

[b] Ethanol used for this procedure should be stored and used from the freezer. Ethanol at room temperature will cause high molecular weight DNA to shear.

[c] If the amount of DNA exceeds 10 μg then the precipitate may be visible after addition of absolute ethanol. In this case the initial centrifugation in step 3 can be for 2 min.

Protocol 10

Spectrophotometric determination of DNA[a]

Equipment and reagents

- UV-visible scanning spectrophotometer
- Quartz cuvettes
- TE buffer pH 7.5: 10 mM Tris–HCl pH 7.5, 1 mM EDTA

Method

1 Transfer 10–100 μl of the DNA to 900–990 μl TE pH 7.5 in a 1 ml quartz cuvette.

2 Mix the contents thoroughly and record a spectrum from 250–320 nm. A smooth peak should be observed with an absorption maximum at around 260 nm and no noticeable shoulder at 280 nm (the latter indicates contamination with protein).

Protocol 10 continued

> The 260:280 nm absorbance ratio should not be more than 1.9. If the ratio is less 1.9 then again contamination with protein should be suspected.
>
> **3** Calculate the quantity of DNA using the guide that 1 ml of a solution with an A_{260} of 1.0 is equivalent to 50 μg of double-stranded or 35 μg of single-stranded DNA.
>
> [a] Taken from ref. 4.

Acknowledgements

Several of the protocols in this chapter are taken from ref. 1. I thank the authors of these protocols for allowing me to use their material.

References

1. Brown, T. A. (ed.) (2000/2001). *Essential molecular biology: a practical approach* (2nd edn). Oxford University Press, Oxford.
2. Ausubel, F. M., Brent, R., Kingston, R. E., Moore, D. D., Seidman, J. G., Smith, J. A., *et al.* (ed.) (2001). *Current protocols in molecular biology*. John Wiley, New York.
3. Boom, R., Sol, C. J. A., Salimans, M. M. M., Jansen, C. L., Wertheim-van Dillen, P. M. E., and van der Noordaa, J. (1990). *J. Clin. Microbiol.*, **28**, 495.
4. Towner, P. (2000). In *Essential molecular biology: a practical approach* (2nd edn) (ed. T. A. Brown), Vol. I, p. 55. Oxford University Press, Oxford.
5. Frohman, M. A., Dush, M. K., and Martin, G. R. (1988). *Proc. Natl. Acad. Sci. USA*, **85**, 8998.
6. Bagasra, O., Seshamma, T., Pomerantz, R., and Hanson, J. (1996). In *Current protocols in molecular biology* (ed. F. M. Ausubel, R. Brent, R. E. Kingston, D. D. Moore, J. G. Seidman, J. A. Smith, *et al.*), Unit 14.8. John Wiley, New York.
7. Kellogg, D. E., Rybalkin, I., Chen, S., Mukhamedova, N., Vlasik, T., Siebert, P. D., *et al.* (1994). *Biotechniques*, **16**, 1134.
8. Brown, T. A. (2001). In *Essential molecular biology: a practical approach* (2nd edn) (ed. T. A. Brown), Vol. II, p. 187. Oxford University Press, Oxford.
9. Robinson, D. H. and Lafleche, G. J. (2000). In *Essential molecular biology: a practical approach* (2nd edn) (ed. T. A. Brown), Vol. I, p. 89. Oxford University Press, Oxford.
10. Wilkinson, M. (2000). In *Essential molecular biology: a practical approach* (2nd edn) (ed. T. A. Brown), Vol. I, p. 69. Oxford University Press, Oxford.
11. Dyson, N. J. (2001). In *Essential molecular biology: a practical approach* (2nd edn) (ed. T. A. Brown), Vol. II, p. 109. Oxford University Press, Oxford.
12. Mundy, C. R., Cunnigham, M. W., and Read, C. A. (2001). In *Essential molecular biology: a practical approach* (2nd edn) (ed. T. A. Brown), Vol. II, p. 63. Oxford University Press, Oxford.
13. Sanger, F., Nicklen, S., and Coulson, A. R. (1977). *Proc. Natl. Acad. Sci. USA*, **74**, 5463.
14. Slatko, B. E., Albright, L. M., Tabor, S., and Ju, J. (1999). In *Current protocols in molecular biology* (ed. F. M. Ausubel, R. Brent, R. E. Kingston, D. D. Moore, J. G. Seidman, J. A. Smith, *et al.*). p. 7.4A.11. John Wiley, New York.
15. Brown, T. A. (2001). In *Essential molecular biology: a practical approach* (2nd edn) (ed. T. A. Brown), Vol. II, p. 157. Oxford University Press, Oxford.
16. Brown, T. A. (2000). In *Essential molecular biology: a practical approach* (2nd edn) (ed. T. A. Brown), Vol. I, p. 151. Oxford University Press, Oxford.

17. Brown, T. A. (2000). In *Essential molecular biology: a practical approach* (2nd edn) (ed. T. A. Brown), Vol. I, p. 175. Oxford University Press, Oxford.

18. Smith, C. P. (2001). In *Essential molecular biology: a practical approach* (2nd edn) (ed. T. A. Brown), Vol. II, p. 223. Oxford University Press, Oxford.

19. Ausubel, F. M., Brent, R., Kingston, R. E., Moore, D. D., Seidman, J. G., Smith, J. A., *et al.* (ed.) (2001). *Current protocols in molecular biology*, Chapter 12. John Wiley, New York.

20. Powell, R. and Gannon, F. (2000). In *Essential molecular biology: a practical approach* (2nd edn) (ed. T. A. Brown), Vol. I, p. 129. Oxford University Press, Oxford.

Chapter 10
Cell lipids: from isolation to functional dynamics

Robert Jan Veldman, Eve-Isabelle Pécheur,
Sven C. D. van IJzendoorn, Jan Willem Kok, and
Dick Hoekstra

University of Groningen, Department of Membrane Cell Biology,
Faculty of Medical Sciences, A. Deusinglaan 1, 9713 AV Groningen,
The Netherlands.

1 Introduction

The boundary of every cell as well as that of intracellular compartments, including intracellular transport vesicles, is provided by a membrane, the major components of which are proteins and lipids. Membrane lipids are amphipathic molecules that consist of a polar head group and usually two non-polar hydrocarbon chains. The hydrocarbon chains are covalently attached to a lipid backbone structure such as glycerol (in the case of phospholipids and glycoglycerolipids) or sphingosine (in the case of (glyco)sphingolipids) via ester, ether, amide, or C–C bonds. In biological membranes, a variety of lipid classes can be distinguished, the major lipids belonging to the class of the phospholipids. Although a minor class, the more complex (glyco)sphingolipids are biologically highly relevant (1).

Lipids vary widely in polymorphic, physicochemical, and biological properties, thus reflecting the diversity of dynamic functions that can be expressed by these molecules or the multitude of processes they can be involved in. Although many of these functions are still poorly understood, some of these functions and processes include their ability to modulate protein functioning, their role in transmembrane signalling events, their ability to regulate intracellular transport, the roles they play in cell recognition, cell differentiation, and cell growth, and their intimate involvement in membrane fusion processes. Accordingly, these examples illustrate that the biological significance of lipids far exceeds that of simply acting as a fence or a matrix for membrane proteins. Indeed, issues dealing with the asymmetric transbilayer distribution of lipids, and the functional significance of their distribution in specific domains in the lateral plane of the bilayer, as is for example the case in so-called 'rafts' that are enriched in sphingolipids (and cholesterol), are only gradually becoming apparent (2, 3). Furthermore, given their non-random distribution among or within various membranes, lipids, like proteins (4), are presumably undergoing sorting events. These sorting events have been most convincingly revealed in the trafficking and distribution of sphingolipids in so-called polarized cells. Thus cells such as epithelial cells and hepatocytes, simultaneously face different extracellular environments, which are functionally tackled by virtue of the maintenance of two different plasma membrane domains, separated by tight junctions. These domains, the apical and basolateral plasma membrane domains, display distinct (protein and) lipid compositions, the apical domain usually being enriched in glycosphingolipids (5, 6). Although intracellular compartments have been defined in which such lipid sorting events may occur (7, 8), the underlying molecular mechanisms are still largely obscure.

Nevertheless, the aforementioned considerations clearly emphasize the cell biological significance of lipids. In this chapter we describe a variety of aspects that are, on the one hand, related to establishing the identity and, on the other, the dynamics of lipids. We will particularly, but not exclusively, focus on glyco-sphingolipids, given the rapid advancements in revealing and understanding their ubiquitous cell biological functions. First we present methods to isolate and detect a variety of complex (glyco)sphingolipids. Subsequently, we outline proto-cols for preparing and applying fluorescently labelled lipid analogues in studies that deal with issues concerning the trafficking of lipids in eukaryotic cells. Such studies are aimed at understanding the cell biology of lipids, including the biogenesis of membrane domains in polarized cells. Lipid membranes can also be reconstituted and their dynamics and biophysical properties studied by employing artificial membranes, such as liposomes. We describe how the purity of the lipids of choice can be determined, how such liposomes can be prepared, and how they can be applied in studies involving peptide- and protein-induced fusion. In this case, protocols are provided for monitoring membrane fusion based upon lipid mixing assays.

2 Analysis of cellular sphingolipids

2.1 Extraction and hydrolysis

The first step of virtually all lipid measurements consists of the extraction of the lipids from the biological sample. Protocol 1 is a modified version of a method which was originally developed by Bligh and Dyer (9), which is very convenient for the extraction of lipids from cultured cells. As noted above, (glyco)sphingo-lipids are usually present in membranes in relatively minor amounts, ranging from 5% to approx. 20% (in apical membrane domains) of the total lipid fraction. Given such minor amounts, it is therefore highly recommended to remove glycerol-based lipids such as phospholipids (with the exception of sphingo-myelin, which is a ceramide-based phospholipid), which make up 70–80% of the total lipid fraction in the extract. Protocol 2 describes a rapid and efficient procedure to do so, without affecting any of the (glyco)sphingolipids. Application of this hydrolysis procedure generally results in cleaner (glyco)sphingolipid samples, which in many cases significantly improves the quality of subsequent analysis.

Protocol 1

Extraction of lipids from cultured cells

Equipment and reagents

- Borosilicate glass tubes
- Bath-sonicator
- Water-bath
- Phosphate-buffered saline (PBS)

- Methanol
- Chloroform
- Nitrogen, argon

Method

1 Rinse the cells twice with 2 ml of phosphate-buffered saline (PBS) and fix by adding 1 ml of ice-cold methanol. Harvest the cells immediately by scraping the flasks with a rubber policeman and transfer the methanol–cell mixtures to borosilicate glass tubes on ice. Rinse the culture flasks with another 1 ml of methanol and pool this wash with the first mixture.

2 Add 1 ml of chloroform to the tubes and bath-sonicate the mixtures until homogeneous solutions are obtained (approx. 30 sec). Note that the sonication step is not strictly necessary, but it will improve the efficiency of the extraction.

3 Add another 3 ml of chloroform and separate phases by the addition of 1.6 ml of water. Vortex, centrifuge (5 min, 1000 g, 20 °C), and collect the lower, organic phase, using a long size (230 mm) Pasteur capillary pipette. Transfer this phase to a new tube.

4 Wash the lower phase by adding 3.6 ml of chloroform/methanol/water (3:48:47, by vol.). Vortex, centrifuge (5 min, 1000 g, 20 °C), and collect the lower phase as in step 3.

5 Transfer the lower phase to a new tube and dry the lipids under a gentle stream of nitrogen in a 37 °C water-bath.

6 For storage (4 °C), saturate the tube's atmosphere with argon, cap the tubes, and seal with Parafilm.

Protocol 2

Mild alkaline hydrolysis of glycerol-based lipids

Equipment and reagents
- Water-bath
- Methanol
- Chloroform
- NaOH

Method

1 Dissolve the dried lipids in 1 ml of chloroform and add 2 ml of methanol containing 0.2 M NaOH.

2 Vortex and incubate the tubes for 2 h in a water-bath at 37 °C. After 30 min tiny 'droplets' will become visible on the walls of the tube.

3 Re-extract the mixture by repeating Protocol 1, steps 3–5.

2.2 Sphingolipid mass measurements

For the quantitation of natural, endogenous (glyco)sphingolipids in biological samples, many different procedures have been described. This chapter presents some of the most commonly used protocols.

The assay for sphingomyelin mass measurement (Protocol 3) is based upon detection of the lipid's phosphate moiety (1 mol of sphingomyelin contains 1 mol of phosphate). Due to the relative insensitivity of the phosphate determination (the lower limit of detection is 5 nmol), it is recommended to start with a larger amount of cells ($>3 \times 10^7$ cells per assay; e.g. a confluent cell layer in a 75 cm^2 flask) than the amounts required for most other assays described in this chapter. Protocol 4 describes the standard procedure for the mass measurement of ceramide, a method originally described by Preiss *et al.* (10), while slight modifications were introduced by van Veldhoven *et al.* (11). This procedure relies on the fact that the enzyme diacylglycerol kinase recognizes ceramide as a substrate, in addition to diacylglycerol, and a detailed account has been presented elsewhere (12). Using ^{32}P-labelled ATP as phosphate donor, this kinase activity results in the formation of easily detectable radiolabelled ceramide-1-phosphate and phosphatidic acid (Figure 1). Since the detection limit is 25 pmol, only small amounts of starting material are required (e.g. semi-confluent cell cultures, grown in 25 cm^2 flasks).

Sphingoid bases are analysed according to the method described by Merrill *et al.* (13) (Protocol 5). After fluorescent derivatization, sphingoid bases are separated by high performance liquid chromatography (HPLC), which not only distinguishes between different types of sphingoid bases, but also between different chain length variants of the same base (Figure 2). The high sensitivity of this assay (a lower limit of detection of 10 pmol) allows small quantities of starting material (e.g. semi-confluent cell cultures from 25 cm^2 flasks).

The procedure for ganglioside analysis (Protocol 6) was originally described by Senn *et al.* (14), and further adapted by Sietsma *et al.* (15). Due to the relative hydrophilic nature of gangliosides, the standard lipid extraction is replaced by a one-phase extraction, followed by an adapted two-phase extraction in which the aqueous phase contains the gangliosides. After further purification by a column washing step, separation of the gangliosides is then performed by high performance thin-layer chromatography (HPTLC). Detection takes place by staining of the lipids with a specific reagent. Since most ganglioside species are present in rather small quantities, at least 6×10^7 cells per extraction are required as starting material.

Protocol 3

Mass measurement of sphingomyelin

Equipment and reagents

- Pyrex glass tubes
- 60 Å silica gel-coated HPTLC plate
- Glass micro-syringe
- Chloroform/methanol (1:1, v/v)

- Nitrogen
- Inorganic phosphate
- Chloroform, methanol, acetic acid, water

Method

1   Add exactly 1 ml of chloroform/methanol (1:1, v/v) to the extract (Protocol 1) and vortex. Transfer an aliquot of exactly 100 μl to a Pyrex glass tube and dry both the sample and the original extract under nitrogen.

2   Perform a phosphate determination (16), on both the aliquot and a series of standard amounts of inorganic phosphate (ranging from 0–200 nmol).

3   Perform a mild alkaline hydrolysis[a] on the remaining lipids (Protocol 2).

4   Dissolve the lipids in a few drops of chloroform/methanol (1:1, v/v) and spot onto a 10 × 10 cm 60 Å silica gel-coated HPTLC plate, using a glass micro-syringe. Wash the tube with some chloroform/methanol (1:1, v/v) and spot onto the first spot.[b]

5   Separate sphingomyelin from all other lipids by developing the TLC plate in chloroform/methanol/acetic acid/water (12:6:1:1, by vol.).

6   Allow the plate to dry and visualize the lipids by staining the plates in an iodine vapour-saturated container (1–3 min). After identification,[c] scrape the sphingomyelin-containing spots from the TLC plate with a razor blade or a surgical scalpel.

7   Transfer the sphingomyelin-containing silica grains to Pyrex glass tubes and perform a phosphate determination (16), on both the silica grains and a series of standard amounts of inorganic phosphate (ranging from 0–200 nmol).

8   Calculate the total phospholipid content (step 3) and the sphingomyelin content (step 7). Express the amount of sphingomyelin as nmol per μmol of total phospholipid.

Protocol 3 continued

[a] The sphingomyelin content of an extract is generally normalized to the total phospholipid content of that particular sample (step 8). The initial extract should therefore not be hydrolysed before the first phosphate determination is performed (step 2).

[b] The spotting procedure is best performed in a stream of warm air, using a hair-drier.

[c] Positive identification of the sphingomyelin spot on the HPTLC plate is best performed by using a purified standard of sphingomyelin, which is run simultaneously on the same plate (though in a separate lane). A confirmative 'negative' identification procedure involves the pretreatment of a separate batch of cells for 2 h with 0.1 U/ml of a bacterial sphingomyelinase (e.g. *B. cereus* or *S. aureus*; Sigma). A lipid sample thus treated should lack any significant iodine staining at the R_f value of sphingomyelin.

Protocol 4

Mass measurement of ceramide

Equipment and reagents

- See Protocols 1–3
- Detergent solution: 1 mM diethylenetriaminopentaacetic acid (DETAPAC) pH 7.0, 12.5 mM 1,2-dioleoyl-*sn*-glycero-3-phosphoglycerol, 3.75% (w/v) octyl-β-glucoside
- Reaction mixture: 100 mM imidazole–HCl pH 7.0, 1 mM DETAPAC, 100 mM LiCl,

25 mM $MgCl_2$, 2 mM EGTA, 2 mM dithiothreitol (add freshly), 0.5 μg/μl diacylglycerol kinase (add freshly)
- ATP mixture: 20 mM imidazole–HCl pH 7.0, 1 mM DETAPAC, 10 mM ATP, 500 μCi/ml [^{32}P]ATP (add freshly)
- $HClO_4$

Method

1   Extract lipids from the cells according to the procedure described in Protocol 1.[a]

2   Redissolve the extracts in exactly 1 ml of chloroform/methanol (1:1, v/v) and perform a phosphate determination on an aliquot (16). Calculate the phospholipid content of the total extract. Transfer an aliquot of 50 nmol to a new tube and dry under N_2.

3   Prepare a series of standards, ranging from 0–1000 pmol of ceramide and diacylglycerol.

4   Dissolve the dried lipids in 40 μl of detergent solution by bath-sonication (1 min).

5   Add 50 μl of reaction mixture and incubate (10 min, 20 °C).

6   Start the diacylglycerol kinase reaction by adding 10 μl of ATP mixture and incubate (30 min, 25 °C).

7   Stop the reaction by adding 0.7 ml of 1% (w/v) $HClO_4$ and 3 ml of chloroform/methanol (1:2, v/v) and vortex.

8   Separate phases by adding 1 ml of chloroform and 1 ml of 1% (w/v) $HClO_4$. Vortex and centrifuge (1000 g, 5 min, 20 °C). Discard the upper phase.

Protocol 4 continued

9   Wash the lower phase twice with 2 ml of 1% (w/v) HClO$_4$/methanol (7:1, v/v).

10 Transfer 1.5 ml of the lower phase to a new tube and dry under N$_2$.

11 Redissolve the lipids in 60 µl of chloroform/methanol (2:1, v/v), apply 20 µl onto a HPTLC plate, and develop the plate with chloroform/methanol/acetic acid (13:3:1, by vol.) as the mobile phase.

12 Localize the labelled lipids by autoradiography (6–18 h), scrape the ceramide-1-phosphate and phosphatidic acid-containing spots, and determine the extent of radiolabelling by scintillation counting (Figure 1).

13 Calculate the amounts of ceramide and diacylglycerol in the extracts with the aid of the external standards and normalize to total phospholipid (step 2).

[a] Note that the assay co-quantifies diacylglycerol, the primary substrate of diacylglycerol kinase. Therefore alkaline hydrolysis should be omitted, as it would destroy the substrate.

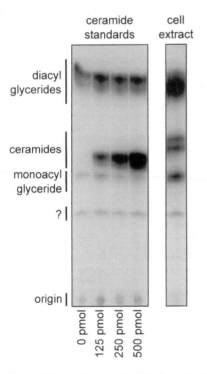

Figure 1 Autoradiogram of a TLC plate on which the ^{32}P-phosphorylated products of the DAG kinase reaction have been separated. The left panel shows lanes that contain different standard amounts of (brain-derived) ceramides. In addition to these ceramides, also mono- and diacylglycerides are present, as well as a yet unidentified lipid. Since these products are present even in the absence of ceramide (blank, left lane), it can be assumed that these lipids originate from the membrane-derived enzyme preparation. The right panel shows the reaction products as detected in a human fibroblast extract. Note that the R$_f$ value of the cellular ceramides differs slightly from that of the standards. This might reflect a difference in fatty acid composition.

Protocol 5

Mass measurement of free sphingoid bases

Equipment and reagents

- See Protocols 1–3
- C:18-coated 4 µm silica HPLC column
- Fluorimeter
- Derivatization mix: 50 mg/ml *o*-phtaldialdehyde and 5% (v/v)
- 2-mercaptoethanol in ethanol (96%); prepare freshly a in a light-protected flask
- Boric acid

Method

1 Extract the cellular lipids according to Protocol 1.

2 Redissolve the extracts in exactly 1 ml of chloroform/methanol (1:1, v/v) and perform a phosphate determination on an aliquot (16). Calculate the phospholipid content of the total extract. Transfer an aliquot of 300 nmol to a new tube and dry under N_2.

3 Subject the lipids to a mild alkaline hydrolysis (Protocol 2) and redissolve in 50 µl of chloroform.

4 Prepare a series of standards of the sphingoid bases to be analysed (ranging from 0–1000 pmol) in 50 µl of chloroform.

5 Dilute the derivatization mix 1:100 in 3% (w/v) boric acid pH 10.5. Vortex and add 50 µl of this mixture to the lipids. Incubate for 5 min at 20 °C.

6 Add 300 µl of methanol/water (9:1, v/v) and vortex.

7 Apply 80 µl[a] to a 250 × 4.6 mm C:18-coated 4 µm silica HPLC column. Run the sample, employing an isocratic mobile phase consisting of methanol/water (9:1, v/v) at a flow rate of 1 ml/min.

8 For detection, connect the column to a 25 µl quartz flow cell, which is placed in a fluorimeter with excitation and emission wavelength settings at 350 and 418 nm, respectively (Figure 2).

9 Measure peak surfaces by on-line integration. Identify and quantify the different sphingoid bases with the aid of the external standards and taking into account their retention times.[b] Normalize the amounts of sphingoid bases to total phospholipid (step 2).

[a] Sample injection is best performed when employing a 100 µl sample loop. However, to avoid sample cut-offs by the injector system, 10 µl of methanol/water (9:1, v/v) should be injected just prior and after injection of the sample to be analysed.

[b] When employing the present technique, the following sphingoid bases can generally be detected in cellular extracts: C:18 phytosphingosine (retention time 7' 30"), C:18 sphingosine (11' 45"), C:18 sphinganine (15' 45"), and C:20 sphingosine (19' 45") (see Figure 2).

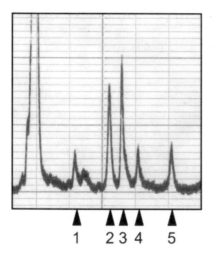

Figure 2 Typical HPLC chromatograph of a hydrolysed and o-phtaldialdehyde-derivatized lipid extract, isolated from HT29 cells. The large peak on the left represents free o-phtaldialdehyde. The other peaks have been identified as phytosphingosine (1), sphingosine (2), sphinganine (4), and C_{20}-sphingosine (5). Peak 3 has not been identified.

Protocol 6

Analysis of gangliosides

Equipment and reagents

- Bath-sonicator
- Sep-Pak C:18 cartridge (Bio-Rad)
- PBS
- Chloroform
- Methanol
- Nitrogen

- Diisopropylether/1-butanol (3:2, v/v)
- NaCl
- $CaCl_2$
- 6 mg/ml dimethyl-aminobenzaldehyde in ethanol/37% HCl (4:1, v/v)

Method

1   Wash the cell cultures twice with PBS and harvest the cells by mild trypsinization. Centrifuge (5 min, 500 g), resuspend the pellets in PBS, and count the cells.

2   Centrifuge 6×10^7 cells (5 min, 500 g, 20 °C) and remove the supernatant. Resuspend the pellet in 0.25 ml of water, and add 5 ml of chloroform/methanol (1:1, v/v). Bath-sonicate the mixtures until homogeneous solutions are obtained. Centrifuge (5 min, 1000 g, 20 °C) and transfer the supernatant to new tubes.[a] Re-extract the pellet with 5 ml of chloroform/methanol (1:1, v/v) and sonicate. Centrifuge (5 min, 1000 g, 20 °C) and pool the supernatants. Re-extract the pellet once more, and pool the supernatants.

3   Dry the pooled supernatants under nitrogen. Redissolve in 2.5 ml of chloroform/methanol (1:1, v/v). Sonicate for 2 min, centrifuge (15 min, 2000 g), and place the tubes at −20 °C for 16 h.

Protocol 6 continued

4 Transfer the supernatant to a new tube, and dry under nitrogen. The fraction thus obtained, represents the total lipid pool.

5 Redissolve the lipids in 4 ml of diisopropylether/1-butanol (3:2, v/v). Vortex and sonicate. Subsequently, add 2 ml of 17 mM NaCl. Vortex and centrifuge (10 min, 2000 g, 20 °C). Discard the upper phase,[b] and wash the lower phase with 4 ml diisopropylether/1-butanol. Vortex and sonicate. Centrifuge for 10 min, 2000 g, at 20 °C.

6 Transfer the lower phase to a new tube. Remove residual organic solvent by a flow of nitrogen (15 min at 37 °C). Dry the lipids by lyophilization.

7 Redissolve the lipids in 5 ml of methanol/water (1:1, v/v). Vortex and sonicate thoroughly.

8 Wash a Sep-Pak C:18 cartridge with 5 ml of methanol, 10 ml of chloroform/methanol (2:1, v/v), 5 ml of methanol, and 10 ml of methanol/water (1:1, v/v), respectively. Load the lipid sample onto the cartridge, and reload the flow-through once.

9 Wash the cartridge with 40 ml of water, and dry by a gentle stream of nitrogen.

10 Elute the gangliosides with 2 ml of methanol and 10 ml of chloroform/methanol (1:1, v/v), respectively. Dry the eluate under nitrogen.

11 Separate the gangliosides by HPTLC, employing chloroform/methanol/0.2% (w/v) $CaCl_2$ (11:9:2, by vol.) as the mobile phase.

12 After drying, spray the plate with 6 mg/ml dimethyl-aminobenzaldehyde in ethanol/37% HCl (4:1, v/v). Cover with a clean glass plate, heat the plates for 15 min at 180 °C, and photograph.[c]

[a] In contrast to other procedures described in this chapter, the present procedure starts with a one-phase extraction.

[b] In contrast to the standard two-phase extraction, the lower phase of this particular (second) extraction procedure is aqueous.

[c] For identification and quantification purposes, apply standard amounts (nmol amounts) of different gangliosides onto the same plate as the samples. After staining, the plate should be subjected to digital scanning and analysed with the aid of appropriate software (e.g. Scion Image software, which yields arbitrary units of lipid). A free version of this program can be downloaded from http://scioncorp.com/frames/fr_scion_products.htm. Obviously, for proper quantitation the sample lipids should be in the linear range of the standards.

2.3 Sphingolipid analysis by metabolic radiolabelling

All radiolabelling studies described below are based upon the fact that proliferating cells in culture readily take up and incorporate precursor molecules into their biosynthetic pathways. This also holds for a variety of precursors for sphingolipids. Since most of these precursor molecules are commercially available in a radiolabelled form, the opportunity is provided to radiolabel cellular sphingolipid pools by the simple addition of these molecules to the culture medium. The highest efficiency of labelling is obtained when radiolabelling is performed during the exponential growth phase, since at least two population doublings, in the presence of the precursors, are generally required to ensure an equilibrium distribution of the label over the entire pool of cellular lipids. Depending on the growth rate of the cell type, 24–48 hours usually suffice to obtain a label distribution which fully reflects the natural, endogenous lipid composition of a cell (17). Although radiolabelling studies do not provide data on the absolute amounts of particular lipids in an extract, they do offer a rapid and very sensitive way to study *relative* changes in the levels of particular sphingo-lipids.

In the following we focus on the two most commonly used precursor molecules, serine and choline, but other precursor molecules, e.g. palmitic acid or dihydrosphingosine, can also be employed, governed by experimental purposes. In addition, we present a protocol for the study of cerebrosides, based upon the use of radiolabelled galactose.

Together with palmitoyl-CoA, serine is the ultimate precursor molecule for the *de novo* synthesis of all sphingolipids. The presence of L-[3-^{3}H]serine in the culture medium therefore eventually results in the radiolabelling of all (glyco)sphingolipids. In contrast, radiolabelled [methyl-^{3}H]choline specifically radiolabels the cellular pool of sphingomyelin. This specificity greatly facilitates the procedure. The original procedure, which is based upon HPTLC analysis, can be further simplified by an alternative protocol, previously published by Andrieu *et al.* (18), which will also be described below. By replacing the HPTLC step by a mild alkaline hydrolysis, sphingomyelin analysis is now even less time-consuming. Growing cells in the presence of [^{3}H]D-galactose, results in the specific radiolabelling of cerebrosides. Cellular epimerase activity ensures a rapid conversion to [^{3}H]D-glucose, which is efficiently incorporated into cerebrosides (and all higher glycosphingolipids) (19). In this particular procedure shorter incubation periods (16 hours) usually suffice.

Protocol 7

Total sphingolipid analysis by L-[3-^{3}H]serine labelling

Equipment and reagents

- See Protocols 1 and 2
- 10 × 10 cm TLC plate
- Scintillation counter
- L-[3-^{3}H]serine

- PBS
- Chloroform, methanol, acetic acid
- Iodine vapour
- H_3BO_3, NH_4OH

Method

1 Culture semi-confluent cells in a 25 cm^2 flask for at least 48 h in the presence of 5 μCi/ml L-[3-^{3}H]serine.

Protocol 7 continued

2 Wash the cells twice with PBS and extract lipids (Protocol 1).

3 Dissolve the extracted lipids in exactly 1 ml of chloroform/methanol (1:1, v/v) and take an aliquot for the determination of total lipid-incorporated radioactivity.[a]

4 Subject lipids to an alkaline hydrolysis procedure (Protocol 2).

5 Spot lipids on the lower left-hand corner of a 10×10 cm TLC plate, together with unlabelled standards of the sphingolipids of interest.

6 Develop the plate in chloroform/methanol/water (14:6:1, by vol.) and dry. Rotate the plate 90° anticlockwise and run the plate in chloroform/acetic acid (9:1, v/v). Dry the plate overnight.

7 Stain lipids in iodine vapour. Identify the ceramide-containing spot with the aid of an unlabelled standard[b] and scrape into scintillation vials, which contain 0.5 ml of water. Vortex and add 4 ml of scintillation cocktail. Vortex again and perform scintillation counting.[a]

8 Carefully wrap the left-hand 2 cm of the plate in aluminium foil and spray the unprotected part of the plate with 2.5% (w/v) H_3BO_3 in methanol. Remove the foil and dry the plate. Develop the plate once more in the second dimension, this time using chloroform/methanol/25% (w/v) NH_4OH (13:7:1, by vol.) as the mobile phase, and dry the plate.

9 Stain the lipids in iodine vapour. Identify the lipids of interest with the aid of the unlabelled standards.[b] For further positive identification and localization of the lipids, treat the plate with En3Hance spray (Dupont) and subject it to auto-radiography for one week at $-80\,°C$. Mark and scrape all lipids of interest into vials, containing 0.5 ml of water. Vortex, add 4 ml of scintillation cocktail and count.[a]

[a] Express lipids as disintegrations per second (d.p.s.) per 1000 d.p.s. in the total (unhydrolysed) extract (step 3).

[b] A positive control for the identification of ceramide and sphingomyelin involves the pre-treatment of a separate batch of cells for 2 h with 0.1 U/ml of a bacterial sphingomyelinase. The extract of these particular cells should lack any significant amount of sphingomyelin and contain excessive amounts of ceramide.

Protocol 8

Sphingomyelin analysis by [methyl-^{3}H]choline labelling

Reagents

- [Methyl-^{3}H]choline
- PBS
- Chloroform, methanol

Protocol 8 continued

Method

1   Culture semi-confluent cells in a 25 cm^2 flask for at least 48 h in the presence of 1 μCi/ml [methyl-^{3}H]choline.

2   Wash the cells twice with PBS and harvest by scraping with a rubber policeman. Resuspend the cells in PBS and perform a protein determination on aliquots (20). Calculate the protein content of the remaining cells.[a]

3   Extract the lipids from the remaining cells (Protocol 1).

4   Subject the lipids to a mild alkaline hydrolysis (Protocol 2) and, after extraction, perform two additional washing steps on the lower phase, using chloroform/methanol/water (3:48:47, by vol.). Pool the upper phases. Take an aliquot and perform a scintillation counting.[b]

5   Transfer the lower phase to a scintillation vial, and dry under nitrogen. Perform a scintillation counting.[b] Express the amount of the lipids as d.p.s./μg of protein.

[a] Express lipids as d.p.s./μg of protein (step 2). Alternatively, a phosphate determination can be carried out (as in Protocol 3) after step 3.

[b] The upper phases of the extraction, after alkaline hydrolysis, contain phosphorylcholine, a radiolabelled degradation product, derived from the original pool of phosphatidylcholine. The lower phase contains sphingomyelin.

Protocol 9

Cerebroside analysis by [^{3}H]D-galactose labelling

Reagents

- [^{3}H]D-galactose
- Glucose-free medium, supplemented with 5% dialysed serum and 0.5 mg/ml glucose
- PBS
- 2.5% (w/v) H$_3$BO$_3$ in methanol
- Unlabelled cerebroside standards
- Chloroform, methanol, NH$_4$OH

Method

1   Culture semi-confluent cells in a 25 cm^2 flask for at least 16 h in the presence of 15 μCi/ml [^{3}H]D-galactose in glucose-free medium, supplemented with 5% dialysed serum and 0.5 mg/ml glucose.

2   Wash the cells twice with PBS and harvest by scraping with a rubber policeman. Resuspend the cells in PBS and perform a protein determination on an aliquot (20). Calculate the protein content of the remaining cells.[a]

3   Extract lipids from the remaining cells (Protocol 1).

Protocol 9 continued

4 Impregnate a TLC plate with borate by spraying with 2.5% (w/v) H_3BO_3 in methanol. Spot the lipids together with unlabelled cerebroside standards. Develop the plate in chloroform/methanol/25% (w/v) NH_4OH (13:7:1, by vol.).

5 Stain the lipids in iodine vapour. Identify the lipids of interest with the aid of the unlabelled standards. For further positive identification and localization of the lipids, treat the plate with En3Hance spray and subject it to autoradiography for one week at $-80\,°C$. Mark and scrape lipids into vials containing 0.5 ml of water. Vortex, add 4 ml of scintillation cocktail, and count.[a]

[a] Express lipids as d.p.s./μg of protein (step 2). Alternatively, a phosphate determination can be carried out (as in Protocol 3) after step 3.

3 Sphingolipid trafficking in living cells

To study lipid metabolism or cellular processes in which the generation of distinct lipid metabolites is of relevance, the employment of radiolabelled approaches, as described above, is the most obvious choice. This is the case, for example, in studies involving the generation of (radiolabelled) ceramide, which may trigger apoptosis, the sphingolipid metabolite arising from activation of sphingomyelinase, which in turn may cause the hydrolysis of sphingomyelin (21, 22). However, in studies that focus on obtaining insight into the fate and mechanisms of intracellular trafficking of distinct lipids or the flow of membranes, marked by distinct lipids, the application of radiolabelled lipids is often less advantageous, as determination of their localization will often require cell fractionation studies. The occurrence of lipid redistribution during such procedures may lead to erroneous interpretations. This has prompted studies in which fluorescent lipid analogues are applied, as they allow the monitoring of lipid flow in living cells, as visualized directly by fluorescence microscopy. In conjunction with image analysis, quantitative determination of compartmentalized pools of such lipids within the cells, thereby avoiding fractionation studies, is also possible. Apart from their fruitful and versatile application in studies dealing with the cell biology of lipids (8, 23, 24), fluorescent lipid analogues have also found a broad application in studies of the biophysical properties of membranes, including their polymorphic properties such as lamellar–hexagonal phase transitions, as well as in work examining lipid phase separations and membrane fusion (24). The molecular nature of the probes may vary, and their choice of application will depend on the purpose of the experiment (e.g. in ref. 25). The most popular ones are those which are derivatized by 7-nitrobenz-2-oxa-1,3-diazol-4-yl (NBD), which is attached via a short carbon chain spacer (six carbon atoms in length: C_6) to either a monoacylglycerol backbone in case of phospholipids, or sphingosine, in case of sphingolipids. Due to their relatively high solubility in water, these C_6-NBD-labelled lipid analogues readily transfer between a donor source, such as micelles or liposomes, and a target membrane, e.g. the plasma membrane, thus allowing their rapid insertion into the target membrane. In membrane fusion studies, *head group*-labelled analogues are employed (see Section 4), because such derivatized lipids appear essentially non-exchangeable, an obvious prerequisite for reliably determining membrane mixing as a result of fusion rather than by lipid exchange.

Here, we focus on the synthesis and application of C_6-NBD-labelled sphingolipids. The synthesis of the phospholipid analogues is largely based on very similar procedures, as described in detail elsewhere (23, 24). In ref. 24, an extensive listing of other probes and applications can be found.

Protocol 10

Synthesis of C_6-NBD-labelled sphingolipids

Equipment and reagents

- HPTLC plates (10 × 10 cm) (Merck, Darmstadt, Darmstadt/Germany)
- Sphingosylphosphorylcholine (lyso-sphingomyelin) (Matrya, Plasant Gap, PA, USA)
- C_6-NBD (Molecular Probes, Eugene, OR, USA)

- Triphenylphosphine
- Dithiodipyridine
- Chloroform, methanol, NH_4OH
- Argon, nitrogen

Method

To prepare 2–4 μmol C_6-NBD-sphingomyelin.[a]

1   Weigh 10 μmol of sphingosylphosphorylcholine in a glass tube.

2   Add C_6-NBD (1:1, mol:mol).

3   Add triphenylphosphine (1:2, mol:mol) and dithiodipyridine (= aldrithiol) (1:2, mol:mol). Add 200 μl chloroform and incubate under argon while continuously stirring, for at least 8 h.

4   Load the suspension onto three HPTLC plates (10 × 10 cm). As a running solvent chloroform/methanol/25% (w/v) NH_4OH (24:6:1, by vol.) is used. Appropriate NBD-labelled reference markers are employed to identify the products.

5   Scrape the band representing C_6-NBD-sphingomyelin from the plate, and stack the silica particles in a self-made glass column ('Pasteur pipette') on top of a 'plug', prepared from a small amount of silica gel 100. Elute the lipid with 10 ml chloroform/methanol (1:1, v/v), and subsequently with 10 ml of methanol.

6   Check the product for purity by HPTLC (see step 4), and repeat step 5 if more than one spot is obtained (see step 8, below).

7   Dry the C_6-NBD-lipid under nitrogen, resuspend in chloroform/methanol (1:1), and determine the concentration spectrofluorometrically (λ_{ex} = 465 nm, λ_{em} = 530 nm) using a C_6-NBD-labelled lipid with a known concentration as a standard.

8 Store the C$_6$-NBD-labelled lipids protected from light under argon at −20°C. Occasionally, run a TLC plate as described in step 4 to monitor purity, which is reflected by the presence of one spot only.

^{a} The same procedure is followed for preparing C$_6$-NBD-labelled ceramide, glucosylceramide, or galactosylceramide from C$_6$-NBD and D-sphingosine, 1-β-D-glucosylsphingosine (glucopsychosine), or 1-β-D-galactosylsphingosine (psychosine), respectively.

For carrying out the cell biological studies of interest, lipid analogue dispersions for cell labelling experiments are made freshly on the day of the experiment. The protocols for preparing the samples and the labelling of the cells, is described in Protocols 11 and 12.

Protocol 11

Preparation of C$_6$-NBD-labelled lipid dispersion for cell labelling

Reagents
- C$_6$-NBD-sphingolipid
- Nitrogen
- Hanks balanced salt solution (HBSS)

Method

1 Take the required amount of C$_6$-NBD-sphingolipid from the stock solution (cell labelling is typically done with 2–4 μM lipid analogue).

2 Dry the lipid analogue under nitrogen.

3 Redissolve the dried lipid analogue in absolute ethanol and inject into the desired volume of HBSS (final concentration of ethanol should not exceed 0.5%), under vigorous vortexing.

3.1 Incorporation of C$_6$-NBD-labelled sphingolipids into the plasma membrane of eukaryotic cells

In order to investigate the endocytotic pathway of C$_6$-NBD-sphingolipids in living cells by pulse-chase labelling, the lipid is incorporated into the cells' plasma membrane. Due to the probe's water soluble properties, the analogue readily partitions from the aqueous phase into the plasma membrane by means of a monomeric transfer process. When labelling is carried out at 4°C, the lipid's localization will be restricted to the outer leaflet of the plasma membrane, since at this temperature, neither endocytic internalization nor transbilayer translocation of the probe occurs.

Protocol 12

Incorporation of C_6-NBD-sphingolipids into the outer leaflet of the plasma membrane

Reagents

- Phosphate-buffered saline (PBS)
- C_6-NBD-labelled sphingolipid

Method

1   Wash the cells (2–4×10^6) three times with ice-cold PBS.

2   Add the lipid suspension, prepared as described in Protocol 11, and incubate the cells with 2–4 μM C_6-NBD-labelled sphingolipid at 4 °C for 30 min. In case of quantitative biochemical analysis of endocytosis and recycling, more than one lipid analogue can be administered simultaneously.

3   To remove the non-incorporated lipid analogue, wash the cells three times with ice-cold PBS.

To verify the exoplasmic orientation of the lipid analogues, the cells can subsequently be subjected to a '*back exchange*' procedure, using albumin (from bovine serum) as a lipid scavenger. Due to its hydrophobic binding pockets, albumin is capable of retrieving the lipid analogues (which are relatively hydrophilic because of the C_6-NBD-chain, which tends to 'loop' back to the bilayer/water interface) from the outer leaflet of the plasma membrane. A back exchange procedure is carried out as follows.

Protocol 13

Back exchange of C_6-NBD-labelled sphingolipid from the plasma membrane

Reagents

- HBSS
- PBS
- Fatty acid-free bovine serum albumin pH 7.4

Method

1   Incubate the cells with a solution of HBSS, supplemented with 5% (w/v) fatty acid-free bovine serum albumin pH 7.4, at 4 °C for 30 min.

2   Repeat this procedure once.

3   Wash the cells three times with ice-cold PBS.

Note that following the back exchange procedure, the entire pool of the initially incorporated lipid analogue should be recovered from the back exchange medium, while none should remain associated with the cells. This can be verified by either fluorescence microscopy (using the appropriate filter blocks for NBD-fluorescence; see above), or by biochemical quantitation of the lipid analogue in the cells and in the back exchange media after lipid extraction as described in Protocol 14.

Protocol 14

Lipid extraction and quantification of C_6-NBD-sphingolipids

Equipment and reagents

- HPTLC (Silica 60) plates (10 × 10 cm)
- Glass microcapillary pipette
- Methanol
- Chloroform
- NH_4OH
- Triton X-100
- PBS

Method

1   Add to 1 vol. of the sample (i.e. cell suspension or back exchange media) in a glass extraction tube, 1 vol. methanol, and 2 vol. chloroform.

2   Vortex and put on ice for 30 min.

3   Centrifuge to obtain chloroform/water phase separation (10 min, 200 g).

4   Collect the (lower) chloroform phase, containing the lipids, with a 'Pasteur pipette'.

5   Evaporate the chloroform and resuspend the dried lipid in 50 μl chloroform/ methanol (2:1).

6   Spot the resuspended lipid onto HPTLC (Silica 60) plates (10 × 10 cm) with a glass microcapillary pipette and run for 20 min, using chloroform/methanol/25% (w/v) NH_4OH (14:6:1, by vol.) as the running solvent.

7   Identify the C_6-NBD-sphingolipid bands with UV light. Mark the spots gently with a soft pencil, and scrape them from the plates.

8   To remove the lipid from the silica particles, shake the silica particles vigorously for at least 1 h in 1% (v/v) Triton X-100 (diluted in PBS).

9   Spin down the silica particles and measure the fluorescence of the C_6-NBD-sphingo-lipids as described in Protocol 10, step 7.

An alternative and faster way to *microscopically* verify the exoplasmic orientation of the lipid analogue following the incorporation step, is to incubate the labelled cells with HBSS, supplemented with 30 mM sodium dithionite (diluted from a 1 M stock solution in 1 M Tris buffer pH 10), prepared freshly (i.e. within

30 min prior to use), at 4 °C for 5–10 min. Sodium dithionite chemically reduces the NBD, thereby irreversibly destroying its fluorescence (26). Due to its very slow permeation rate across membranes, selective quenching of the fraction of NBD that is directly accessible to the reductant, i.e. in the outer leaflet of the plasma membrane, can be accomplished in this manner.

3.2 Processing of membrane-inserted sphingolipid by endocytosis

Following incorporation of the lipid analogue into the plasma membrane, endocytosis and recycling to the plasma membrane can be conveniently examined by warming the cells to a transport permissive temperature (typically above 15 °C). The cells are then incubated for the desired time interval in culture medium (without serum) or HBSS, commonly 30–60 min. Following such an incubation, the cells are subjected to a back exchange procedure as described in Protocol 13, either to allow visualization of the intracellular compartments that comprise the transport routes by microscopy (Figure 3), or to quantitate the kinetics of endocytosis. Note that the presence of plasma membrane associated fluorescence often makes it difficult to clearly discern intracellularly labelled compartments, which display fluorescence with a (much) lower intensity. It is therefore recommendable to remove residual label from the plasma membrane by back exchange, prior to microscopic examination. However, such a back exchange step is not necessary when the sample is examined by confocal laser scanning or deconvolution microscopy.

For determination of endocytosis and its kinetics, the cells are scraped following the back exchange (Protocol 13) and the amount of C_6-NBD-sphingolipid, present in the cell suspension, incubation medium (containing any expelled lipid analogues), and back exchange media, is quantitated as described in Protocol 14. The fraction of endocytosed lipid is calculated from the fraction of lipid that is non-back exchangeable ('intracellular fraction') relative to the total fluorescence measured in all fractions (27, 28).

3.3 Determination of recycling of internalized sphingolipids

To investigate the kinetics of sphingolipid recycling, after endocytosis, the fluorescent lipids are incorporated into the plasma membrane, and allowed to internalize at elevated temperatures for the desired time. Note that 3 min at 37 °C is appropriate to label early endocytotic compartments; to label the recycling endosomes, incubate at 37 °C for 10–15 min or at 18 °C for 60 min (28). Next, the residual pool of sphingolipid analogues from the plasma membrane is depleted by subjecting the cells to a back exchange procedure (Protocol 13) or to a sodium dithionite treatment at 4 °C, as described above. Recycling can then be monitored by a continued incubation of the cells at 37 °C for different time intervals in back exchange medium. After each time point, the amount of lipid analogue present in the back exchange medium as a percentage of the total (cell-associated + medium) NBD-fluorescence reflects the amount of recycled lipid

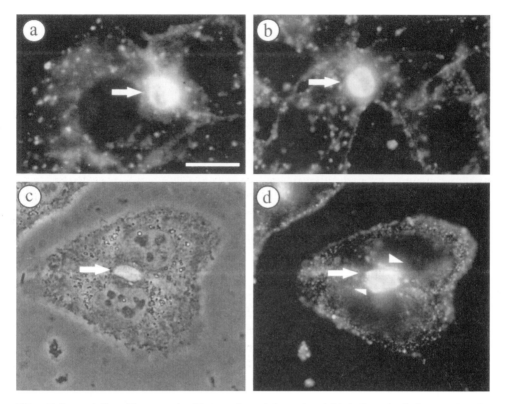

Figure 3 Accumulation of fluorescent sphingomyelin and glucosylceramide in the subapical compartment, SAC, a sorting centre in polarized HepG2 cells. The cells were labelled with C_6-NBD-labelled sphingomyelin (a, d) and -glucosylceramide (b) as described in Section 3.2. After back exchange, the lipids are seen in the bile canalicular membranes (a, b), located in between two apposed cells. When the incubation is continued at 18 °C, the lipids transfer from the bile canalicular membrane to a compartment, SAC (arrow heads), that plays a crucial role in the maintenance of membrane polarity (c, phase contrast to d). For further details, see ref. 8.

analogue. By fluorescence microscopy the occurrence of recycling can be visualized by the reappearance of fluorescence at the plasma membrane (27).

3.4 Biosynthetic transport of C_6-NBD-labelled sphingolipids to the plasma membrane

The synthesis of (glyco)sphingolipids occurs at the endoplasmic reticulum and the Golgi apparatus. Specifically, the ceramide backbone is synthesized at the endoplasmic reticulum, which is then transferred to the Golgi complex where the head group (phosphorylcholine, in the case of sphingomyelin and carbohydrate for all other sphingolipid species) is attached. When C_6-NBD-ceramide is incubated with mammalian cells at 4 °C, this lipid analogue (in contrast to the C_6-NBD-derivatives of sphingomyelin and glucosylceramide, as noted above) acquires rapid access to the cell interior as a free monomer and accumulates specifically in the Golgi apparatus, during metabolism to the sphingolipid

336

products C₆-NBD-sphingomyelin and C₆-NBD-glucosylceramide (29). In that sense, the NBD-labelled ceramide is a highly useful tool as a vital stain of the Golgi complex (23, 29). More complex sphingolipids are usually not observed, implying that the NBD-labelled analogues are apparently less suitable substrates for the enzymes involved in the synthesis of complex glycosphingolipids. The advantage is that this 'limitation' allows detailed studies of the exclusive fate of glucosylceramide and sphingomyelin in the biosynthetic transport pathway by either microscopical or biochemical means (30, 31).

Protocol 15

Monitoring of NBD-labelled sphingomyelin and glucosylceramide in the biosynthetic pathway

Equipment and reagents

- TLC equipment
- PBS
- C₆-NBD-Cer
- HBSS, supplemented with 5% fatty acid-free bovine serum albumin

Method

1 Wash the cells three times with ice-cold PBS.

2 Incubate the cells with 2–4 μM C₆-NBD-Cer at 4 °C for 60 min (see Protocol 11).

3 Wash the cells three times with ice-cold PBS.

4 Incubate the cells with HBSS, supplemented with 5% fatty acid-free bovine serum albumin ('back exchange medium'), at 4 °C for 2 × 30 min, to remove all lipid analogues present at the plasma membrane.

5 Incubate the cells in back exchange medium (step 4) at 37 °C.

6 Collect back exchange fraction.

7 Wash cells and collect wash buffer.

8 Scrape cells in ice-cold PBS.

9 Extract lipids from the pooled back exchange media, wash buffers, and from the cell fraction, as described in Protocol 14. Separate C₆-NBD-sphingomyelin and C₆-NBD-glucosylceramide by thin-layer chromatography and quantify by fluorescence measurement (Protocol 14).

Following Protocol 15, insight can thus be obtained in the kinetics of synthesis of glucosylceramide and sphingomyelin and the fate of these lipids in the cell when travelling along the biosynthetic pathway. For example, if carried out in polarized cells, cultured on filters, their selective appearance at either the basolateral or apical plasma membrane domain can be determined by specifically bathing either surface with back exchange medium (32).

During vesicle-mediated transport processes, as occur in the endocytic and biosynthetic transport pathways, the delivery of intravesicular or membrane-associated cargo is ultimately accomplished via a membrane fusion process, involving the merging of two initially separated membrane containers. It has become increasingly clear that these fusion events are mediated by proteins (33, 34). However, lipids, constituting the basic core structure of the membrane, are intimately involved as they will have to depart from a stable bilayer structure and undergo bilayer-to-nonbilayer transitions, a prerequisite for fusion to take place. In the next section we describe how membrane vesicles of defined lipid compositions can be prepared for the purpose of such studies and how protein or peptide-induced membrane fusion can be studied in such model systems. In particular, we will also describe the protocol for a lipid mixing assay that conveniently allows one to monitor the kinetics of membrane fusion in a sensitive manner.

4 Lipid vesicles and their use in studies of membrane fusion

Given the complexity of biological membranes, lipid vesicles or liposomes are still a highly useful and valid tool in studies aimed at revealing insight into mechanisms of membrane fusion. For example, to characterize the fusogenic features of biological membranes, like viruses or membrane vesicles derived from organelles, lipid vesicles of defined lipid composition can be employed as one of the fusion partners in the system, harbouring probes to monitor and quantify the fusion event in a convenient manner. Protein-induced fusion events can be reconstituted. In such studies, peptides are often employed that correspond to the putative fusion domains of fusogenic proteins, which are added to the liposome suspension as free monomers. Fusion is then monitored by following the mixing of membranes and of liposomal aqueous contents, using a variety of fluorescence assays. These peptides usually induce efficient lipid mixing but cause extensive contents leakage, reflecting poor control of the fusion event. An alternative, and more closely matching the molecular environment of membrane-bound fusion proteins, is a simplified membrane system, consisting of a liposome-anchored short fusogenic peptide, which allows the study of fusion in a more controlled fashion. Apart from the nature of the fusion protein, the lipid composition of the membrane also appears to be a regulating factor in governing membrane fusion. Since relative minor amounts of 'lipid impurities' can affect membrane fusion properties, it is imperative to routinely analyse the purity of the phospholipids, used to prepare artificial membranes.

In the following sections we will describe assays for analysing lipid purity, the preparation of lipid vesicles, coupling of peptides, and how fusion can be monitored, based upon lipid mixing.

4.1 Analysis of phospholipid purity

Lipids are checked regularly by UV spectrophotometry and thin-layer chromatography on silica gel plates for oxidation products and purity. Phospholipid
standards and molybdenum blue spray reagent are obtained from Sigma.
Natural phospholipids contain non-adjacent double bonds that give rise to UV
absorbance peaks at short wavelengths in ethanol (200–210 nm). Oxidation
products, as a result of a free radical chain mechanism (to which double bonds
are most susceptible), appear as additional peaks in the UV spectrum: the
presence of dienes is indicated by a peak around 230 nm, and trienes give rise to
an additional peak around 270 nm. The degree of oxidation can be calculated
from the OD reading at 233 nm (molar extinction coefficient for dienes = 30 000
$M^{-1}cm^{-1}$), according to the following formula:

$$\% \text{ oxidation} = ([\text{diene}] / [\text{phosphate}]) \times 100$$

The phosphate concentration is determined by a phosphate determination (16).

Minor amounts (*c.* 1%) of oxidation products (dienes) are routinely found in
freshly-opened ampules of natural phospholipids (e.g. egg yolk phosphatidylcholine, phosphatidylserine or phosphatidylethanolamine from bovine brain,
phosphatidylinositol from bovine liver); however, heavily oxidized batches (>5%)
are discarded, since these oxidation products can affect the fusion properties of
liposomes (see below).

The purity of lipid batches is also readily revealed by HPTLC on silica-coated
sheets (cf. Protocol 14). By this technique, any contamination by other lipid
species and the presence of hydrolysis products can be detected. Lipid standards
are used as reference spots, and lipid samples are spotted from chloroform
or chloroform/methanol solutions (>5 μmol lipid). The plates are run with a
solvent such as chloroform/methanol/water (65:25:4, by vol.), until the solvent
front is *c.* 0.5 cm from the top of the plate. Phospholipids and lysophospholipids
can be revealed as blue spots by spraying a molybdenum blue reagent on the
dried plate. Under these conditions, lysolipids typically run to positions below
the phospholipid spots, and quantitation can be performed after scanning of the
plate and densitometric analysis (for example with Scion Image software).

Purified lipids are stored at −20 °C under argon and routinely (monthly)
checked for impurities.

4.2 Preparation of liposomes, anchor lipid, and peptide-coupled vesicles

A variety of techniques has been presented in the literature over the last decades
(35), describing the preparation of large unilamellar lipid vesicles (diameter
approx. 150 nm). Here, we briefly describe one of these procedures, which is conveniently used in our laboratory.

Protocol 16

Preparation of large unilamellar vesicles by freeze thaw and extrusion

Equipment and reagents

- Polycarbonate membrane with a pore size of 0.1 mm (Nucleopore Corp.)
- Chloroform
- Nitrogen

Method

1 Mix appropriate amounts of lipids in chloroform and remove the organic solvent by evaporation under a stream of nitrogen.

2 Resuspend the lipid film by vigorous vortexing in 0.5–1.0 ml of the desired aqueous buffer.

3 Submit the suspension to ten cycles of freezings in liquid nitrogen followed by thawing in a 40 °C water-bath.

4 Extrude the liposomes through a polycarbonate membrane with a pore size of 0.1 μm. This will yield particles with a diameter of about 150 nm, as determined by electron microscopy and dynamic light scattering in a Coulter N4S submicron particle analyser.

Note: The liposome suspension obtained by vortexing (step 2) can also be submitted to sonication for 30 min with a 50% active cycle on a VibraCell (Bioblock) at 4 °C. The sonicated suspension is then extruded as described (step 4).

To couple a peptide to these liposomal bilayers, advantage is taken of the efficient coupling that can be accomplished via establishment of a disulfide bond, formed between the free sulfhydryl groups provided by a C terminal cysteine residue in the peptide of choice (36) and that present in the anchor lipid, PE-PDP (37). This derivative is synthesized from L-α-dipalmitoylphosphatidylethanolamine (DPPE) and N-succinimidyl-3-(2-pyridyldithio)propionate (SPDP), as described in Protocol 17.

Protocol 17

Synthesis of PE-PDP

Equipment and reagents

- HPTLC equipment
- L-a-dipalmitoylphosphatidylethanolamine (DPPE)
- N-succinimidyl-3-(2-pyridyldithio)propionate (SPDP)
- Triethylamine
- Chloroform, methanol, acetic acid
- Tris-buffered saline pH 7.4
- Nitrogen

Method

1   Dissolve 20 μmol DPPE in chloroform at T ≥ 50 °C.

2   Mix with 30 μmol SPDP in ethanol and 30 μmol triethylamine.

3   Incubate the mixture (1 ml final volume, in a sealed tube) in the dark for 3 h at room temperature under agitation and an argon atmosphere.

4   Analyse an aliquot of the reaction mixture by HPTLC, with chloroform/methanol/acetic acid as running solvent (60:30:3, by vol.). Two spots will be apparent, if the reaction has not yet run to completion. PE-PDP appears as a spot, which is located ahead of DPPE (i.e. use the latter also as a reference). For completion of the reaction, add a few more drops of triethylamine, leave the sample for another hour, and verify by HPTLC.

5   When the reaction is completed, add Tris-buffered saline pH 7.4 to the PE-PDP-containing chloroform solution (ratio organic (O)/aqueous (A) phase 1:4. v/v), and centrifuge for 10 min at 900 g at room temperature. Collect the organic phase and interface.

6   Add water (ratio O/A 1:3, v/v) and centrifuge for 5 min at 900 g. Collect the organic phase and if necessary, add methanol (i.e. until a clear solution is obtained).

7   Concentrate the PE-PDP sample by evaporation under a stream of nitrogen at room temperature, and assay for phosphate. The lipid derivative is stable up to six months upon storage in glass ampules under argon at −50 °C. Before use, the presence of hydrolysis products should be checked by HPTLC (step 4).

The PE-PDP is incorporated into the liposomal bilayer by mixing an appropriate amount of the lipid analogue with additional lipids, as described above for preparation of the liposomes, which are subsequently produced as described in Protocol 16. The peptide-coupled liposomes are then prepared according to Protocol 18, based upon the coupling principle as indicated below. Note that the peptide of choice should contain a C terminal cysteine residue.

Protocol 18

Covalent coupling of peptides to liposomes

Equipment and reagents

- Spectrophotometer
- Sephadex G-25 column (PD-10, Pharmacia, Sweden)
- Liposomes
- Peptide

Method

1 Incubate a desired amount of liposomes (e.g. 1 μmol) of a given composition (e.g. egg yolk PC/cholesterol/PE-PDP, 3.5:1.5:0.25) overnight at 4°C on a gently swirling device. The peptide is added to this suspension in a PE-PDP/peptide molar ratio of 1:5.

The following reaction takes place:

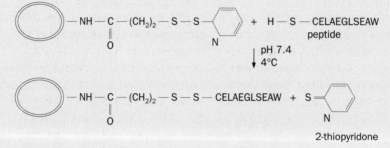

2 Monitor the efficiency of the coupling reaction by measuring the released 2-thiopyridone (see ref. 30) spectrophotometrically at 343 nm.[a]

3 Uncoupled peptide is removed by gel filtration on a Sephadex G-25 column (PD-10), and lipid phosphorus is measured (16) to determine the vesicle concentration.

[a] Typically, at the conditions as described, the coupling efficiency of the added peptide fraction is 10–20%, implying that at a density of 5 mol% PE-PDP and at a PE-PDP/peptide ratio of 1:5, at least half of the PE-PDP molecules contain coupled peptide (38).

The stability of the coupling is excellent over a period of one week, even in diluted suspensions of liposomes (lipid concentration <1 mM), provided that the peptide-coupled vesicles are stored in the dark at 4°C. This stability allows concentration of the liposome sample after gel filtration by Airfuge™ or Speedvac™ (although this latter technique may lead to an undesired increase in the salt concentration, which may result in the swelling of liposomes when placed into a large volume of isotonic buffer).

4.3 Fusion assay based upon lipid mixing

When employing liposomal model systems or semi-artificial systems, in which the fusion of liposomes with biological target membranes is examined, lipid mixing can be monitored as a convenient assay for the detection of membrane fusion. When possible however, it is recommended if not imperative to verify the occurrence by another criterion for membrane merging, e.g. by monitoring the mixing of contents (refs 34, 39; see also below) and/or the detection of fusion by electron microscopic criteria. This is dictated by the notion that lipid movement as a result of exchange or monomeric transfer should be excluded in the assay as described below. For a detailed discussion of such hazards, the reader is referred to ref. 39.

The membrane mixing assay that we describe here relies on the use of resonance energy transfer between two fluorescent lipids, N-(7-nitrobenz-2-oxa-1,3-diazol-4-yl) phosphatidylethanolamine (N-NBD-PE), the energy donor and, N-(Lissamine rhodamine B sulfonyl)-phosphatidyl ethanolamine (N-Rh-PE), the energy acceptor (39, 40). These lipid analogues can be obtained from either Avanti Polar Lipids (Alabaster, AL) or Molecular Probes (Eugene, OR). When both are present in vesicles at a ratio not exceeding 1 mol% of the total lipids, the efficiency of energy transfer, between N-NBD-PE and N-Rh-PE, changes in a linear fashion when their density changes as a result of fusion, thus giving a direct measure of the extent of fusion. In practice, a density of approx. 1 mol% each of the fluorescent lipid analogues corresponds to a minimal fluorescence intensity of the energy donor, N-NBD-PE when monitored at its excitation and emission wavelengths (460 and 535 nm, respectively).

During the process of membrane merging between fluorescently labelled and unlabelled vesicles or target membranes, the distance separating both fluorophores will increase as their surface density decreases, which translates into a linear decrease in the efficiency of the energy transfer, and consequently an increase in the fluorescence intensity of the NBD moiety.

In a membrane fusion assay, two populations of membranes are defined: donor membranes and target membranes. In case of a pure liposomal system, the probes can be inserted into either membrane. In case of a semi-artificial system, e.g. upon studying fusion between liposomes and viruses or Golgi membranes, the lipid analogues are located in the liposomal membrane, since the non-exchangeability of the probes excludes the possibility of their insertion into intact (as opposed to reconstituted) biological membranes at appropriate ratios. Finally, in systems in which we studied peptide-induced fusion, donor liposomes are defined as the liposomes to which the peptide is coupled and which contain the fluorescence probes. Typically, they are composed of egg yolk phosphatidylcholine (EYPC), cholesterol (chol), PE-PDP, N-NBD-PE, and N-Rh-PE (molar ratio 3.5:1.5:0.25:0.05:0.05, respectively). Target liposomes consist of natural or synthetic phosphatidylcholines, cholesterol, and lysine-coupled PE (PE-lys) (molar ratio 11:6:3). PE-lys or, alternatively, a cationic lipid, act as lipids that facilitate the (electrostatic) interaction between donor and acceptor membranes (see ref. 38), aggregation representing a necessary step in the overall fusion reaction. However, to obtain efficient fusion with the peptide described here (see Protocol 18, step 1), the target membranes can also be composed of natural or synthetic phosphatidylserine (PS) and phosphatidylethanolamine (PE). Liposomes are freshly prepared and used within three days. Liposomes prepared with polyunsaturated PC species are used within one day and kept in the dark at any time.

The membrane fusion procedure described below is based upon the use of peptide-coupled liposomes. Obviously the same procedure is applied when using labelled vesicles, devoid of the coupled peptide and biological or artificial target membranes.

Protocol 19

Membrane fusion assay based on lipid mixing

Reagents

- Donor liposomes
- Target liposomes
- Triton X-100

Method

1 Donor liposomes, e.g. EYPC/chol/PE-PDP/N-NBD-PE/N-Rh-PE (3.5:1.5:0.25:0.05:0.05) containing a covalently-coupled peptide (see Protocol 18), are equilibrated under agitation at the desired temperature in a quartz cuvette for 1 min. Their fluorescence intensity is taken as 0% fluorescence (λ_{exc} = 460 nm; λ_{em} = 535 nm).

2 Add an aliquot of target liposomes ('acceptors'; PC/chol/PE-lys, 11:6:3) in a molar ratio donor/acceptor of at least 1:4 (final lipid concentration not exceeding 100 μM to avoid light scattering and inner filter effects), and monitor the kinetics of NBD fluorescence increase in a continuous fashion (see Figure 4).[a]

3 After a plateau is reached, lyse the vesicles by adding Triton X-100 (0.1% final concentration) to the suspension. The value thus obtained is multiplied by 1.54 (to take into account the effect of the detergent on NBD fluorescence), and corrected for sample dilution. This value is set to 100% fluorescence (24, 39). To convert this number to the actual level of *fusion*, the ratio of labelled to unlabelled vesicles should also be taken into account. Thus in case of a 1:4 ratio, 100% lipid mixing would correspond to 80% fluorescence. This value can also be obtained by measuring the fluorescence intensity of a mock fusion product in which the probes are incorporated at a final density that would be reached when all vesicles would fuse, i.e. in the case of a 1:4 ratio (labelled vs. unlabelled vesicles), the fluorecence is measured of a vesicle sample in which the concentration of the probes is five-fold diluted. The percentage of fusion at any time f(t) can then be calculated according to:

$$f(t) = [(F(t) - F_o) / (F_{max} - F_o)] \times 100$$

where F(t) is the fluorescence intensity at time t, F_o the initial fluorescence of the labelled liposomes, and F_{max} the maximal level of fluorescence reached at a given ratio of labelled/unlabelled vesicles (= 80% at a 1:4 ratio, respectively). Full details of this technique can be found in refs 39 and 40.

[a] Note that the extent of fluorescence increase will depend on the ratio of labelled to unlabelled vesicles. Thus the smaller this ratio, the larger the surface area for lipid dilution, and the higher the net increase in fluorescence when fusion occurs.

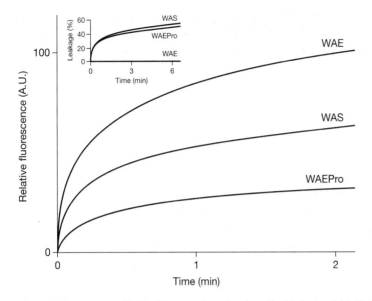

Figure 4 Time course of lipid mixing, monitored as described in Protocol 19. Lipid mixing is induced upon the interaction of target vesicles and fluorescently labelled donor vesicles to which different peptides (11-mers) had been attached by covalent coupling via PE-PDP (Protocol 18). Note that in contrast to WAS and WAEPro, the peptide denoted as WAE does not induce the release of vesicle contents (insert). Indeed, WAE-induced fusion measured by lipid mixing closely matches that measured by contents mixing. For leakage-inducing peptides, lipid mixing provides a convenient alternative for the detection of membrane fusion. For further details, see ref. 36.

Upon fusion, both leaflets should mix. To exclude therefore that mixing occurs as a result of the exclusive mixing of outer leaflets only (hemi-fusion), the fluorescently labelled liposomes can be treated with sodium dithionite (20 mM for 20 min at 37 °C), which, as described above only slowly permeates membrane bilayers. Hence, such a treatment will extinguish specifically the fluorescence of the outer leaflet. In this way, any increase in the fluorescence signal of NBD, occurring when such asymmetrically labelled vesicles are mixed with non-labelled membranes, will imply that the inner leaflet of the bilayer participates in the membrane mixing process, thus likely reflecting a genuine membrane merging process (41). This approach is sometimes taken, e.g. when contents mixing cannot be carried out, to support the occurrence of genuine fusion, reported by the lipid mixing assay, rather than lipid exchange or transfer. It is evident though that, if possible, contents mixing is preferred as an additional means to rigorously demonstrate the fusion event.

References

1. Hoekstra, D. (ed.) (1994). *Current topics in membranes*, Vol. 40. Academic Press, San Diego.
2. Simons, K. and Ikonen, E. (1997). *Nature*, **38**, 569.
3. Brown, D. A. and London, E. (1998). *Annu. Rev. Cell Biol. Dev.*, **14**, 111.

4. Matter, K. and Mellman, I. (1994). *Curr. Opin. Cell Biol.*, **6**, 545.

5. Mostov, K. E. and Cardone, M. H. (1995). *Bioessays*, **17**, 129.

6. Zegers, M. M. P. and Hoekstra, D. (1998). *Biochem. J.*, **336**, 257.

7. Rodriguez-Boulan, E. and Powell, S. K. (1992). *Annu. Rev. Cell Biol.*, **8**, 395.

8. van IJzendoorn, S. C. D. and Hoekstra, D. (1999). *Trends Cell Biol.*, **9**, 144.

9. Bligh, E. J. and Dyer, W. J. (1959). *Can. J. Biochem. Physiol.*, **37**, 911.

10. Preiss, J. E., Loomis, C. R., Bell, R. M., and Niedel, J. E. (1987). In *Methods in enzymology* (ed. A. R. Means and M. Conn), Vol. 141, pp. 294–300. Academic Press, Orlando.

11. Van Veldhoven, P. P., Matthews, T. J., Bolognesi, D. P., and Bell, R. M. (1992). *Biochem. Biophys. Res. Commun.*, **187**, 209.

12. Perry, D. K. and Hannun, J. (1999). *Trends Biochem. Sci.*, **24**, 226.

13. Merrill, Jr., A. H., Wang, E., Mullins, R. E., Jamison, W. C., Nimkar, S., and Liotta, D. C. (1988). *Anal. Biochem.*, **171**, 373.

14. Senn, H. J., Orth, M., Fitzke, E., Wieland, H., and Gerok, W. (1989). *Eur. J. Biochem.*, **181**, 657.

15. Sietsma, H., Nijhof, W., Dontje, B., Vellenga, E., Kamps, W. A., and Kok, J. W. (1999). *Cancer Res.*, **58**, 4840.

16. Böttcher, C. J. F., van Gent, C. M., and Pries, C. A. (1961). *Anal. Chim. Acta*, **24**, 203.

17. Babia, T., Veldman, R. J., Hoekstra, D., and Kok, J. W. (1998). *Eur. J. Biochem.*, **258**, 233.

18. Andrieu, N., Salvayre, R., and Levade, T. (1994). *Biochem. J.*, **303**, 341.

19. Van Echten, G. and Sandhoff, K. J. (1989). *Neurochemistry*, **52**, 207.

20. Smith, P. K., Krohn, R. I., Hermanson, G. T., Mallia, A. K., Gartner, F. H., Provenzano, M. D., *et al.* (1985). *Anal. Chem.*, **150**, 76.

21. Hoekstra, D. (1999). *J. Hepatol.*, **31**, 161.

22. Veldman, R. J., Klappe, K., Hoekstra, D., and Kok, J. W. (1998). *Biochem. J.*, **331**, 563.

23. Pagano, R. E. and Martin, O. C. (1997). In *Cell biology: a laboratory handbook* (ed. J. E. C. Celis), pp. 507–12. Academic Press, San Diego.

24. Kok, J. W. and Hoekstra, D. (1999). In *Fluorescent and luminescent probes for biological activity* (2nd edn) (ed. W. T. Mason), pp. 136–55. Academic Press, London.

25. Chen, C. S., Martin, O. C., and Pagano, R. E. (1997). *Biophys. J.*, **72**, 37.

26. McIntyre, J. C. and Sleight, R. G. (1991). *Biochemistry*, **30**, 11819.

27. Kok, J. W., Hoekstra, K., Eskelinen, S., and Hoekstra, D. (1992). *J. Cell Sci.*, **103**, 1139.

28. van IJzendoorn, S. C. D. and Hoekstra, D. (1998). *J. Cell Biol.*, **142**, 683.

29. Kok, J. W., Nikolova-Karakashian, M., Klappe, K., Alexander, C., and Merrill, A. H. Jr. (1997). *J. Biol. Chem.*, **272**, 21128

30. Zegers, M. M. P. and Hoekstra, D. (1997). *J. Cell Biol.*, **138**, 307.

31. Kok, J. W., Babia, T., Filipeanu, C. M., Nelemans, A., Egea, G., and Hoekstra, D. (1998). *J. Cell Biol.*, **142**, 25.

32. Van Meer, G., Stelzer, E. H. K., Wijnaendts-van Resandt, R. W., and Simons, K. (1987). *J. Cell Biol.*, **105**, 1623.

33. White, J. M. (1992). *Science*, **258**, 917.

34. Pécheur, E. I., Sainte-Marie, J., Bienvenüe, A., and Hoekstra, D. (1999). *J. Membr. Biol.*, **167**, 1.

35. New, R. R. C. (1990). In *Liposomes: a practical approach* (ed. R. R. C. New), pp. 105–62. IRL Press, Oxford.

36. Pécheur, E. I., Martin, I., Ruysschaert, J. M., Bienvenüe, A., and Hoekstra, D. (1998). *Biochemistry*, **37**, 2361.

37. Martin, F. J., Hubbell, W. L., and Papahadjopoulos, D. (1981). *Biochemistry*, **20**, 4229.

38. Pécheur, E. I., Hoekstra, D., Sainte-Marie, J., Maurin, L., Bienvenüe, A., and Philippot, J. R. (1997). *Biochemistry*, **36**, 3773.

39. Hoekstra, D. and Düzgüneş, N. (1993). In *Methods in enzymology* (ed. N. Düzgüneş), Vol. 220, pp. 15–32. Academic Press, San Diego.

40. Struck, D., Hoekstra, D., and Pagano, R. E. (1981). *Biochemistry*, **20**, 4093.

41. Hoekstra, D., Buist-Arkema, R., Klappe, K., and Reutelingsperger, C. P. M. (1993). *Biochemistry*, **32**, 14194.

Chapter 11
Extracellular matrix protocols for the study of complex phenotypes

Calvin D. Roskelley
Department of Anatomy, University of British Columbia,
Vancouver BC V6T 1Z3, Canada.

Shoukat Dedhar
British Columbia Cancer Agency, Jack Bell Research Center,
Vancouver BC V6H 3Z6, Canada.

1 Introduction

Cell adhesion assays are most often carried out on rigid planar substrata coated with ligands that facilitate cell binding and cell spreading. These ligands include: complex extracellular matrices (ECM; i.e. basement membrane), defined extra-cellular matrix proteins (i.e. laminin), short peptide fragments of ECM proteins that bind integrins (i.e. contain RGD sequences) or antibodies that engage, and in some cases cluster, cell surface integrins. Under these conditions integrin-dependent signalling events and changes to the cytoskeleton can be assessed, often concomitantly. These classical 'stick-and-spread' assays have proven par-ticularly useful for identifying the mechanisms responsible for basic adhesion-dependent phenotypes such as anchorage-dependent proliferation, membrane ruffling, or the initiation of cell motility (1–3). However, it is rarely technically possible to examine more complex differentiative phenotypes using these assays. For example, mammary epithelial cells spread on a rigid basement membrane substratum and begin to proliferate. In contrast, when these cells are placed on a malleable gel of the same matrix, morphogenesis and differentiation occur. Given the complexity inherent in the emergence of the latter phenotypes it is very difficult to determine the molecular mediators responsible. To overcome this problem we have developed specialized assays in which the entire process can be deconstructed such that individual morphogenic and differentiative events can be assayed and manipulated in isolation (4–6). As detailed below, the data generated from these assays have shed considerable light on the ECM-mediated signals, both biochemical and physical, that regulate development in the mammary epithelium. It is clear, however, that these assays can be modified to assess complex ECM-dependent phenotypes in other cell types (7).

2 Mammary gland development *in vivo*

Mammary gland development is marked by numerous cellular changes that begin in the embryo, continue postnatally, are modified at puberty, and come to full fruition during adult cycles of pregnancy, lactation, and involution. These changes are regulated by hormones produced at distant sites and by short-range factors within the glandular microenvironment. A critical component in this microenvironment is the complex basement membrane ECM (8).

2.1 Embryonic development

In the embryo, mammary gland morphogenesis begins when epidermal protrusions branch within an underlying adipocytic mesenchyme to form the preliminary ductal tree. At birth, the epithelial and mesenchymal compartments of the developing gland are clearly separated by a continuous basement membrane. The deposition of laminin and heparan sulfate proteoglycan into the basement membrane by the adipocytic mesenchyme of the gland is crucial for further epithelial development. In males, androgen-mediated conversion to a fibro-elastic mesenchyme results in fibronectin and tenascin deposition which sparks ductal regression (9, 10). Thus, the specific molecular composition of the basement membrane and, presumably differential integrin activation, is an important mediator of early ductal outgrowth.

2.2 Postnatal development

Postnatally, a second phase of mammary morphogenesis occurs as growth and branching of the ductal tree accelerates. Epithelial proliferation at puberty is directed by ovarian hormones that trigger the mesenchymal production of locally acting growth factors such as TGF-α (11). Other factors produced by the mesenchyme alter ECM-dependent branching decisions. For example, localized production of TGF-β causes the deposition of a fibroelastic ECM and this prevents branching along the walls of the ducts (12). In contrast, site-specific side branching is induced by HGF which acts, at least in part, by regulating epithelial interactions with the basement membrane (13).

2.3 Lobulo-alveolar development

Lobulo-alveolar development takes place during pregnancy. Lactogenic hormones such as oestrogens, progestins, insulin, insulin-like growth factors, and prolactin all act to stimulate the massive epithelial proliferation that takes place early in pregnancy. This short-lived hyperplasia is kinetically coupled to the formation of small alveolar outpocketings that bud from the lateral margins of the terminal portions of the ductal tree. The latter process is partially regulated by the local production of the paracrine factor neuregulin (14) and by transient discontinuities that develop in the basement membrane. As alveolar budding nears completion a continuous basement membrane is restored and proliferation ends. Lactogenic hormones then initiate alveolar morphogenesis, differentiative milk protein production, and lactational secretion, all of which absolutely require an intimate association of the alveolar mammary epithelium with the basement membrane (15).

2.4 Involution

At the end of lactation, during weaning, the great majority of the alveolar mammary epithelium degenerates and the gland returns to its resting state. This process is mediated by a wave of epithelial apoptosis that is initiated, in large part, by metalloprotease-mediated destruction of the basement membrane which interrupts integrin-dependent survival signals (16).

3 Mammary epithelial cell morphogenesis, differentiation, and apoptosis *in vitro*

When mammary epithelial cells are isolated from mid-pregnant mice they rapidly de-differentiate in monolayer culture, even when they are maintained in the presence of a full spectrum of lactogenic hormones. In contrast, when they are cultured on a reconstituted basement membrane gel these cells undergo a complex morphogenic reorganization. They round-up, aggregate, undergo an apical/basal polarization, and the clusters cavitate to form a central lumen (17, 18). The resulting three-dimensional spheroids closely resemble lactational alveoli *in vivo*. These structures also differentiate. Specifically, the cells express milk protein genes, package the products in secretory vesicles, transport the vesicles to the apical cell surface, and release the products into the central lumen (17, 19–23). If cellular interactions with the basement membrane are disrupted differentiation is prevented and apoptosis is initiated (24, 25). Therefore, dynamic and reciprocal interactions between the cells and the surrounding basement membrane are key regulators of complex mammary epithelial phenotypes (26, 27). We have long been interested in the role that integrin signalling and other ECM-dependent events play in these processes. Toward this end we and others have developed functional cell lines and specialized culture models to examine individual aspects of basement membrane-dependent morphogenic, differentiative, and apoptotic events in isolation.

4 Experimental models and approaches

4.1 Functional cell lines

The cell lines that can be used in these studies are numerous. We routinely utilize a functional, homogeneous mouse mammary epithelial cell line that cannot deposit an endogenous basement membrane such that the effects of exogenously added ECM can be assayed with certainty. This line, designated scp2, was isolated by limited dilution cloning (28) from the functional, but heterogeneous CID-9 cell line (29). The CID-9 line was originally isolated by differential trypsinization of the spontaneously immortal COMMA-1D line (30). The HC-11 (31), IM-2 (32), and EpH4 (14) mouse mammary epithelial cell lines can also be used to assay basement membrane-dependent induction of differentiative milk protein gene expression. The usefulness of each line for morphogenic studies varies widely. In our hands CID-9 cells consistently form spheres with uniformly large lumina, perhaps because they contain a mixture of epithelial and stromal-like cells. scp2 cells are very responsive to changes in cell shape both in terms of morphogenesis and differentiation. EpH4 cells form compact spheres, with small lumina. As an alternative to cell lines, primary mammary epithelial cell cultures derived from mid-pregnant mice (17) can also be used (25). Cell lines are advantageous because they can be used to generate relatively consistent data between experiments and because they can be used to produce stable transfectants. Thus, their phenotype can be easily manipulated by the expression of transgenes and the transactivation state of milk protein gene promoters can be determined using reporter genes (21, 29, 33), although primary cultures from transgenic mice can also be used for these purposes (34, 35).

4.2 Monolayer culture

Functional mammary epithelial cell lines are routinely maintained in monolayer culture in medium containing insulin to protect against apoptosis. Except in special circumstances (see Section 4.7) the cells should be subcultured at subconfluence to reduce phenotypic drift and loss of function (i.e. ability to induce milk protein gene expression in response to exogenously added ECM). To decrease the loss of cell surface integrin and cell–cell adhesion molecules at passaging we routinely use trypsin at low concentrations (i.e. 0.05%) for cell dissociation. In addition, we find that prolonged passaging in high serum leads to post-confluent proliferation that ultimately contributes to loss of function. Therefore, we routinely use 2% FBS as a medium supplement that supports controlled proliferative growth in stock monolayer culture.

Protocol 1

Routine culture and maintenance of functionality

Equipment and reagents

- 100 mm diameter tissue culture dish
- Functional scp2 mouse mammary epithelial cell line (28)
- Growth medium: Dulbecco's modified Eagles medium (DMEM)/F12 medium (1:1) (Sigma) containing 2% FBS (Hyclone, Logan UT), 5 μg/ml insulin (Sigma), and 50 μg/ml gentamycin sulfate (Sigma)
- Trypsin solution: 0.05% trypsin, 0.02% EDTA in Hanks balanced salt solution (HBSS) (Sigma)

Protocol 1 continued

Method

1 Remove and discard culture medium from subconfluent cultures in a 100 mm diameter tissue culture dish.

2 Wash the cells with 10 ml of serum-free DMEM/F12 media, three times for 10 min each, with gentle agitation. This ensures that serum is completely removed which greatly reduces the time required for trypsinization.

3 Add 2 ml of 0.05% trypsin/0.02% EDTA in HBSS at room temperature and leave until the cells begin to retract and round on the plate. This takes approx. 5 min; aspirate.

4 Add 2 ml fresh trypsin, maintain at 37°C until all cells detach. This takes approx. 10 min. Using this low concentration of trypsin, the cells often detach as small (three to five cell) clusters, and, unless a specific experiment requires single cells, this is optimal as it maintains homogeneity and functionality in the cell line stocks.

5 Add 8 ml of DMEM/F12 medium containing 10% FBS to quench the trypsin. Centrifuge at 200 g for 5 min.

6 For routine monolayer culture and cell line stock expansion the cell pellet is split at a ratio of 1:4 in growth medium and maintained until subconfluent in the following passage. For freezing of stocks the cells are split at a ratio of 1:4 in growth medium supplemented with 10% DMSO.

7 To further ensure the maintenance of scp2 cell line homogeneity and functionality we perform a differential adhesion protocol every five passages. To do this, freshly trypsinized cells from one 100 mm dish are diluted in 12 ml of growth medium and plated directly into a fresh 100 mm dish. After 20 min unattached cells are gently removed and plated into four 100 mm dishes (3 ml per dish, 1:4 split) followed by the addition of 7 ml of fresh growth medium to each dish. These cells will then attach to the dish overnight and form small, tight epithelial cell clusters. These less adherent cells remain completely dependent on exogenously added basement membrane for differentiation. In contrast, the more adherent cells left behind in the original culture dish are a heterogeneous mixture of epithelial and fibroblastic cells. They tend to lose their strict dependence on exogenously added basement membrane and thus act like the original CID-9 parental line. For example, when they are forced to round and cluster they are capable of depositing some endogenous basement membrane and differentiating. This is particularly problematic for the type of experiments outlined in Section 4.5.

4.3 Three-dimensional spheroidogenesis and differentiation

When they are placed on a reconstituted basement membrane gel (i.e. Matrigel), mammary epithelial cells round-up, engage integrins, and cluster to form small spherical structures that are approx. 50 μm in diameter. These spheroids have a smooth basal surface where integrins (i.e. α6, β1, and β4 subunits) are localized at the interface between the cells and the basement membrane. These spheroids also form adherens junctions which contain E-cadherin and β-catenin that local-ize to the lateral surfaces of the cells. They also form apical tight junctions which contain occludin and ZO-1. The majority of F-actin localizes to the apical domain of the cells and forms a ring around the spheroid's central lumen. In the presence of lactogenic hormones the cells within the spheroids differentiate and express milk proteins. The milk protein genes induced include lactoferrin, β-casein, and whey acidic protein (WAP). Complete, polarized spheroidal morphogenesis is required prior to the induction of WAP (36); β-casein induction requires both cell rounding and integrin engagement (20); lactoferrin induction requires only cell rounding and is, therefore, integrin-independent (21).

All of the morphological changes and differentiative responses described above occur within three days after plating of the cells on the basement mem-brane gel. Therefore, it is possible to assess these end-points when the cells have been transiently transfected, for example with signal transduction modifiers. Alternatively, stable transfectants can be generated prior to carrying out the basement membrane assays. However, producing the stable transfectants is time-consuming, laborious, and it is often subject to clonal variation with respect to functionality. Therefore, as a compromise, we now routinely generate hetero-geneous pools of 'rapid stables'. To do this we introduce the transgene by retro-viral infection, genetically select for five days, and carry out the experiments immediately (37). When assessing the effects of pharmacological agents or blocking antibodies we routinely dilute stock solutions of these reagents to their final working concentration in cold, liquefied Matrigel and in the culture medium. However, because it is very difficult to rapidly extract the spheroids from the basement membrane gel, these three-dimensional gel assays are not conducive to signal transduction studies.

Protocol 2

Basement membrane gel culture

Equipment and reagents

- Cell lines (see Protocol 1)

- Basement membrane gel (Matrigel; Collaborative Research)

- Attachment medium: DMEM/F12 containing 1% FBS and 5 μg/ml insulin

- Differentiation medium: serum-free DMEM/F12 medium containing a complete lactogenic hormone cocktail of 5 μg/ml insulin, 1 μg/ml hydrocortisone (Sigma), and 3 μg/ml prolactin (ovine luteotropic hormone, Sigma)

Method

1   Thaw Matrigel overnight on ice to ensure that it remains liquid. Dilute the Matrigel 1:1 with differentiation medium.

2   Pre-cool tissue culture plates on ice. Add Matrigel to the dishes on ice (15 μl/cm² surface area; for example 120 μl per 35 mm dish). Spread the liquefied Matrigel on

Protocol 2 continued

the bottom of the dish using the blunt end of a sterile P-200 pipette tip. It is important to work quickly and keep everything cool such that the Matrigel remains liquid until it is completely and evenly spread on the bottom of the dish.

3   Incubate the Matrigel for 2 h at 37°C to ensure complete gelling.

4   Dilute freshly trypsinized cells in attachment medium to 5×10^4 cells/cm^2 (i.e. 2×10^5/ml and plate 2 ml in a 35 mm dish), plate on the basement membrane gel, and incubate for 6 h to allow for cell attachment and cell clustering.

5   Gently wash the cells attached to the Matrigel three times with differentiation medium. It is important to thoroughly remove the serum to ensure complete responses to the basement membrane. Change medium daily and maintain the cultures for 72 h prior to assessing morphogenic and differentiative end-points.

6   To assess morphology and to carry out immunostaining for integrins, junctional proteins, and cytoskeletal elements we initially spread the Matrigel on glass cover-slips placed in the tissue culture dishes prior to cell plating. The cultures can be fixed in paraformaldehyde or methanol prior to staining. We routinely assess immunolocalization by confocal microscopy (23). However, sectioning can be carried out on frozen cultures (19, 38).

7   Differentiative milk protein expression can be assessed in a number of ways. For mRNA analysis (i.e. RT-PCR or Northern blotting) we routinely extract directly into guanidinium thiocyanate. This is preferable to less chaotropic treatments such as Trizol as the latter tends cause the Matrigel to clump and become very sticky. For protein and reporter gene analysis (i.e. transcriptional activation of milk protein promoters) we first remove the Matrigel prior to lysis by incubating the cultures in full-strength Dispase (Collaborative Research) for 60 min. This releases intact cell spheroids that can be lysed and analysed (23, 29). Milk protein expression can also be assessed by immunofluorescence of paraformaldehyde fixed spheres within the matrix (36).

4.4 Extracellular matrix overlays

Matrix overlays are accomplished by diluting ECM molecules in the culture medium and adding it to cells that are pre-attached to a rigid substratum. We utilize matrix overlays for three reasons:

(a) It allows us to assess the effects of the matrix independent of an associated requirement for adhesion.

(b) It allows us to utilize purified matrix components where required.

(c) It is much easier to assess signal transduction changes, and modify them, under overlay conditions than it is when the cells are cultured on basement membrane gels.

When flat mammary epithelial cell monolayers pre-attached to tissue culture plastic are overlaid with a 1% solution of Matrigel they differentiate, but in an incomplete manner. Lactoferrin and β-casein milk protein gene induction occurs, but the cells do not express WAP (20). Additionally, the differentiative induction that does take place is slower than that observed when the cells are plated on basement membrane gels; it takes five to seven days. With respect to morphology, the cells round-up in response to the overlay and form clusters in an adherens junction-dependent manner. Tight junction formation is also initiated but the clusters do not cavitate to form a central lumen. Similar differentiative and morphogenic changes are also induced by a purified laminin overlay (39). Antibody blocking experiments indicate that α6 and β1 subunit containing integrins are required for the differentiative induction of β-casein expression. The signals initiated by these integrins include focal adhesion kinase (FAK) and MAP kinase activation. Activation of the latter and induction of β-casein is, at least in part, FAK-independent as expression of a dominant negative Shc mutant prevents both. Both basement membrane and purified laminin overlays also activate the integrin-linked kinase (ILK). Expression of a dominant negative ILK does not prevent β-casein induction, but it does inhibit cell clustering. This suggests that ILK may function to link integrin signalling to ECM-mediated changes in cell–cell adhesion. This notion is supported by the fact that overexpression of wild-type ILK leads to the dissolution of adherens junctions by down-regulating the expression of the cell–cell adhesion molecule E-cadherin (37).

In the overlay assays, the differentiative induction of lactoferrin and morphologic cell rounding are not integrin-dependent (i.e. neither can be prevented by integrin function-blocking antibodies). The same is true for the recruitment of F-actin to the plasma membrane and exit from the cell cycle that are associated with overlay-mediated cell rounding and clustering. Peptide blocking experiments indicate that non-integrin receptors that bind to the E3 fragment of laminin is responsible for these functions (22).

Protocol 3

Matrix overlay of monolayer cultures

Equipment and reagents

- Cell lines and growth medium (see Protocol 1)
- Basement membrane overlay medium: dilute cold, thawed Matrigel to a final concentration of 1% (v/v; final protein concentration ~150 μg/ml) in cold

- differentiation medium (see Protocol 2)
- Laminin overlay medium: dilute cold, thawed laminin (Sigma, stock 1 mg/ml) to a final concentration of 50 μg/ml in cold differentiation medium

Method

1 Dilute freshly trypsinized cells in growth medium to 1×10^5 cells/cm^2 (i.e. 4×10^5/ml and plate 2 ml in a 35 mm dish) and plate in uncoated regular tissue culture plastic dishes.

Protocol 3 continued

2 Incubate cells on uncoated tissue culture plastic dishes overnight during which time they will form flat, subconfluent monolayers.

3 Wash cells three times with differentiation medium.

4 Prepare overlay medium immediately before use. Mix by thorough but gentle vortexing to prevent foaming. Add cold overlay medium to the cells to prevent precipitation of the matrix proteins prior to contact with the cells. When this is done correctly the matrix will form a fine, floccular precipitate on the top of the cells that can be seen approx. 4 h after the addition of the overlay.

5 Incubate for five days changing the overlay medium every two days. At the end of the five day treatment period end-points can be assayed as described in Protocol 2, except that there is no need to use Dispase prior to lysis for protein analysis (i.e. the cells can be lysed directly without difficulty).

4.5 Cell shape

ECM overlays of flat monolayers suggested that cell rounding and clustering, which also occurs in basement membrane gel culture, may be an important functional component of the differentiative response. To assess this directly we plated the cells on tissue culture plastic coated with a low concentration of the anti-adhesive polymer poly 2-hydroxyethlmethacrylate (polyHEMA). Under these conditions the cells attach to the rigid polyHEMA coated dishes, but they do not spread. Thus, they form rounded clusters independent of exogenous ECM addition. Cells within these 'naked' clusters respond poorly to growth factor induction of distal (i.e. transcription factor) outputs and they exit the cell cycle. The cells within the clusters also form adherens junctions. Tight junction proteins are recruited to the plasma membrane in association with cortical F-actin but there is no evidence of junctional polarization. This indicates that cell rounding can initiate, but not complete, changes in cell junctions which are required for alveolar morphogenesis (i.e. tight junction barrier formation at the apical surface of the cells facing the central lumen). Differentiation in these naked clusters is very limited. The cells do not express β-casein but lactoferrin is induced. These findings, together with those obtained in the monolayer overlay assays, indicate that β-casein, but not lactoferrin induction, requires integrin signalling initiated by cell contact with laminin. polyHEMA experiments in low calcium containing medium, which disrupts cell–cell junctions, indicate that the induction of lactoferrin is entirely cell shape-dependent (21).

Protocol 4

'Naked' polyHEMA cell clusters

Equipment and reagents

- Cell lines, attachment medium, and differentiation medium (see Protocols 1 and 2)
- Poly 2-hydroxyethylmethacrylate (polyHEMA, Sigma)

Method

1   PolyHEMA stock solution: Dissolve polyHEMA to a final concentration of 50 mg/ml in 95% ethanol (it is important not to use dehydrated ethanol as this will lead to cracking of the polyHEMA on the surface of the tissue culture dish upon drying). Heat the stock solution in a 37 °C water-bath overnight to dissolve completely. Store at room temperature.

2   Dilute stock polyHEMA solution 1:100 with 95% ethanol (0.5 mg/ml final concentration) and coat regular tissue culture dishes with 125 μl/cm^2 surface area (i.e. 1.0 ml/35 mm dish). Evaporate the ethanol overnight in a dry, sterile incubator at 37 C. This will leave a clear, low concentration polyHEMA coating on the surface of the dish.

3   Dilute freshly trypsinized cells in attachment medium to 1×10^5 cells/cm^2 (i.e. 4×10^5/ml and plate 2 ml in a 35 mm dish) and plate directly on low concentration polyHEMA coated dishes. Incubate overnight. The cells will attach, but not spread. Thus, they will form rounded 'naked' cell clusters (i.e. clustered in the absence of exogenously added ECM).

4   Wash cells three times with differentiation medium and incubate for up to five days changing differentiation medium every two days. Throughout the five day treatment period differentiative and morphological end-points can be assayed as described in Protocols 2 and 3 (i.e. there is no need for Dispase treatment prior to lysis for protein analysis).

4.6 Differentiative integrin signalling

When mammary epithelial cell monolayers are overlaid with basement membrane or purified laminin, integrin-dependent induction of β-casein is slow (i.e. requires five day treatment for full induction) and it is associated with cell rounding and clustering. Therefore, we reasoned that the changes in cell shape, which are initiated by the overlay in an integrin-independent manner, may be a prerequisite for the differentiative response to integrin signalling. To address this directly we first formed 'naked' cell clusters on polyHEMA prior to adding the overlay. Under these conditions, both basement membrane and laminin overlays rapidly induced the transcriptional activation of β-casein expression (i.e. within 4 h) in both an α6 and β1 integrin subunit-dependent manner. This indicates that cell rounding 'primes' the cells to respond appropriately to differentiative integrin signals.

Importantly, the nature of laminin-mediated integrin signalling is not altered by cell rounding. For example, the overlay rapidly initiates FAK and MAP kinase activation in flat monolayers and in 'naked' clusters. However, cell rounding does alter the output response to these signals by regulating downstream signalling 'set points'. For example, cell rounding dampens basal AP1 transcription factor activity in naked clusters. As a result, laminin-mediated integrin signalling initiates a transient AP1 activation of considerable amplitude in the pre-clustered cells. This high amplitude response is not observed in flat monolayers, presumably because of the high resting set point which others have demonstrated is initiated by cell spreading. Therefore, the artefact of cell spreading on rigid substrata must be overcome when assessing complex differentiative responses to integrin-mediated signals. This is precisely the opposite response of classical 'stick-and-spread' integrin signalling assessed where the end-points assessed are often proliferative rather than differentiative.

Protocol 5

ECM overlay of pre-clustered cells

Equipment and reagents

- Cell lines, polyHEMA, attachment medium, differentiation medium, and overlay medium (see Protocols 1–4)

Method

1   Prepare naked cell clusters on low concentration 0.5 mg/ml polyHEMA coated dishes as described in Protocol 4 and incubate overnight in attachment medium.

2   Wash cells very gently three times with differentiation medium and incubate for 24 h. These cultures will consist of 'naked' rounded cell clusters that will be expressing lactoferrin but not β-casein. They will also have dampened 'set point' responses to growth factors and be out of the cell cycle (i.e. they will be 'primed' to respond to the differentiative integrin signalling).

3   Treat the 'naked' cell clusters with either basement membrane or purified laminin overlay media as described in Protocol 3.

4   Integrin signalling in the cell clusters will be initiated within 5–15 min after addition of the overlay, β-casein mRNA and promoter activation are observable within 4 h. Significant accumulation of β-casein protein accumulation is observable within 12 h (20, 27). The signalling responses are similar when flat monolayers are overlaid with either basement membrane or laminin but rapid induction of β-casein does not occur.

4.7 Suspension-mediated apoptosis

When epithelial cells are detached from the extracellular matrix they undergo apoptosis. This has been termed 'anoikis' by Frisch and Francis, which is a Greek term meaning 'homelessness' (40). Induction of anoikis has been closely linked to the disruption integrin signalling and both *in vivo* and *in vitro* evidence strongly suggests that a similar disruption is responsible for initiating the wave of apoptosis that is a prerequisite for the involution of the mammary epithelium after weaning (16, 24, 25).

Various methodologies for releasing cells from extracellular matrix substrata to induce anoikic apoptosis have been developed, all of which involve maintaining the cells in suspension culture. The pre-formation of adherens junctions in confluent monolayers prior to maintenance of the cells in suspension enhances anoikis. In addition, disrupting adherens junctions in the suspended cultures also facilitates anoikis. These protocols, as well as direct experimental evidence (41, 42), strongly suggest that cell–cell adhesion also plays a role in protecting epithelial cells from anoikic apoptosis. It appears that the common protective effect of integrin and adherens junction signalling may be mediated by their common ability to maintain baseline levels of protein kinase B (PKB) activity. As discussed above, the integrin-linked kinase appears to influence both cell–substratum and cell–cell adhesion in mammary epithelial cells. ILK is also a potent co-activator of PKB (43).

By coating tissue culture dishes with high concentrations of polyHEMA (10 mg/ml) we produced 'naked' mammary epithelial cell clusters that do not attach to the substratum. Under regular high calcium (i.e. 1.2 mM) conditions these cells begin to apoptose within 24 h. This process can be inhibited by adding overlay medium to the clusters or (i.e. by initiating integrin signalling) and it can be enhanced by decreasing the calcium concentration of the medium (i.e. 50 μM) which disrupts cell–cell junctions and thus produces single cells in suspension. Interestingly, forced expression of ILK in these single cell cultures inhibits anoikic apoptosis (44).

Protocol 6

Suspension culture

Equipment and reagents

- Cell lines, differentiation medium, and stock polyHEMA solution (see Protocols 1, 2, and 4)
- Low calcium differentiation medium: dilute stock $CaCl_2$ solution (1 M in HBSS) to a final concentration of 50 μM in calcium-free DMEM/F12 medium (Sigma). Supplement with a complete cocktail of lactogenic hormones as described above to produce low calcium differentiation medium.

Protocol 6 continued

Method

1 Prepare polyHEMA plates as described in Protocol 4, except dilute the stock solution 1:10 (5 µg/ml) to form 'high' concentration polyHEMA plates. We find that this high concentration polyHEMA coating is more efficient in keeping mammary epithelial cells in suspension than are non-tissue culture plastic (i.e. bacteriological) dishes. In addition, we find that high concentration polyHEMA coated dishes are preferable to rotational suspension culture in roller bottles or plastic tubes as the latter protocols tend to initiate the formation of very large aggregates that are difficult to disperse, even in low calcium differentiation medium.

2 Allow stock monolayers of cells in regular growth medium on regular tissue culture plastic to reach confluence. Maintain these stocks in differentiation medium (i.e. serum-free) for a further 48 h to ensure complete adherens junction formation prior to the initiation of the assays. Formation of these junctions can be assayed by immunofluorescence for E-cadherin and β-catenin.

3 Dilute freshly trypsinized cells in small clusters (see Protocol 1) in regular differentiation medium to 2×10^5 cells/cm^2 (i.e. 8×10^5/ml and plate 2 ml in a 35 mm dish) and place directly in high concentration polyHEMA coated tissue culture dishes. Incubate for up to 48 h. Cells will form floating clusters of approx. 100 µM in diameter. Significant apoptosis will be initiated within 24 h.

4 Alternatively, dilute the cells as described in step 3 in low calcium differentiation medium and incubate for up to 48 h. Cells will remain mostly single and apoptosis will be initiated within 8 h. We prefer to use low calcium medium (i.e. 50 µM) rather than calcium-free medium to carry out these rapid anoikis assays as integrins are still functional under these conditions (i.e. when overlay medium is added to the suspensions integrin signalling can be initiated to inhibit the anoikic response).

5 Cells are collected from the medium without trypsinization or Dispase treatment and apoptosis can be assessed directly by fluorescent annexin-5 binding or by nucleosomal DNA laddering (44). Alternatively, the cells can be cytospun onto coverslips and apoptosis can be assessed by nuclear fragmentation analysis after staining with routine DNA binding dyes (i.e. DAPI).

References

1. Hynes, R. O. (1992). *Cell*, **69**, 11.
2. Giancotti, F. G. and Rouslahti, E. (1999). *Science*, **285**, 1028.
3. Dedhar, S. (2000). *Curr. Opin. Cell Biol.*, **12**, 250.
4. Lin, C. Q. and Bissell, M. J. (1993). *FASEB J.*, **7**, 737.
5. Roskelley, C. D., Srebrow, A., and Bissell, M. J. (1995). *Curr. Opin. Cell Biol.*, **7**, 736.
6. Roskelley, C. D., Wu, C., and Somasiri, A. M. (2000). *Methods Mol. Biol.*, **136**, 27.
7. Boudreau, N. J. and Jones, P. L. (1999). *Biochem. J.*, **339**, 481.
8. Schmeichel, K. L., Weaver, V. M., and Bissell, M. J. (1998). *J. Mammary Gland Biol. Neoplasia*, **3**, 201.

9. Kimata, K., Sakakura, T., Inaguma, K., Kato, M., and Nishizuka, Y. (1985). *J. Embryol. Exp. Morphol.*, **89**, 243.

10. Chiquet-Ehrismann, R., Mackie, E. J., Pearson, C. A., and Sakakura, T. (1986). *Cell*, **47**, 131.

11. Snedeker, S. M., Brown, C. F., and DiAugustine, R. P. (1991). *Proc. Natl. Acad. Sci. USA*, **88**, 276.

12. Silberstein, G. B., Flanders, K. C., Roberts, A. B., and Daniel, C. W. (1992). *Dev. Biol.*, **152**, 354.

13. Brinkmann, V., Foroutan, H., Sachs, M., Weidner, K. M., and Birchmeier, W. (1995). *J. Cell Biol.*, **131**, 1573.

14. Yang, Y., Spitzer, E., Meyer, D., Sachs, M., Niemann, C., Hartmann, G., *et al.* (1995). *J. Cell Biol.*, **131**, 215.

15. Howlett, A. R. and Bissell, M. J. (1993). *Epithelial Cell Biol.*, **2**, 79.

16. Talhouk, R. S., Bissell, M. J., and Werb, Z. (1992). *J. Cell Biol.*, **118**, 1271.

17. Barcellos-Hoff, M. H., Aggeler, J., Ram, T. G., and Bissell, M. J. (1989). *Development*, **105**, 223.

18. Aggeler, J., Ward, J., Blackie, L. M., Barcellos-Hoff, M. H., Streuli, C. H., and Bissell, M. J. (1991). *J. Cell Sci.*, **99**, 407.

19. Streuli, C. H., Bailey, N., and Bissell, M. J. (1991). *J. Cell Biol.*, **115**, 1383.

20. Roskelley, C. D., Desprez, P.-Y., and Bissell, M. J. (1994). *Proc. Natl. Acad. Sci. USA*, **91**, 12378.

21. Close, M. J., Howlett, A. R., Roskelley, C. D., Desprez, P. Y., Bailey, N., Rowning, B., *et al.* (1997). *J. Cell Sci.*, **110**, 2861.

22. Muschler, J., Lochter, A., Roskelley, C. D., Yurchenco, P., and Bissell, M. J. (1999). *Mol. Biol. Cell*, **10**, 2817.

23. Somasiri, A. M., Wu, C., Ellchuk, T., Turley, S., and Roskelley, C. D. (2000). *Differentiation*, **66**, 116.

24. Boudreau, N., Sympson, C. J., Werb, Z., and Bissell, M. J. (1995). *Science*, **267**, 891.

25. Farrelly, N., Lee, Y. J., Oliver, J., Dive, C., and Streuli, C. H. (1999). *J. Cell Biol.*, **144**, 1337.

26. Bissell, M. J., Hall, H. G., and Parry, G. (1982). *J. Theor. Biol.*, **99**, 31.

27. Roskelley, C. D. and Bissell, M. J. (1995). *Biochem. Cell Biol.*, **73**, 391.

28. Desprez, P.-Y., Roskelley, C., Campisi, J., and Bissell, M. J. (1993). *Mol. Cell. Differ.*, **1**, 99.

29. Schmidhauser, C., Bissell, M. J., Myers, C. A., and Casperson, G. F. (1990). *Proc. Natl. Acad. Sci. USA*, **87**, 9118.

30. Danielson, K. G., Osborn, C. J., Durban, E. M., Butel, J. S., and Medina, D. (1984). *Proc. Natl. Acad. Sci. USA*, **81**, 3756.

31. Ball, R. K., Friis, R. R., Schoenenberger, C. A., Doppler, W., and Groner, B. (1988). *EMBO J.*, **7**, 2089.

32. Reichmann, E., Ball, R., Groner, B., and Friis, R. R. (1989). *J. Cell Biol.*, **108**, 1127.

33. Doppler, W., Welte, T., and Phillipp, S. (1995). *J. Biol. Chem.*, **270**, 17962.

34. Streuli, C. H., Edwards, G. M., Delcommenne, M., Whitelaw, C. B., Burdon, T. G., Schindler, C., *et al.* (1995). *J. Biol. Chem.*, **270**, 21639.

35. Gordon, K. E., Binas, B., Chapman, R. S., Kurian, K. M., Clarkson, R. W., Jhon Clark, A., *et al.* (2000). *Breast Cancer Res.*, **2**, 222.

36. Lin, C. Q., Dempsey, P. J., Coffey, R. J., and Bissell, M. J. (1995). *J. Cell Biol.*, **129**, 1115.

37. Somasiri, A., Howarth, A., Goswami, D., Dedhar, S., and Roskelley, C. D. (2001). *J. Cell Sci.*, **114**,

38. Weaver, V. M., Petersen, O. W., Wang, F., Larabell, C. A., Briand, P., Damsky, C., *et al.* (1997). *J. Cell Biol.*, **137**, 231.

39. Streuli, C. H., Schmidhauser, C., Bailey, N., Yurchenco, P., Skubitz, A. N., Roskelley, C. D., *et al.* (1995). *J. Cell Biol.*, **129**, 591.

40. Frisch, S. M. and Francis, H. (1994). *J. Cell Biol.*, **124**, 619.
41. Kantak, S. S. and Kramer, R. H. (1998). *J. Biol. Chem.*, **273**, 16953.
42. Pece, S., Chiariell, M., Murga, C., and Gutkind, J. S. (1999). *J. Biol. Chem.*, **274**, 19347.
43. Persad, S., Attwell, S., Gray, V., Delcommenne, M., Troussard, A., Sanghera, J., *et al.* (2000). *Proc. Natl. Acad. Sci. USA*, **97**, 3207.
44. Attwell, S., Roskelley, C. D., and Dedhar, S. (2000). *Oncogene*, **19**, 3811.

Chapter 12
The cytoskeleton

Theresia B. Stradal, Antonio S. Sechi,
Jürgen Wehland, and Klemens Rottner
GBF, National Research Center for Biotechnology, Department of Cell Biology,
Mascheroder Weg 1, D-38124 Braunschweig, Germany.

1 Introduction

The cytoskeleton is a complex fibrillar system required for mechanical stability and phenomena as diverse and essential as cell division, anchorage, and cell locomotion. This system further provides highways for the directed transport of organelles or vesicles, driven by molecular motors. The tight spatial compartmentalization of cytoplasmic components by the cytoskeleton is a prerequisite for the high order of signalling that discriminates eukaryotic cells from prokaryotes.

The cytoskeleton is grossly divided into three distinct filamentous systems: actin filaments, microtubules, and the intermediate filament system. All three filament systems are formed by monomers from protein families that can each polymerize into filaments in a non-covalent manner. The dynamic turnover of these filaments via polymerization and depolymerization creates the high flexibility of the cell that is needed to respond to extracellular signals or to follow endogenous rhythms like the cell cycle. Each cytoskeletal filament system is composed of various subcompartments that are defined by:

(a) The different isoforms or family members of the polymer-forming proteins.

(b) A set of binding/interacting proteins, which can be restricted to tissues and/or subcompartments of the cell, like the Arp2/3 complex, myosin IIb, and MAPs.

1.1 Intermediate filaments

Intermediate filaments (IFs) are stable, rope-like filaments that most probably provide mechanical stability to cells or cell sheets (1). For example, in prismatic epithelial cell layers, IFs spread into neighbouring cells through desmosomes and are anchored to the basal lamina via hemidesmosomes, thereby causing excellent resistance to shear stress (2). The turnover of IFs is slow during interphase, but these filaments are rapidly depolymerized during mitosis due to hyperphosphorylation (3). They can be divided into four major groups that are highly tissue- and cell type-specific. For example, desmin is muscle-specific, glial fibrillar acidic protein (GFAP) is expressed in the neuroglia, and peripherin and the neurofilament proteins (NFs) occur only in neurons. The set of IFs expressed in a cell can define its tissue origin and the state of differentiation, making these proteins widespread markers for tumour cells. In addition, a typical feature of eukaryotic cells, the nuclear lamina, is formed by a highly conserved group of IF proteins, the lamins (4).

1.2 The actin cytoskeleton

Actin is highly conserved through evolution and for most actin binding proteins, there are little or no differences in binding affinity for actins from different species (5). Filamentous actin shows a rapid turnover *in vitro* and *in vivo*, and therefore provides the structural framework for the most dynamic network in the cell (6). Actin filaments can be associated with myosin filaments, both of which form the contractile apparatus of eukaryotic cells (7).

Besides the establishment of contractile force, actin filaments are required for the maintenance of cell shape, motility, and adhesion of interphase cells. During mitosis, actin and myosin filaments form the contractile ring required for the formation of the two daughter cells in a process called cytokinesis. In addition to myosin, a large number of actin binding proteins interacts with filamentous or monomeric actin, thereby regulating the polymerization, bundling, and/or turnover of actin filaments (5). The polymerization of monomeric actin (G-actin) into filamentous actin (F-actin) leads to a polar filament with a 'barbed' (plus) end and a 'pointed' (minus) end. The *de novo* formation of an actin filament is called nucleation and requires the stable association of a minimum of three actin monomers. The addition and removal of actin monomers can occur on both ends, albeit with different velocities, causing the continuous turnover (treadmilling) in the filament (8). In living cells, the amount of F-actin is highly regulated by the various types of interacting proteins mentioned above, whereas *in vitro*, pure actin can polymerize spontaneously. Major sites of actin polymerization in the cell, where plus ends are concentrated, are membrane protrusions like lamellipodia (9) and filopodia or focal adhesions (10).

1.3 The microtubule system

Microtubules (MTs) are cylindrical structures that very often emerge from one centre in the cell body, the centrosome. The diameter of a microtubule (25 nm) is three times higher than that of F-actin and twice that of IFs (11). The microtubule is also a polar filament, but unlike actin is formed by heterodimers of alpha- and beta-tubulin (12). Microtubules are anchored at the centrosome with their minus ends, whereas the plus ends reach out into the cytoplasm. The nucleation of a new microtubule requires gamma-tubulin which is restricted to the centrosomal region (13). A subset of microtubules is not connected to the centrosome, but emerges from breakage of anchored microtubules (14). The addition and removal of tubulin dimers predominantly occurs at the free plus ends. The poly-

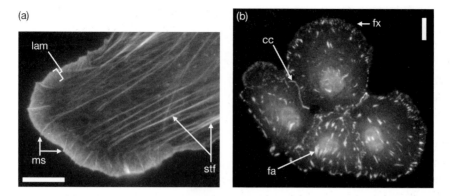

Figure 1 The actin cytoskeleton in interphase cells. (a) 'Subcompartments' of the actin cytoskeleton: Swiss 3T3 fibroblasts (ATCC No. CCL-92) were allowed to spread on FN coated coverslips for 24 h. Cells were fixed with a mixture of 0.25% GA in 4% PFA/PBS for 15 min and then permeabilized with 0.1% Triton X-100 in 4% PFA/PBS. F-actin was visualized with Alexa™-594 phalloidin. The actin cytoskeleton can be divided into different structural arrays (24), the most prominent of which are strong, contractile bundles termed stress fibres (stf) or the protrusive structures called lamellipodia (lam) and microspikes (ms). Parallel bundles of actin filaments protruding independently of lamellipodia are termed filopodia (not shown). Stress fibre bundles are anchored with the substrate in specialized junctions, the focal adhesions, while the protrusive structures are frequently associated with so-called focal complexes (see below). Bar represents 5 μm. (b) Types of cell–substrate and cell–cell adhesions established by the actin cytoskeleton: MDBK cells (ATCC No. CRL-6071) were plated onto FN coated coverslips 24 h prior to fixation. They were extracted with a mixture of 0.1% Triton X-100 and 4% PFA in PBS for 1 min and fixed with 4% PFA/PBS for 20 min. Adhesion sites are visualized employing monoclonal anti-vinculin antibodies (hVin-1, Sigma) followed by Alexa™-594 labelled goat anti-mouse antibodies (Molecular Probes). Vinculin is just one of a plethora of cytoplasmic proteins specifically recruited to cell–substrate adhesions upon ECM-triggered clustering of integrin-transmembrane receptors. Different types of substrate adhesions are the arrow-shaped focal adhesions (fa) and dot-shaped focal complexes (fx) at the cell periphery, some of which can develop into focal adhesions (25). In addition, vinculin can also be found in cell–cell contacts (cc). Bar represents 10 μm.

merization and depolymerization events follow each other frequently leading to a steady turnover of the filaments. When depolymerization is very rapid, it is termed chaotic breakdown or catastrophy. Post-translational enzymatic modifications of tubulin that can also occur after polymerization into tubules, can lead to variations in microtubule stability and altered affinity for microtubule binding proteins (15, 16). In interphase cells, microtubules provide important tracks for the directed, long-distance transport of vesicles and organelles. This transport is mainly driven by two families of microtubule-based motor proteins, the dynein and kinesin-families, which differ in the direction they 'walk' along microtubules (17). In addition, MTs are involved in the determination of cell polarity (18) in a mechanism involving the actin cytoskeleton most presumably at the sites of cell anchorage (19). MTs are essential for cell division, which is initiated by the depolymerization of all interphase microtubules and by division of the centrosome. The two halves of the centrosome separate and nucleate the

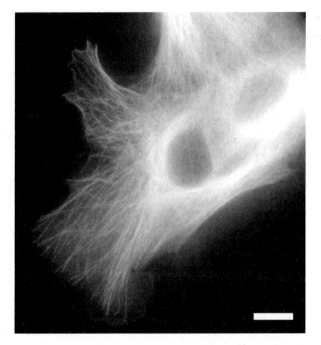

Figure 2 Interphase microtubules in LLC-PK1 cells. LLC-PK1 (ATCC No. CL-101) cells were plated onto FN coated coverslips 24 h prior to fixation. Cells were fixed with a mixture of 0.25% GA in 4% PFA/PBS for 15 min and then permeabilized with 0.1% Triton X-100 in 4% PFA/PBS. To remove free aldehyde groups, a 5 min treatment with NaBH$_4$ was followed. Microtubules were stained with anti-tyrosine-tubulin antibodies directly coupled to Alexa™-488 as described in Protocol 6A. Tyrosine tubulin is the nascent form of alpha-tubulin. Once polymerized into microtubules, the C terminal tyrosine of alpha-tubulin is slowly removed by a carboxypeptidase, producing the glu-form. After depolymerization of the microtubules, the degraded alpha-tubulin is recycled in the cytoplasm from the glu- to the tyr-form by an enzyme called TTL (tubulin-tyrosine-ligase) (15, 26). In rapidly growing cultured cells such as LLC-PK1, the majority of microtubules consists of tubulin in the tyr-form. Bar represents 5 μm.

polymerization of new microtubules to form the mitotic spindle apparatus, which then sorts and separates the two sets of chromosomes.

2 Visualization of the cytoskeleton

The following protocols are designed to provide an overview of standard methods to visualize the different cytoskeletal filament systems and their associated proteins, both in fixed and live cells. The visualization of two or three different components in the same cell sometimes requires compromises, since the optimal method for one component may not give an ideal result for the other(s). We try to provide aid in developing new protocols that combine various methods leading to images of satisfying spatial resolution and signal intensity for each component tested. Buffer compositions and suppliers of the compounds are given in Section 4. For additional techniques such as electron microscopy refer to: *The cytoskeleton: a practical approach* (ed. K. L. Carraway and C. A. C. Carraway) (1992).

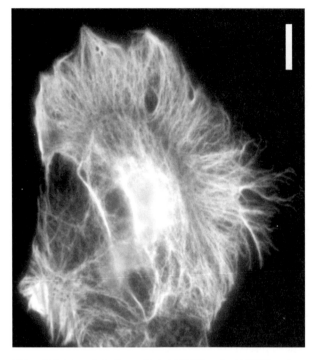

Figure 3 Intermediate filaments in LLC-PK1 cells. LLC-PK1 cells were plated onto FN coated coverslips 24 h prior to fixation. They were first extracted with 0.1% Triton X-100 in 4% PFA/PBS for 1 min and then fixed with 4% PFA/PBS for 20 min. The primary reagent was a polyclonal goat-antiserum specific for vimentin (kindly provided by P. Traub, Goettingen), which was followed by Cy3-labelled donkey anti-goat antibodies (Jackson Laboratories). Note the dense arrangement of intermediate filaments in these cells. Bar represents 10 μm.

2.1 Plating of cells on different substrates

The visualization of subcellular structures in the light microscope requires high resolution optics employing lenses with high numerical aperture and short working distances. In order to optimally exploit the advantages of high resolution light microscopy, cells are usually viewed through thin glass coverslips. In order to guarantee cell viability, the coverslips are extensively washed and frequently coated with various extracellular matrix proteins in order to enhance adhesion to the glass surface or to specifically stimulate the formation of various cytoskeletal structures, e.g. lamellipodia. The most common extracellular matrix (ECM) proteins used for coating coverslips are fibronectin (FN), laminin (Lam), and collagens (Co). The behaviour of cells on different substrates or on different concentrations of one substrate can be highly variable for different cell types, and tests should be performed to find the optimal conditions for the cells of interest. In general, FN is the preferred substrate for fibroblastic cells, since it is a well-established ligand of integrin receptors that link the substrate to the cytoskeleton in focal adhesions, while Lam and Co enhance the spreading of most epithelial cell types. The rate of cell spreading, motility, or cell division can be highly variable according to the ECM chosen (20, 21). The efficiency of ECM-coating on glass can significantly be enhanced by previous coating of coverslips with PLL (poly-L-lysine), although in addition to this effect, PLL alone is occasionally used to promote the 'non-specific' spreading of certain cell types (22).

Protocol 1
Preparation of coverslips

Equipment and reagents

- Coverslips
- Glass Petri dish
- UV light or a dry heat sterilizer
- Parafilm
- Ethanol
- Hydrochloric acid
- 0.1 mg/ml poly-L-lysine in water
- 25 μg/ml ECM in sterile PBS
- DMEM

A.   Sterilization

1   Wash 100 coverslips (12–15 mm Ø, 0.17 mm thick) in 50 ml of 60% ethanol/ 40% hydrochloric acid (v/v) in a beaker for 20 min.

2   Rinse the coverslips ten times with deionized water and dry between two layers of lint-free paper (lens paper).

3   Transfer coverslips to a glass Petri dish and sterilize by exposure to UV light for 2–3 h or in a dry heat sterilizer at 220 °C overnight.

B.   PLL coating

1   Prepare 0.1 mg/ml poly-L-lysine in water and place drops of 100 μl on a sheet of Parafilm. Allow the coverslips to float on these drops for 1 h.

2   Wash by dipping them ten times into a large volume of water (200 ml). After washing, air dry and sterilize the coverslips by UV light.

3   PLL-coated, dried coverslips can be stored in a Petri dish at room temperature.

C.   ECM coating

1   Prepare a dilution of approx. 25 μg/ml ECM in sterile PBS.

2   Place drops of 150 μl (for 15 mm Ø-coverslips) or 100 μl (for 12 mm Ø-coverslips) on UV sterilized Parafilm.

3   Incubate coverslips on the drops for 1 h at room temperature.

4   Transfer coverslips to a 3 cm Ø cell culture dish upside up and add 2 ml of plain DMEM for washing. Aspirate DMEM directly before plating the cells.

ECM-coated coverslips should not be allowed to dry out and should be prepared shortly before re-plating the cells. After coating, they can be stored in the dish in DMEM for several hours.

For most types of analysis, cells are plated subconfluently and allowed to spread for between 3 and 24 hours, depending on cell type and ECM component used.

2.2 Localization of cytoskeletal components by indirect immunofluorescence microscopy

Immunolocalization studies can in principle be performed in living (after micro-injection of tagged antibodies) or fixed cells, and either with fluorescently-coupled primary antibodies, or indirectly making use of coupled secondary reagents. Since the indirect immunolabelling of fixed cells is the most frequently used method, we will describe the latter method in detail. Indirect immuno-fluorescence is preceded by the fixation of cells and their extraction to enable the penetration of antibodies. Proteins are most commonly fixed by chemical crosslinking with formaldehyde or glutaraldehyde, or alternatively by dehydra-tion with organic solvents such as methanol, ethanol, or acetone. In general, the choice of fixation and extraction procedure is dependent on the structure labelled and on the antibody to be used. For most applications, we recommend the use of aldehyde-based fixation procedures. However, the optimal fixation procedure for a given antibody can only be tested experimentally. The following parameters have to be considered: (a) structural preservation, (b) non-specific background and antibody accessibility. While the structural preservation is best using strong fixatives (such as glutaraldehyde) or gentle extraction procedures, better antibody accessibility and a decrease of non-specific background can be achieved with weaker fixatives or stronger extraction. Therefore, it is necessary to find compromises between these parameters. Aldehyde-based fixatives are most frequently dissolved in isotonic buffers such as PBS (phosphate-buffered saline) or CB (cytoskeleton buffer).

Protocol 2

Fixation of cells

Equipment and reagents

- Cells cultured on coverslips
- DMEM or PBS
- Aldehyde-based fixative
- Triton X-100

Method

1 Wash the cells cultured on coverslips once with warm (37 °C) plain DMEM or PBS.

2 Add 3 ml of warm aldehyde-based fixative (at least room temperature) for 20 min.

3 Remove the fixative and add fixative or PBS (CB) containing 0.1–0.5% (v/v) Triton X-100 for 1–5 min.

4 Wash the cells three times with the buffer selected for the next procedure.

Note that the extraction is more intense at higher temperature (20→37 °C). The visualization of more resistant structures like focal adhesions or intermediate filaments may give better results when the extraction step is carried out for 0.5–1 min before fixation. Then, the majority of soluble, cytoplasmic proteins is removed, which reduces non-specific background. Formaldehyde (PFA, Merck) is commonly used at concentrations of 3–4%, while glutaraldehyde (GA) is employed at concentrations between 0.25–1%. In case antibodies recognize GA-fixed epitopes, free aldehyde groups must be 'neutralized' after fixation, which helps to prevent high fluorescent background resulting from autofluorescence and from non-specific fixing of primary antibodies.

For deactivation of aldehyde groups:

- dissolve sodium borohydride ($NaBH_4$) at a final concentration of 0.5 mg/ml in ice-cold PBS (or CB) and apply immediately to the fixed cells for 5 min on ice. Most fixed cells can be stored in buffer at 4 °C overnight or for several days.

Protocol 3A

Immunolabelling of fixed cells

The easiest way to incubate fixed cells without wasting antibodies is to place them upside down on 20 µl drops that are placed on Parafilm in a moist chamber (i.e. Petri dish with wet Whatman paper).

Equipment and reagents

- Petri dish
- Whatman paper
- PBS (or TBS) containing 1 g/litre BSA and 5% donor horse serum

- Mounting medium

Method

1 Block in PBS (or TBS) containing 1 g/litre BSA and 5% donor horse serum for at least 15 min.

2 Dilute antibodies in PBS (or TBS) containing 1 g/litre BSA and incubate cells for 1 h (don't wash the cells between blocking and the primary antibody).

3 Wash the cells up to three times for 5 min in buffer on top of large drops or in a 24-well plate.

4 Incubate the cells in the secondary antibody for 1 h diluted in buffer containing 1 g/litre BSA.

5 Wash three times, 5 min each.

Protocol 3A continued

6   Remove the last drop of washing buffer by touching a Whatman paper with the very edge of the coverslip.

7   Mount the coverslips on slides in small drops of mounting medium such as Gelvatol or Moviol containing 2.5 mg/ml *n*-propylgallate or an alternative anti-bleach reagent.

8   Allow to dry for several hours at room temperature.

9   Mounted coverslips can be stored at 4 °C for several weeks.

Protocol 3B
Double and triple stainings

1   When different components are visualized in the same cell (23), for example microtubules and focal adhesions, the antibodies raised against the respective components either have to be coupled directly or they have to be of different origin/isotype in order to specifically distinguish between them with the fluorescently labelled secondary antibodies. The latter possibility is employed most frequently and a simple example is given in Table 1.

2   An alternative for one of the components is the use of biotinylated primary or secondary antibodies in combination with fluorescently-coupled avidin—or the tetravalent streptavidin—reagents, which specifically interact with biotin.

3   For triple stainings, investigators most frequently use blue dyes excited by very short wavelengths, for instance coumarin derivatives of antibodies or DAPI, a chromatin-specific dye. More recently, the development of novel dyes excitable by wavelengths towards the infrared end of the visible light spectrum (for instance Cy5) enables the use of an additional colour, but the clear separation of these colours is not easy and requires a careful choice of filter sets. In general, the most common problem when using different dyes in one preparation, is the non-specific excitation of additional dyes in a given light channel, the so-called 'bleed-through', which has to be kept at a minimum. We recommend to test the lack of non-specific excitation of a given dye before its use in double- and triple-labelling experiments.

Table 1 Typical antibody combination for double immunofluorescence

Primary antibodies	Secondary antibodies
Monoclonal mouse anti-vinculin	Sheep/goat/donkey (not rabbit) anti-mouse conjugated with a green dye such as FITC, Cy2, or Alexa488
Polyclonal rabbit anti-tubulin	Sheep/goat/donkey anti-rabbit conjugated with a red dye like the rhodamine-derivative TRITC, Cy3, or Alexa594

Protocol 3C

Staining of F-actin with phalloidin derivatives

1   The most common method to visualize F-actin is fluorescently labelled phalloidin, a low molecular weight compound from the fungus *Amantia phalloidea*. Phalloidin specifically interacts with polymerized actin, but not with monomeric G-actin. Affinity and specificity of this staining cannot be achieved by antibody labelling.

2   Fluorescently labelled phalloidin derivatives are commercially available and can be diluted and incubated together with the secondary antibody for 1 h at room temperature at an appropriate concentration (between 0.1 and 1 µg/ml depending of the fluorescent dye). Stock solutions are stored in methanol. After two short washes, the preparation should be free of background. All antibody—and phalloidin—incubations can also be performed at 4 °C overnight.

2.3 Cytoskeletal dynamics in living cells

The visualization of proteins and their dynamics in live cells is used to obtain information on subcellular localizations that are frequently lost after fixation and extraction procedures. In addition, live cell dynamics of proteins allows for the co-ordination of information on protein localization with a certain activity of the whole cell or a given subcellular structure the protein is localizing to. A specific example is given below. Video microscopy requires the establishment of conditions of cell growth on the microscope identical or at least highly similar to the conditions during regular cultivation of the cells in the incubator. This includes the regulation of temperature and pH of the culture medium. Further, the illumination with light causes heat and can lead to photodamage. These requirements lead to the development of heat stages, culture media that do not need CO_2 for pH stabilization, highly sensitive cameras, and computer-controlled shutter systems. A high sensitivity of the camera helps to reduce illumination time. The shutters restrict the illumination of the cells to the time of image acquisition. They also enable visualization of cells by both trans-illumination (such as phase contrast or differential interference microscopy) and fluorescence optics. In addition, a filter wheel can be used to simultaneously visualize two different fluorescent proteins in the same cell. The improvement of camera sensitivity over the last two decades was based on the transformation and multiplication of light into electronic signals directly read out into computers. Software controlling the camera and shutter systems has to allow for flexibility to set various exposure times (dependent on signal intensity) and picture frequency. The confocal or two-photon microscopes are used to extract three-dimensional information.

Two principal methods can be used to study the dynamics of cytoskeletal proteins *in vivo*, the cytoplasmic injection of chemically-modified cytoskeletal

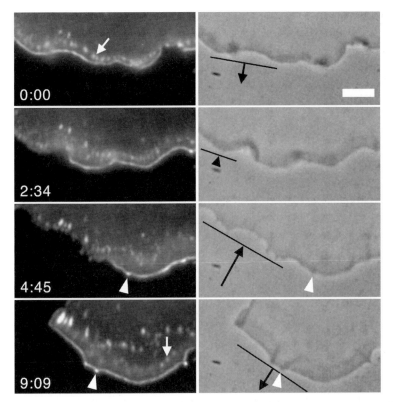

Figure 4 Correlating VASP recruitment with peripheral protrusion/retraction events. Panels correspond to selected time points of a movie shot in fluorescence and phase contrast of a motile B16-F1 melanoma cell (ATCC No. CRL-6323) expressing GFP-tagged VASP (vasodilator-stimulated phosphoprotein). VASP is recruited to the very periphery of protruding lamellipodia and to cell–substrate contacts. The black arrows in the phase contrast images indicate the motile history of the zone marked by the black line, and as measured approx. 1 min before each frame. This analysis demonstrates that the GFP construct disappears from peripheral zones that have stopped their protrusion (2:34) or that have retracted (4:45), while the fluorescent signal is retained upon re-initiation of protrusion (9:09). In addition to the continuous line of GFP-VASP at the periphery of protruding lamellipodia, we find bright spots, which move laterally along their tip. These bright spots correspond to the tips of protruding microspikes (compare to phase contrast images and to Figure 1a), which are embedded in lamellipodia (white arrowheads). Further, the focal complexes at the base of lamellipodia are rapidly remodelled during protrusion of the cell periphery (white arrows). Time is given in minutes and seconds and the bar corresponds to 5 μm. A more detailed analysis of B16-F1 cells stably expressing GFP-VASP revealed a linear correlation between GFP intensity and lamellipodial protrusion rate (27).

protein or the ectopic expression of mutants fused to an autofluorescent protein tag such as GFP (green fluorescent protein).

2.3.1 Microinjection of fluorescently labelled probes

Microinjection is preceded by purification of a given cytoskeletal component or an appropriate antibody and their chemical modification with a fluorescent dye (for a protocol see below). Protein coupling is followed by extensive dialysis into an appropriate microinjection buffer. The choice of injection buffer depends on both the stability of injected protein and on cell viability following injection. Cell viability is improved with pH levels close to neutral and buffers containing relatively low salt. For most cytoskeletal proteins, 5 mM Tris pH 7 containing 50 mM KCl may be appropriate. However, some proteins like actin, tubulin, or myosin require different buffers. While KCl induces the polymerization of actin, this salt is used to prevent the spontaneous polymerization of myosin. Further, actin or tubulin require cyclic nucleotides for protein stability. The injection volume is dependent on injection pressure, needle diameter, injection time, and viscosity of the protein solution. The best results are obtained with needles that have a tip diameter of 1 μm or less and with injection pressures of 10–100 hPa. The injection times should not exceed 0.5 sec and for most proteins, concentrations of around 1 mg/ml are sufficient. Injections require micromanipulators and pressure supplies and are performed manually or by making use of half- or fully-automated systems. Cytoplasmic injections of adherent cells are most easily performed in close proximity to the nucleus. Since microinjection can cause severe effects on cell morphology and viability, control experiments with buffers alone or with inactive probes are required.

2.3.2 Transfection of cells with plasmid DNA

The use of GFP-tagged cytoskeletal components opened new horizons in the visualization of cytoskeletal dynamics (28). In addition, the tag on the protein

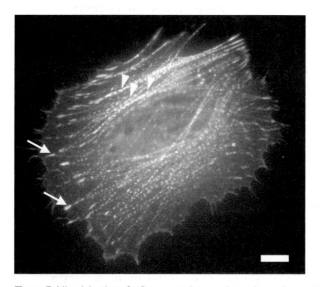

Figure 5 Microinjection of a fluorescently tagged protein can be used to study its localization and dynamics. Alpha-actinin from turkey gizzard, kindly provided by Mario Gimona, was coupled to 5-TAMRA (carboxytetramethyrhodamine succinimidyl ester; Molecular Probes) and microinjected into a CAR goldfish fibroblast (ATCC No. CCL-71). The fluorescent derivative localizes to both focal adhesions (arrows) and periodically along stress (arrowheads), which is consistent with immunolocalization of the endogenous protein (58). Bar represents 10 μm.

(GFP, myc, Flag, etc.) can be elegantly used for pull down experiments and immunoprecipitations employing tag-specific antibodies instead of protein antibodies that might interfere with binding partners. Common procedures for transfection of cells with DNA are lipofection and electroporation, which replaced calcium phosphate precipitation or the microinjection of DNA for most applications. The transfection efficiency is not only dependent on the method and the type or concentration of construct used, but seems to positively correlate with cell growth. In general, optimal conditions have to be tested for each given cell type.

Protocol 4A

Transfection using lipophilic reagents

The delivery of DNA into mammalian cells using lipophilic reagents usually results in high efficiency and great reproducibility (29). Several different lipofection reagents have recently been developed, which give variable results with different cell types. The principle is that plasmid DNA is complexed to liposomes in serum-free medium and then added to the cells. Volumes and concentrations of the following protocol are for exponentially growing cells in a 3 cm diameter dish

Equipment and reagents

- Coverslips
- Superfect™ (Qiagen)
- DNA
- Medium with and without serum

Method

1 Dilute 1 µg DNA in 300 µl serum-free medium.

2 Add 7 µl Superfect™ reagent and mix by gently flipping the tube.

3 Incubate at room temperature for 15–20 min to allow complex formation.

4 Add 1.5 ml serum-containing medium.

5 Aspirate medium from the cells and add DNA/liposome complex directly to the cells.

6 Incubate overnight at regular growth conditions.

7 Re-plate the transfected cells onto coverslips.

8 After spreading, the cells can be fixed or subjected to video microscopy.

Slowly growing cells like rat A7r5 smooth muscle cells or WI38 human fibroblasts show a higher transfection efficiency when transfected for 24 hours. Some cell types give the best results when transfected in serum-free medium. In this case the cells need a recovery from lipofection in serum-containing medium for one hour to overnight before re-plating. The lipofection of cells on glass coverslips is not recommended due to lower efficiencies and high autofluorescence of most lipophilic reagents. In addition, re-plating of transfected cells removes dead and selects for viable cells capable of spreading and cytoskeletal reorganization.

Disadvantages:

(a) Most lipophilic reagents become toxic after extended transfection times or at high concentrations.

(b) Time periods of several hours following lipofection may therefore be required for sufficient cell recovery.

Protocol 4B

Transfection using electroporation

During electroporation, a high voltage electric shock is employed to deliver the DNA into living cells. This technique makes use of the fact that plasma membranes behave as an electric capacitor when subjected to an electric field. The application of high voltages causes temporary and localized ruptures and formation of pores in the cell membrane that are large enough to allow the penetration of DNA.

Equipment and reagents

- Electroporation apparatus
- Cell culture dish with coverslips
- Incubator
- Serum-free medium
- DNA

Method

1 Trypsinize cells and wash once with serum-free medium.

2 Spin and resuspend the cell pellet in serum-free medium supplemented with 20 mM HEPES to a density of between 5×10^6 and 2×10^7 cells/ml.

3 Add DNA (20–40 µg) to 0.4 ml of the cell suspension.

4 Transfer the cell/DNA suspension to an electroporation cuvette.

5 Electroporate cells using a pulse with voltage and capacitance parameters according to distributor's recommendations and depending on cell type and size.

6 Transfer transfected cells to cell culture dish with coverslips.

7 Incubate at 37 °C and 7% CO_2.

8 Change medium after 3 h to remove dead cells.

Disadvantages:

(a) Electroporation requires high amounts of DNA and a high number of cells compared to the liposome-mediated methods.

(b) It can further lead to extensive cell death, although for some cell types, it is the only applicable method.

2.4 Cytoskeletal drugs

Bioactive compounds cause specific cytoskeletal changes, which can be used to interfere with various aspects of assembly or disassembly of specific filamentous systems. Some of these compounds have proved useful both as experimental tools, for a better functional understanding of specific filamentous system, and as tools in clinical therapy. The actin cytoskeleton often shows prominent responses to changes in the cell environment or upon addition of drugs. Most compounds specifically interfere with the assembly/disassembly of actin filaments or of microtubules, although actin cytoskeleton- or microtubule-disrupting drugs also change the appearance of the IF system, since IFs are passively spread by both microtubules and the actin system. A summary of some compounds that are of special interest for cell biological experiments and their effects is given in Table 2.

3 Basic biochemical analysis of actin and actin binding proteins

3.1 Purification of actin from smooth muscle

Many cytoskeletal components such as actin, myosin, or tubulin can easily be purified from muscle or brain due to their relative abundance in these tissues. These purification methods are still used today, because, as in the case of actin, purification from recombinant sources does not give acceptable results, since bacteria lack the chaperones required for proper folding of this protein. The following protocol describes how to obtain actin or actin binding proteins from smooth muscle, since we describe actin co-sedimentation and polymerization assays in the subsequent sections.

Table 2 Overview of drugs most commonly used for interfering with cytoskeletal assembly

Compound	Effect	Reversible	Cell permeable	Reference
Nocodazole	Depolymerization of microtubules	+	+	30, 31
Vinblastin	Depolymerization of microtubules	+	+	32
Taxol	Stabilization of microtubules	+	+	33, 34
Cytocalasin B and D	Depolymerization of F-actin by plus end capping	+	+	35, 36
Phalloidin	Stabilization of F-actin by intercalation within the filament	–	–	36–38
Jasplakinolide	Stabilization of F-actin	–	+	39, 40
Latrunculin	Depolymerization of F-actin by sequestering G-actin	+	+	41

Protocol 5A

Obtaining the raw material

1 Take 3–5 kg of pig stomach (or turkey gizzard) fresh from the slaughter house on ice.

2 Dissect the smooth muscle tissue with scalpels and scissors (avoid contact of the muscle tissue with the inner epithelia and/or content of the stomach).

3 Mince the pure smooth muscle tissue (yield about 1 kg).

This material can serve as source for the preparation of most proteins related to actomyosin (actin, myosin, tropomyosin, alpha-actinin, vinculin, ...) and further treatment of the mince depends on the required proteins.

Protocol 5B

Preparation of acetone powder (based on ref. 42)

Equipment and reagents

- 100 μm filter cloth
- Waring blender
- Minced tissue
- KCl, NaHCO$_3$, acetone

Method

1 Extract the minced tissue with 3 vol. of ice-cold 0.1 M KCl, and slowly stir for 15 min at 4 °C.

2 Filter the mince through several layers of 100 μm filter cloth and discard the filtrate.

3 Extract the mince again with 3 vol. of 0.05 M NaHCO$_3$ at 4 °C under gentle stirring for 15 min.

4 Filter the mince through several layers of 100 μm filter cloth.

5 Wash the mince with 10 vol. of water.

6 Filter again.

7 Wash the mince twice with 2 vol. of acetone.

8 Filter the mince through a fresh filter cloth and mix with 1 vol. of fresh cold acetone in a Waring blender for 15 sec.

9 Filter again and spread the now fibrous mince on fresh cloth and air dry at room temperature overnight.

10 The acetone powder can be stored at −20 °C for months and at −70 °C for years.

Protocol 5C

Purification of actin from acetone powder

Equipment and reagents

- Centrifuge
- Superose S-200, 130 cm $\times$ 3 cm
- Acetone powder

- G-actin buffer
- $MgCl_2$, KCl

Method

1 Extract 1 g of acetone powder in 10 ml of G-actin buffer by stirring for 30 min at 4 °C.

2 Re-extract with another 10 ml of G-actin buffer.

3 Clear the extract by centrifugation (to remove denatured protein) for 20 min at 10 000 g and 4 °C, and combine the supernatants.

4 Enrich the actin by two to three cycles of polymerization and depolymerization (see below).

5 Subject the G-actin to a gel filtration step (Superose S-200, 130 cm $\times$ 3 cm) to remove residual actin binding proteins (i.e. tropomyosin).

Polymerization and depolymerization cycles

6 Induce polymerization by addition of 1.5 mM $MgCl_2$ and 100–150 mM KCl.

7 Incubate for 45–90 min on ice to reach equilibrium (the solution becomes highly viscous).

8 Collect F-actin by ultracentrifugation for at least 1 h at 100 000 g and 4 °C.

9 Resuspend the F-actin pellet in 5–10 ml G-actin buffer and dialyse against G-actin buffer for at least 3 h, or better overnight.

The first polymerization step after extraction is less efficient due to impurities and can be performed longer (3 h to overnight) or at higher temperature (10–18 °C). For resuspension of F-actin pellets, buffer volumes are not critical, as long as the actin concentration is above 0.2 mg/ml and below 5 mg/ml. Below 0.2 mg/ml, *in vitro* polymerization is very inefficient, whereas at actin concentrations above 5 mg/ml, the depolymerization is inefficient. The optimal working concentration is between 1–3 g/litre. During gel filtration, tropomyosin, which is a dimer, clearly separates from monomeric G-actin. Pure actin can be stored for days under steady dialysis with buffer changes every day (degassed with fresh DTT and ATP). For long-term storage at −70 °C, the actin is depolymerized and sucrose is added to the G-actin buffer at 1 mg/mg actin.

Acetone powder from muscle tissue as well as readily purified actin are commercially available. It is recommended to probe the purity on a Coomassie stained SDS–polyacrylamide gel in order to decide which steps from this protocol are needed to obtain pure actin. In our hands, it was always useful to subject the actin at least to one polymerization/depolymerization cycle.

3.2 Fluorescent labelling of purified proteins

This technique can be applied to all pure protein and peptide preparations. Most frequently it is used to label primary and secondary antibodies (see above). After the coupling reaction, the dye is chemically inert and covalently bound to the protein. Fluorescent dyes are low molecular weight compounds and can be advantageous to protein tags like GFP, since they may less frequently cause sterical problems. On the other hand, labelling of a protein with fluorescent dyes leads to derivatization of a subset of amino acid residues, which can affect its biochemical characteristics and/or its interaction with other proteins. Different coupling chemistries are available and one should carefully read the literature about the protein before selecting a dye with a specific chemistry (43, 44).

Coupling with amino reactive groups is most commonly used, since it is easy to perform and leads to highly efficient labelling of the protein. For this purpose, succinimidyl ester- and isothiocyanate-activated dyes are used. The dye can potentially react with all lysine and arginine residues in the protein, but over-labelling can lead to non-functionality or precipitation of the protein and should be avoided.

Possible alternatives are thiol reactive fluorescent probes, which target the cysteine residues in the protein. For this purpose, iodoacetamide- and maleimide-coupled dyes are available. Since sulfhydryl groups are less abundant in proteins than amines, thiol modification often means to selectively label a protein at a defined site. The technique is therefore frequently used to address the activity of a protein or to monitor conformational changes. This is the case for pyrene labelling of actin that is used in polymerization assays (see below). For the fluorescent labelling of a new protein or for establishing a new fluorescence-based assay, extensive experimentation may be required.

Protocol 6

Basic protocol for antibody labelling with amino reactive dyes

Purified antibody is required. When the antibody is obtained with stabilizing agents like gelatin or BSA, it should be purified on a protein G column (Pharmacia) according to the suppliers protocol. Azide, amine-containing chemicals like Tris, or other low molecular weight compounds have to be removed by dialysis against PBS overnight.

Equipment and reagents

- Desalting column (PD10 Pharmacia)
- Immunoglobulin in PBS
- 0.1 M NaHCO$_3$
- Dye dissolved in DMSO or DMF
- 1 M hydroxylamine pH 8.5

Method

1   Mix 0.5–5 mg/ml immunoglobulin in PBS with 0.1 vol. of 0.1 M NaHCO$_3$.
2   Add a 20-fold molar excess of the dye, usually dissolved in DMSO or DMF.

Protocol 6 continued

3   Incubate on ice for 1 h.

4   Stop the reaction by the addition of 0.1 vol. of 1 M hydroxylamine pH 8.5.

5   Incubate for 45 min on ice to remove dye from secondary and tertiary amines.

6   Remove the excess of uncoupled dye, labelling buffer, and blocking reagent by a gel filtration step on a desalting column (PD10) followed by extensive dialysis.

3.2.1 Labelling of actin with thiol reactive dyes

A prerequisite for labelling proteins with thiol reactive dyes is the presence of free sulfhydryl (SH–) residues. Proteins harbouring those are kept in DTT (dithio-threitol)-containing buffers and have to be transferred into a SH-free reaction buffer. This is achieved most easily with a desalting column (PD10) immediately before labelling. In case of actin, one free cysteine residue per molecule is available in the polymerized form. The dye is normally added in two-fold to five-fold molar excess over the reactive thiols of the protein. The labelling procedure itself is similar to the one described above except for the blocking, that is carried out with a 10-fold molar excess of a thiol reagent like DTT. Independent of the label, actin is modified in the F-actin form, to prevent reaction with residues essential for interaction of the monomers in the filament. Actin is first poly-merized, then the filaments are labelled, collected by ultracentrifugation, and dialysed against G-actin buffer to depolymerize it again. Note that actin cannot be subjected to chromatographic steps in its filamentous form.

Actin, pyrenyl labelled at the reactive cysteine residue has already proven to be a useful tool for polymerization assays (see below). Rhodamine and fluorescein labelled actins have successfully been microinjected in classical experiments that yielded insights into F-actin dynamics and revealed the primary sites of actin polymerization in cells. Several companies offer complete protein labelling kits with easy to follow protocols. All dyes from the above described protocols were obtained from Molecular Probes.

3.3 Co-sedimentation of actin binding and bundling proteins with F-actin

This technique allows for the analysis of the actin binding or bundling activity of proteins or the putative actin binding domains. Co-sedimentation assays are also suitable to titrate the molar ratio between actin and the protein of interest that is reached upon saturation of the filament with the binding protein. Addition-ally, the binding site on actin can be identified in competition assays with known actin binding proteins (see ref. 45). In case a protein exhibits actin bundling activity, the aggregates can be pelleted at lower centrifugal forces than non-bundled F-actin.

> ## Protocol 7
> ## Co-sedimentation for actin binding and bundling proteins
>
> ### Equipment
> - Beckman airfuge or ultracentrifuge
> - SDS–polyacrylamide gel electrophoresis
>   equipment
>
> ### A.   Actin binding proteins (high speed assay)
>
> **1**   Incubate 2.5 μm actin plus increasing amounts of protein of interest for 10–30 min in a volume of 200 μl.
>
> **2**   Centrifuge 30 min at 100 000 g in Beckman airfuge or ultracentrifuge.
>
> **3**   Wash the pellet carefully.
>
> **4**   Resuspend the F-actin pellet in a volume equal to the reaction volume.
>
> **5**   Compare samples of the supernatant and the resuspended pellet by SDS–polyacrylamide gel electrophoresis.
>
> ### B.   Actin bundling proteins (low speed assay)
>
> **1**   This assay is performed as described above except that the centrifugation step is carried out at 10 000 g.

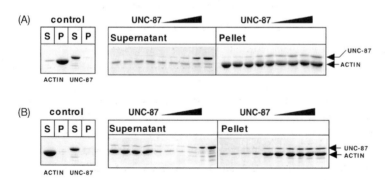

Figure 6 Actin co-sedimentation assay with the actin binding and bundling protein UNC-87 from the body wall muscle of *Caenorhabditis elegans* (45). (A) High speed assay. To constant amounts of F-actin (3 μM) increasing amounts of recombinant UNC-87 were added and after incubation co-sedimented at 100 000 *g* with filamentous actin. The assay demonstrates that UNC-87 is an F-actin binding protein. Occurrence of UNC-87 in the supernatant is due to saturation of actin filaments with the protein. (B) Low speed assay. Constant amounts of F-actin and increasing amounts of UNC-87 are sedimented at 10 000 *g*. The bundling of F-actin by UNC-87 leads to the formation of aggregates that can be pelleted at 10 000 *g* in contrast to F-actin alone, which remains in the supernatant. This assay demonstrates the actin bundling activity of UNC-87. Also compare the control supernatants (S) and pellets (P) that contain F-actin or UNC-87 alone.

3.4 Actin polymerization assay (based on refs 46 and 47)

This standard protocol is employed to test proteins or other compounds with respect to their influence on actin polymerization *in vitro* (48). In the last decade, this method yielded valuable insights into the kinetics of actin polymerization influenced by cellular regulators such as the Arp2/3 complex (49, 50), profilin (51), members of the Scar/WASP family (52, 53), the Rho family GTPase Cdc42 (54), phosphatidyl-4,5-bisphosphate (PIP2) (56), or adaptor proteins like GRB2 (55) or by *the Listeria* spp. protein ActA (see Figure 7).

Protocol 8

Actin polymerization assay

Equipment and reagents

- UV spectrofluorimeter
- Quartz glass cuvette
- Pyrenyl labelled and unlabelled G-actin

- F-actin buffer
- Polymerization promoting or inhibiting agent

Method

1 Mix 10% pyrenyl labelled and 90% unlabelled G-actin at a final concentration of 2.5 μM in F-actin buffer.

2 Place the mixture into a UV spectrofluorimeter in a quartz glass cuvette.

3 Set the excitation wavelength to 366 nm and the emission to 407 nm.

4 Add different amounts of the polymerization promoting or inhibiting agent.

5 Monitor the increase in fluorescence over a time period of 5–120 min.

The pyrenyl fluorescence increases 25-fold upon formation of the actin filament. The maximum slope of the obtained curves are calculated and taken as a measure for the polymerization rates. At any time of the polymerization process, the rate of polymerization V, derived from the measured slope is expressed as a function of the number of filaments and the concentration of G-actin present at the given time, as follows:

$$V = k_+ \cdot [F] \cdot (C - Cc)$$

where k_+ is the rate constant for association of G-actin to filament ends (essentially barbed ends contribute under standard conditions); [F] is the concentration of filament ends present in solution at the time of the measure; C and Cc are the G-actin concentration (at the time of the measure) and the critical concentration respectively. The fluorescence units can be converted in concentration of F-actin (in μM) using a critical concentration plot made with the same actin solution.

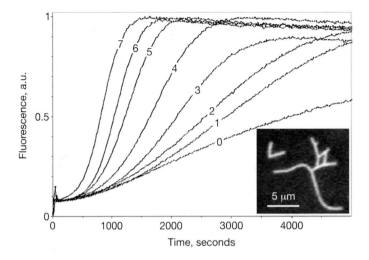

Figure 7 The *Listeria* protein ActA mimics WASp proteins: it activates Arp2/3 complex to initiate the formation of branched filaments. MgATP–G-actin (2.5 µM, 10% pyrenyl labelled) is polymerized in the presence of 0.1 M KCl, 1 mM MgCl₂, and the following additions: 0; none or 20 nM Arp2/3 complex alone or 0.1 µM ActA alone. 1 to 7; 20 nM Arp2/3 complex (from bovine brain) and full-length ActA at 5.4 nM, 11 nM, 27 nM, 54 nM, 135 nM, 500 nM, and 1000 nM (1 to 7). The accelerated polymerization time courses are due to continuous filament branching by Arp2/3 complex activated by ActA. Inset: branched filaments are formed in the polymerization medium containing actin (5 µM), Arp2/3 complex (40 nM), and ActA (100 nM). At the end of the polymerization process, filaments are supplemented with one molar equivalent rhodamine labelled phalloidin and diluted 400-fold for fluorescence microscopy observation. (Courtesy of R. Boujemaa, S. Samarin, C. Kocks, and M.-F. Carlier.)

4 Buffers and stock solutions

All chemicals were obtained from Sigma except otherwise indicated.

ECM stocks

FN (Roche) and Lam (Sigma) are obtained in lyophilized form. After dissolving them, they tend to polymerize and form aggregates, which has to be prevented before coating of the coverslips. To circumvent this problem, FN is dissolved at a concentration of 1 mg/ml in 2 M urea and can be stored at 4°C for months (57). Immediately before coating, it is diluted 1:40 in sterile PBS. It is essential to wash FN coated coverslips as described above, since urea is toxic to cells. Lam is dissolved in 10 mM Tris pH 7.5, 50 mM NaCl at a concentration of 1 mg/ml. 25 µl aliquots are snap-frozen in liquid nitrogen and stored below −70°C. The aliquots are thawed by the addition of 1 ml of 10 mM Tris pH 7.5, 50 mM NaCl immediately before coating the coverslips. Diluted ECM solutions cannot be stored and the coating should not be prolonged extensively due to aggregation.

Fixatives

4% PFA is dissolved in PBS or CB at 65°C. The solution should be cooled down immediately after the PFA has dissolved (5 min). It might be necessary to filtrate the fixative to remove residual particles. The fixative can be stored in frozen aliquots for months or at 4 °C for one to two weeks. GA is obtained as 25% or 50% stock solution and diluted to a concentration of 0.05–1%.

Buffers

PBS: 10 mM phosphate buffer pH 7.2, 135 mM NaCl, 15 mM KCl, 3 mM $MgCl_2$

CB: 10 mM MES pH 6.1, 150 mM NaCl, 5 mM EGTA, 5 mM $MgCl_2$

Triton X-100 stock: 20% (v/v) in water (store at 4 °C)

Microinjection buffer: 2 mM Tris pH 7, 50 mM KCl, other components may be needed depending on the requirements of the respective protein

G-actin buffer: 20 mM imidazole pH 7.0, 0.2 mM $CaCl_2$, 0.1 mM DTT, 0.2 mM ATP
F-actin buffer: G-actin buffer plus 2 mM $MgCl_2$ and 100 mM KCl

Acknowledgements

We thank Marie-France Carlier for providing Figure 7 and for her invaluable help with Section 3.4. We also thank Mario Gimona for generously providing Figure 6 and J. Victor Small for the permission to use images from work performed in his laboratory. In addition, we would like to thank Peter Traub for kindly providing the anti-vimentin antiserum and Helen Morrison for critically reading the manuscript.

References

1. Goldman, R. D. and Chou, Y.-H. (1999). In *Guidebook to the cytoskeletal and motor proteins*, 2nd edn (ed. T. Kreis and R. Vale), p. 281. Oxford University Press, Oxford.
2. Janmey, P. A. (1991). *Curr. Opin. Cell Biol.*, **3**, 4.
3. Celis, J. E., Fey, S. J., Larsen, P. M., and Celis, A. (1985). *Ann. N. Y. Acad. Sci.*, **455**, 268.
4. Nigg, E. A. (1992). *Semin. Cell Biol.*, **3**, 245.
5. Pollard, T. (1999). In *Guidebook to the cytoskeletal and motor proteins*, 2nd edn (ed. T. Kreis and R. Vale), p. 3. Oxford University Press, Oxford.
6. Ampe, C. and Vandekerckhove, J. (1999). In *Guidebook to the cytoskeletal and motor proteins*, 2nd edn (ed. T. Kreis and R. Vale), p. 11. Oxford University Press, Oxford.
7. Gregorio, C. C. and Antin, P. B. (2000). *Trends Cell Biol.*, **10**, 355.
8. Carlier, M. F. (1998). *Curr. Opin. Cell Biol.*, **10** (1), 45.
9. Small, J. V., Isenberg, G., and Celis, J. E. (1978). *Nature*, **272**, 638.
10. Wang, Y. L. (1985). *J. Cell Biol.*, **101**, 597.
11. Cleveland, D. W. (1999). In *Guidebook to the cytoskeletal and motor proteins*, 2nd edn (ed. T. Kreis and R. Vale), p. 189. Oxford University Press, Oxford.
12. Caplow, M., Nogales, E., and Downing, K. H. (1999). In *Guidebook to the cytoskeletal and motor proteins*, 2nd edn (ed. T. Kreis and R. Vale), p. 241. Oxford University Press, Oxford.
13. Oakley, B. R. (1999). In *Guidebook to the cytoskeletal and motor proteins*, 2nd edn (ed. T. Kreis and R. Vale), p. 271. Oxford University Press, Oxford.
14. Waterman-Storer, C. M. and Salmon, E. D. (1997). *J. Cell Biol.*, **139**, 417.
15. Wehland, J. and Rüdiger, M. (1999). In *Guidebook to the cytoskeletal and motor proteins*, 2nd edn (ed. T. Kreis and R. Vale), p. 249. Oxford University Press, Oxford.
16. Rosenbaum, J. (2000). *Curr. Biol.*, **10**, R801.

17. Kamal, A. and Goldstein, L. S. (2000). *Curr. Opin. Cell Biol.*, **12**, 503.

18. Vasiliev, J. M. and Gelfand, I. M. (1976). In *Cell motility* (ed. R. Goldman, T. Pollard, and J. Rosenbaum), p. 279. Cold Spring Harbor Laboratory, Cold Spring Harbor, NY.

19. Small, J. V., Kaverina, I., Krylyshkina, O., and Rottner, K. (1999). *FEBS Lett.*, **452**, 96.

20. Reichardt, L. F. and Tomaselli, K. J. (1991). *Annu. Rev. Neurosci.*, **14**, 531.

21. Schwarzbauer, J. (1999). *Curr. Biol.*, **9**, R242.

22. Hotchin, N. A. and Hall, A. (1995). *J. Cell Biol.*, **131**, 1857.

23. Herzog, M., Draeger, A., Ehler, E., and Small, J. V. (1994). In *Cell biology: a laboratory handbook*, 1st edn (ed. J. E. Celis), Vol. 3, p. 355. Academic Press, San Diego.

24. Small, J. V., Rottner, K., Kaverina, I., and Anderson, K. I. (1998). *Biochim. Biophys. Acta*, **1404**, 271.

25. Rottner, K., Hall, A., and Small, J. V. (1999). *Curr. Biol.*, **9**, 640.

26. Erck, C., Frank, R., and Wehland, J. (2000). *Neurochem. Res.*, **25**, 5.

27. Rottner, K., Behrendt, B., Small, J. V., and Wehland, J. (1999). *Nature Cell Biol.*, **1**, 321.

28. Misteli, T. and Spector, D. L. (1997). *Nature Biotechnol.*, **15**, 961.

29. Gimona, M. (1998). In *Cell biology: a laboratory handbook*, 2nd edn (ed. J. E. Celis), Vol. 4, p. 265. Academic Press, San Diego.

30. Sentein, P. (1977). *Cell Biol. Int. Rep.*, **1**, 503.

31. De Brabander, M., De May, J., Joniau, M., and Geuens, G. (1977). *Cell Biol. Int. Rep.*, **1**, 177.

32. Sajo, I. (1977). *Acta Biochim. Biophys. Acad. Sci. Hung.*, **12**, 259.

33. De Brabander, M., Geuens, G., Nuydens, R., Willebrords, R., and De Mey, J. (1981). *Proc. Natl. Acad. Sci. USA*, **78**, 5608.

34. Kumar, N. (1981). *J. Biol. Chem.*, **256**, 10435.

35. Flanagan, M. D. and Lin, S. (1980). *J. Biol. Chem.*, **255**, 835.

36. Nickola, I. and Frimmer, M. (1986). *Cell Tissue Res.*, **245**, 635.

37. Dancker, P., Low, I., Hasselbach, W., and Wieland, T. (1975). *Biochim. Biophys. Acta*, **400**, 407.

38. Wehland, J., Osborn, M., and Weber, K. (1980). *Eur. J. Cell Biol.*, **21**, 188.

39. Bubb, M. R., Senderowicz, A. M., Sausville, E. A., Duncan, K. L., and Korn, E. D. (1994). *J. Biol. Chem.*, **269**, 14869.

40. Bubb, M. R., Spector, I., Beyer, B. B., and Fosen, K. M. (2000). *J. Biol. Chem.*, **275**, 5163.

41. Spector, I., Shochet, N. R., Blasberger, D., and Kashman, Y. (1989). *Cell Motil. Cytoskeleton*, **13**, 127.

42. Carsten, M. E. and Mommaerts, W. F. (1963). *Biochemistry*, **2**, 28.

43. Kreis, T. E. (1986). In *Methods in enzymology* (ed. R. B. Vallee), Vol. 134, p. 507. Academic Press, London.

44. Wang, Y. L. (1991). In *Methods in enzymology* (ed. R. B. Vallee), Vol. 196, p. 497. Academic Press, London.

45. Kranewitter, W. J., Ylanne, J., and Gimona, M. (2001). *J. Biol. Chem.*, **276**, 6306.

46. Kouyama, T. and Mihashi, K. (1980). *Eur. J. Biochem.*, **105**, 279.

47. Kouyama, T. and Mihashi, K. (1981). *Eur. J. Biochem.*, **114**, 33.

48. Cooper, J. A., Walker, S. B., and Pollard, T. D. (1983). *J. Muscle Res. Cell Motil.*, **4**, 253.

49. Welch, M. D., Rosenblatt, J., Skoble, J., Portnoy, D. A., and Mitchison, T. J. (1998). *Science*, **281**, 105.

50. Pantaloni, D., Boujemaa, R., Didry, D., Gounon, P., and Carlier, M. F. (2000). *Nature Cell Biol.*, **2**, 385.

51. Pantaloni, D. and Carlier, M. F. (1993). *Cell*, **75**, 1007.

52. Machesky, L. M., Mullins, R. D., Higgs, H. N., Kaiser, D. A., Blanchoin, L., May, R. C., *et al.* (1999). *Proc. Natl. Acad. Sci. USA*, **96**, 3739.

53. Yarar, D., To, W., Abo, A., and Welch, M. D. (1999). *Curr. Biol.*, **9**, 555.

54. Ma, L., Rohatgi, R., and Kirschner, M. W. (1998). *Proc. Natl. Acad. Sci. USA*, **95**, 15362.

55. Rohatgi, R., Ma, L., Miki, H., Lopez, M., Kirchhausen, T., Takenawa, T., *et al.* (1999). *Cell*, **97**, 221.

56. Carlier, M. F., Nioche, P., Broutin-L'Hermite, I., Boujemaa, R., Le Clainche, C., Egile, C., *et al.* (2000). *J. Biol. Chem.*, **275**, 21946.

57. Avnur, Z. and Geiger, B. (1981). *Cell*, **25**, 121.

58. Lazarides, E. and Burridge, K. (1975). *Cell*, **6**, 289.

List of suppliers

Amersham Biosciences UK Ltd,
Amersham Place, Little Chalfont,
Buckinghamshire HP7 9NA,
UK
(see also Nycomed Amersham
Imaging UK; Pharmacia)
Tel: 0800 515313
Fax: 0800 616927
URL: http//www.amershambiosciences.
com/

Anderman and Co. Ltd, 145 London
Road, Kingston-upon-Thames,
Surrey KT2 6NH, UK
Tel: 0181 5410035
Fax: 0181 5410623

**American Type Culture Collection
(ATCC)**, PO Box 1549, Manassas,
VA 20108, USA
Tel: 1 703 365 2700
Fax: 1 703 365 2701
URL: http//www.atcc.org

Aurion, Costerweg 5, NL-6702 AA
Wageningen, Netherlands
Tel: +31 317 497 676
Fax: +31 317 415 955
URL: http://www.aurion.nl/

Bal-Tec, Foehrenweg 16,
FL-9496 Balzers, Liechtenstein
Tel: +423 388 1212
Fax: +423 388 1260
URL: http//www.bal-tec.com/

Beckman Coulter (UK) Ltd, Oakley
Court, Kingsmead Business Park,
London Road, High Wycombe,
Buckinghamshire HP11 1JU, UK
Tel: 01494 441181
Fax: 01494 447558
URL: http://www.beckman.com/
Beckman Coulter Inc., 4300 N. Harbor
Boulevard, PO Box 3100, Fullerton,
CA 92834–3100, USA
Tel: 001 714 8714848
Fax: 001 714 7738283
URL: http://www.beckman.com/

Becton Dickinson and Co., 21 Between
Towns Road, Cowley, Oxford OX4 3LY,
UK
Tel: 01865 748844   Fax: 01865 781627
URL: http://www.bd.com/
Becton Dickinson and Co., 1 Becton
Drive, Franklin Lakes, NJ 07417–1883,
USA
Tel: 001 201 8476800
URL: http://www.bd.com/

Bio 101 Inc., c/o Anachem Ltd,
Anachem House, 20 Charles Street,
Luton, Bedfordshire LU2 0EB, UK
Tel: 01582 456666
Fax: 01582 391768
URL: http://www.anachem.co.uk/
Bio 101 Inc., PO Box 2284, La Jolla,
CA 92038–2284, USA
Tel: 001 760 5987299
Fax: 001 760 5980116
URL: http://www.bio101.com/

Bio-Rad Laboratories Ltd, Bio-Rad House, Maylands Avenue, Hemel Hempstead, Hertfordshire HP2 7TD, UK
Tel: 0181 3282000
Fax: 0181 3282550
URL: http://www.bio-rad.com/
Bio-Rad Laboratories, Inc., Life Science Division, 2000 Alfred Nobel Drive, Hercules, CA 94547, USA
Tel: 001 510 741 1000
Fax: 001 510 741 5811
URL: http://www.bio-rad.com/

Boehringer Mannheim (see Roche)

Braun Melsungen AG, Carl Braun Strasse 1, D-34212 Melsungen, Germany
Tel: +49 5661 710
Fax: +49 5661 714567
URL: http://www.bbraun.de

British BioCell International, Golden Gate, Ty Glas Avenue, Cardiff CF14 5DX, UK
Tel: +44 (0) 29 20747232
Fax: +44 (0) 29 20747242
URL: http://www.british-biocell.co.uk

CN Biosciences (UK) Ltd, Boulevard Industrial Park, Padge Road, Beeston, Nottingham NG9 2JR, UK
Tel: 0115 943 0840
Fax: 0115 943 0951
URL: http://www.calbiochem.com
Calbiochem-Novabiochem International, PO Box 12087, La Jolla, California 92039-2087, USA
Tel: (800) 854-3417
Fax: (800) 776-0999

CP Instrument Co. Ltd, PO Box 22, Bishops Stortford, Hertfordshire CM23 3DX, UK
Tel: 01279 757711
Fax: 01279 755785
URL: http//:www.cpinstrument.co.uk/

Diatome (see Leica)

Dionex Corporation, 1228 Titan Way, PO Box 3603, Sunnyvale, CA 94088, USA
Tel: 408 737 0700
Fax: 408 730 9403

Duchefa Biochemie BV, PO Box 2281, 2002 CG Haarlem, The Netherlands
Tel: +31 (0)23 531 9093
Fax: +31 (0)23 531 8027
URL: http://www.duchefa.com

Dupont (UK) Ltd, Industrial Products Division, Wedgwood Way, Stevenage, Hertfordshire SG1 4QN, UK
Tel: 01438 734000
Fax: 01438 734382
URL: http://www.dupont.com/
Dupont Co. (Biotechnology Systems Division), PO Box 80024, Wilmington, DE 19880–002, USA
Tel: 001 302 7741000
Fax: 001 302 7747321
URL: http://www.dupont.com/

Eastman Chemical Co., 100 North Eastman Road, PO Box 511, Kingsport, TN 37662–5075, USA
Tel: 001 423 2292000
URL: http//:www.eastman.com/

European Collection of Cell Cultures (ECACC), CAMR, Salisbury, Wiltshire SP4 0JG, UK
Tel: 44 1980 612512
Fax: 44 1980 611315
URL: http//www.ecacc.co.uk

Fisher Scientific UK Ltd, Bishop Meadow Road, Loughborough, Leicestershire LE11 5RG, UK
Tel: 01509 231166
Fax: 01509 231893
URL: http://www.fisher.co.uk/

Fisher Scientific, Fisher Research,
2761 Walnut Avenue, Tustin,
CA 92780, USA
Tel: 001 714 6694600
Fax: 001 714 6691613
URL: http://www.fishersci.com/

Fluka, PO Box 2060, Milwaukee,
WI 53201, USA
Tel: 001 414 2735013
Fax: 001 414 2734979
URL: http://www.sigma-aldrich.com/
Fluka Chemical Co. Ltd, PO Box 260,
CH-9471 Buchs, Switzerland
Tel: 0041 81 7452828
Fax: 0041 81 7565449
URL: http://www.sigma-aldrich.com/

Gibco-BRL (see Life Technologies)

Hybaid Ltd, Action Court, Ashford Road,
Ashford, Middlesex TW15 1XB, UK
Tel: 01784 425000
Fax: 01784 248085
URL: http://www.hybaid.com/
Hybaid US, 8 East Forge Parkway,
Franklin, MA 02038, USA
Tel: 001 508 5416918
Fax: 001 508 5413041
URL: http://www.hybaid.com/

HyClone Laboratories, 1725 South
HyClone Road, Logan, UT 84321, USA
Tel: 001 435 7534584
Fax: 001 435 7534589
URL: http://www.hyclone.com/

Hycor Biomedical, Pentlands Science
Park, Bush Loan, Penicuik, Edinburgh
EH26 OPL, UK
Tel: +44 131 4457111
Fax: +44 1314457112

Invitrogen Corp., 1600 Faraday
Avenue, Carlsbad, CA 92008, USA
Tel: 001 760 6037200
Fax: 001 760 6037201
URL: http://www.invitrogen.com/

Invitrogen BV, PO Box 2312, 9704 CH
Groningen, The Netherlands
Tel: 00800 53455345
Fax: 00800 78907890
URL: http://www.invitrogen.com/

Jackson Immunoresearch, 872 West
Baltimore Pike, PO Box 9,
West Grove,
PE 19390, USA
Tel: 610 8694024
Fax: 610 8690171
URL: http://www.jacksonimmuno.com/

Leica, Lilienthalsstr. 39-45, D-64625
Bensheim, Germany
Tel: +49 6251 1360
Fax: +49 6251 136155
URL:
http://www.leica-microsystems.com

Life Technologies Ltd, PO Box 35,
3 Free Fountain Drive, Inchinnan
Business Park, Paisley PA4 9RF,
UK
Tel: 0800 269210
Fax: 0800 243485
URL: http://www.lifetech.com/
Life Technologies Inc., 9800 Medical
Center Drive, Rockville, MD 20850,
USA
Tel: 001 301 6108000
URL: http://www.lifetech.com/

**Merck Sharp & Dohme Research
Laboratories**, Neuroscience Research
Centre, Terlings Park, Harlow,
Essex CM20 2QR, UK
URL: http://www.msd-nrc.co.uk/
Frankfurter Straße 250,
D-64293 Darmstadt,
Germany
Tel: +49 (0) 6151 720
Fax: +49 (0) 6151 722000
URL: http://www.merck.de/

Millipore (UK) Ltd, The Boulevard, Blackmoor Lane, Watford, Hertfordshire WD1 8YW, UK
Tel: 01923 816375   Fax: 01923 818297
URL: http://www.millipore.com/local/UKhtm/
Millipore Corp., 80 Ashby Road, Bedford, MA 01730, USA
Tel: 001 800 6455476
Fax: 001 800 6455439
URL: http://www.millipore.com/

New England Biolabs, 32 Tozer Road, Beverley, MA 01915–5510, USA
Tel: 001 978 9275054

Nikon Inc., 1300 Walt Whitman Road, Melville, NY 11747–3064, USA
Tel: 001 516 5474200
Fax: 001 516 5470299
URL: http://www.nikonusa.com/
Nikon Corp., Fuji Building, 2–3, 3-chome, Marunouchi, Chiyoda-ku, Tokyo 100, Japan
Tel: 00813 32145311
Fax: 00813 32015856
URL: http://www.nikon.co.jp/main/index_e.htm/

Nycomed Amersham Imaging, Amersham Labs, White Lion Road, Amersham, Buckinghamshire HP7 9LL, UK
Tel: 0800 558822 (or 01494 544000)
Fax: 0800 669933 (or 01494 542266)
URL: http://:www.amersham.co.uk/
Nycomed Amersham, 101 Carnegie Center, Princeton, NJ 08540, USA
Tel: 001 609 5146000
URL: http://www.amersham.co.uk/

Perbio Science UK Ltd (formerly Pierce & Warriner Ltd), Century House, High Street, Tattenhall, Cheshire CH3 9RJ, UK
Tel: 01829 771744   Fax: 01829 771644

Perkin Elmer Ltd, Post Office Lane, Beaconsfield, Buckinghamshire HP9 1QA, UK
Tel: 01494 676161
URL: http//:www.perkin-elmer.com/

Pharmacia, Davy Avenue, Knowlhill, Milton Keynes, Buckinghamshire MK5 8PH, UK (also see Amersham Pharmacia Biotech)
Tel: 01908 661101
Fax: 01908 690091
URL: http//www.eu.pnu.com/

Plano, Ernst-Befort-Straße 12, D-35578 Wetzlar, Germany
Tel: +49 6441 97650
Fax: +49 6441 976565
URL: http://www.plano-em.de/

Polysciences, Handelsstraße 3, D-69214 Eppelheim, Germany
Tel: +49 6221 765767
Fax: +49 6221 764620
URL: http//www.polysciences.com/
Polysciences, 400 Valley Road, Warrington, PA 18976, USA

Promega UK Ltd, Delta House, Chilworth Research Centre, Southampton SO16 7NS, UK
Tel: 0800 378994
Fax: 0800 181037
URL: http://www.promega.com/
Promega Corp., 2800 Woods Hollow Road, Madison, WI 53711–5399, USA
Tel: 001 608 2744330
Fax: 001 608 2772516
URL: http://www.promega.com/

Qiagen UK Ltd, Boundary Court, Gatwick Road, Crawley, West Sussex RH10 2AX, UK
Tel: 01293 422911
Fax: 01293 422922
URL: http://www.qiagen.com/

Qiagen Inc., 28159 Avenue Stanford,
Valencia, CA 91355, USA
Tel: 001 800 4268157
Fax: 001 800 7182056
URL: http://www.qiagen.com/

Roche Diagnostics Ltd, Bell Lane,
Lewes, East Sussex BN7 1LG, UK
Tel: 0808 1009998 (or 01273 480044)
Fax: 0808 1001920 (01273 480266)
URL: http://www.roche.com/
Roche Diagnostics Corp., 9115 Hague
Road, PO Box 50457, Indianapolis,
IN 46256, USA
Tel: 001 317 8452000
Fax: 001 317 8452221
URL: http://www.roche.com/
Roche Diagnostics GmbH,
Sandhoferstrasse 116,
68305 Mannheim, Germany
Tel: 0049 621 7594747
Fax: 0049 621 7594002
URL: http://www.roche.com/

Sartorius, Weender Landstraße
94-108, D-37075 Göttingen, Germany
Tel: +49 551 3080
Fax: +49 551 308 3289
URL: http://www.sartorius.com/

Schleicher and Schuell Inc., Keene,
NH 03431A, USA
Tel: 001 603 3572398

Shandon Scientific Ltd, 93–96
Chadwick Road, Astmoor,
Runcorn, Cheshire WA7 1PR, UK

Tel: 01928 566611
URL: http://www.shandon.com/

Sigma–Aldrich Co. Ltd, The Old
Brickyard, New Road, Gillingham,
Dorset SP8 4XT, UK
Tel: 0800 717181 (or 01747 822211)
Fax: 0800 378538 (or 01747 823779)
URL: http://www.sigma-aldrich.com/
Sigma Chemical Co., PO Box 14508,
St Louis, MO 63178, USA
Tel: 001 314 7715765
Fax: 001 314 7715757
URL: http://www.sigma-aldrich.com/

Stratagene Inc., 11011 North Torrey
Pines Road, La Jolla, CA 92037, USA
Tel: 001 858 5355400
URL: http://www.stratagene.com/
Stratagene Europe, Gebouw
California, Hogehilweg 15,
1101 CB Amsterdam Zuidoost,
The Netherlands
Tel: 00800 91009100
URL: http://www.stratagene.com/

TOSOH BIOSEP, 156 Keystone Drive,
Montgomeryville, PA 18936,
USA
Tel: 215 283 5000
Fax: 215 283 5035

United States Biochemical (USB),
PO Box 22400, Cleveland, OH 44122,
USA
Tel: 001 216 4649277

Index